# HISTORICAL
# BUILDING
# CONSTRUCTION

A NORTON PROFESSIONAL BOOK

# HISTORICAL
# BUILDING
# CONSTRUCTION

## DESIGN, MATERIALS, AND TECHNOLOGY

### DONALD FRIEDMAN

W•W•NORTON & COMPANY

NEW YORK • LONDON

This book is dedicated to
Robert and Constance Friedman
for their support and encouragement.

Copyright © 1995 by Donald Friedman
All rights reserved
Printed in the United States of America
First Edition

The text of this book is composed in Berkeley Book
with the display set in Gill Sans
Printed and bound by Edwards Brothers
Book design by Gilda Hannah

Library of Congress Cataloging-in-Publication Data

Friedman, Donald.
    Historical building construction / Donald Friedman. — 1st ed.
       p.    cm.
    Includes bibliographical references and index.
    ISBN 0-393-70200-6
        1. Historic buildings—maintenance and repair. 2. Historic buildings—Conservation and restoration. 3. Building—History.
I. Title.
TH3361.F751995                      94-23906
690'.24—dc2094-23906           CIP

ISBN 0-393-70200-6
W. W. Norton & Company, Inc., 500 Fifth Avenue, New York NY 10110
W. W. Norton & Company, Ltd., 10 Coptic Street, London WC1A 1PU
0 9 8 7 6 5 4 3 2

# CONTENTS

# 1. INTRODUCTION

The current concept of historic preservation and rehabilitation is a recent one. In 1963, a building as widely recognized for the quality of its architecture as New York's Pennsylvania Station could be demolished by its owner without the necessity of an established forum to present public opposition. Although architects and other people interested in preserving the station protested, and newspapers published editorials about the loss, the idea that buildings should be preserved—and if need be, modified for new uses—had not firmly taken hold. In the thirty years since, public interest in preservation has increased greatly, best indicated by the continual pressure from community groups to protect from change buildings of local prominence that are unknown to the general public. This interest in preservation extends from the remnants of the American colonial past to buildings instrumental in the development of modern architecture.

The growth of interest in building preservation has led to the creation of a new branch of architecture and engineering. In the past, old buildings were modified without any special attention given to preserving specific construction details or materials. The techniques used in design and construction of old-building modifications were the same as those used in new buildings. This has changed dramatically, with sources of information about ways of combining new construction with old becoming more popular. In addition, original sources of detailing information now are being reprinted for use in restoration.

Most of the information currently available concerns architectural issues, such as the detailing of windows and trim or catalogs of interior finishes. The reason for this is the overriding concern for the appearance of historic buildings. In perfect preservation, the entire building would be kept intact, but this is rarely, if ever, possible. The very changes in use that make possible the extended life of old buildings often require that these buildings be functionally upgraded. The installation of ventilation equipment or sprinklers, and the strengthening of floors for use as public assembly spaces, are among the most frequent changes required. When such changes are made, preserving unchanged the appearance of the original building is often possible, but preserving unchanged its physical structure is not. Structural modification can also be required by the effect of aging on the building structure and deficiencies in its original design. The close tie between the amount of structural retrofitting required to complete a project and the project's economic viability makes the lack of source material on historic construction crucial. Many projects are delayed for extensive probing to determine every aspect of the structure of a building, or are delayed during construction because of framing layouts or construction types not expected after limited probing.

Engineers practicing renovation or rehabilitation of buildings need to know the specifics of obsolete construction in order to work in a nondestructive and unobtrusive manner. Other than a couple of pages giving broad history in design textbooks, there are no contemporary sources of this information. Given the limited acceptance by preservationists of modern structure in a historic building, it becomes crucial for structural engineers engaged in renovation of older buildings to understand previously used construction and design techniques. Modifying the structure will affect the restoration or preservation of architectural finishes and can put constraints on the architectural use of space. Without knowledge of the existing structure, it may be possible to design a renovation or restoration, but the work will almost certainly be more intrusive than necessary.

This guide is meant to permit the design of "elegant" solutions to the structural problems caused by changes in use, changes in architectural design, or changes in the mechanical plant by enabling modern engineers to understand what assumptions their predecessors made, what materials they used, how they designed structures, and how their designs were built. It is not an attempt to trace the development of structural theory.

Architects renovating older buildings, while not usually responsible for structural design, need to understand the design restrictions forced on them by the original structure. Many common modern alterations, such as the installation of cooling towers on building roofs, can be difficult with old buildings if the structural capacity is not clear. Unlike steel frame buildings, which can be *reinforced* in straightforward ways, buildings with wood joist floors or cast-iron columns may require *replacement* of structural members to meet new load requirements. The more architects know of the basic structural systems in the existing building stock, the earlier in the design process decisions about such issues can be addressed.

There are certain obvious pitfalls in producing such a guide. The history of structural design is far less documented than the history of architecture. This is true despite the relative youth of the profession: All of the currently used frame-analysis techniques are less than 120 years old; the use of structural steel in buildings is less than 110 years old, and large-scale use of reinforced concrete in buildings is less than 100 years old. The advances in analysis and design were so rapid, especially before the twentieth century, that a few years' difference in the date of construction could make a tremendous difference in a building's structure. In addition, engineers in the past were no less liable to make mistakes than engineers today. Just because the records indicate that something should be true, such as the design live load

for a floor, this is not proof that the floor can support that weight. It is also possible that an original floor did meet the required live load capacity when first built, but was modified at some time and weakened.

The lack of documentation of built engineering history can also be traced to a general attitude towards the hidden building structure that is no less true today than it was in 1896, when an article in the *Engineering News* stated that

> [t]he construction of the exceptionally high buildings with steel framework, which now form such a notable feature of the larger American cities, has led not only to the development of a new and large market for structural steel and iron work, but also to new problems in structural design. . . . These difficulties are probably appreciated not at all by the owner, and only to a limited extent by the architect (unless he is himself an engineer experienced in structural work), and the designing engineer can expect but little credit for the successful results of his labor, since the work is all built in and enclosed as soon as erected, so that the casual observer never gives it a thought, and even if anyone were interested in it, it is hidden from examination.[1]

Taking into account the restrictions on the usefulness of written records, this guide can act as one step in the process of analyzing existing conditions in a building. The information here is meant to represent the state of common knowledge at various times in the past. Based on the age and original use of a building, it can suggest possibilities for the construction hidden behind floor, wall, and ceiling finishes. Examples are given of various developments, because any material or technique that existed in one building of a certain age almost certainly was used in other buildings of that era. These possibilities are based on general trends in construction. There is always overlap between new and old technology—one example being the continued use of built-up column sections after rolled, wide-flange Bethlehem columns were commercially available. Based on these possibilities, actual design will have to follow verification of the type of construction and the value of the materials used. Especially before the American Society of Testing and Materials (ASTM) began standardizing construction materials, the quality of irons, steels, cements, and concretes varied greatly. Testing samples of existing materials is a prerequisite for alterations that cannot be eliminated for buildings that predate the ASTM specifications, and should be considered for all buildings.

The same structural element will often have two different capacities, that found using original codes and analysis techniques, and that found using the current codes and techniques. These differences in older design are based largely upon the expansion of knowledge in the field of structural engineering. It is always viable to analyze an older structure under the new codes, as long as the properties of the materials are known. Often this will show that the older design was conservative—with a larger safety factor than is now required—either in the applied loads, the material resistance, or both.

The design of buildings has always been influenced by local conditions, economics, and laws. It is difficult at best to compare the influences on construction in different countries, even countries with conditions as similar as Canada and the United States. In order to tell one story clearly, rather than several inadequately, the information here is limited to the United States. This is not to imply that important developments did not occur elsewhere, but simply to limit the discussion of foreign technology to its effect within the United States. On an even more local level, construction varies with geography within the United States, now and in the past. The bulk of the discussion here concerns iron and steel frame construction, which has been used nationally in large cities since at least 1880. The forms of the construction types described vary little with location, since modern construction is related to both mass production and scientific analysis. In order to show a consistent and well-documented set of examples evolving in chronological order, New York City is used for the majority of the descriptions. During the period of structural form's greatest development, between 1890 and 1920, New York experienced explosive growth, ensuring that many examples of obsolete forms are still available for examination and that the drive for innovation was increased by the economics of rapid development. In addition, since the engineering press coverage of building

development necessarily followed the most intense and most innovative activity, New York is better represented than any other city in nationally available source material. The general information given here applies, with few exceptions, to buildings of the same types built all over the country during the same period. Examples from other cities could easily be substituted: the brick row-houses of Boston's Back Bay for New York's brown-stones; Chicago's Monadnock Building for the Chelsea Hotel; Philadelphia's original Bank of the United States for the Merchant's Exchange; St. Louis's Telephone Building, Los Angeles's City Hall Tower, Minneapolis's Metropolitan Building, or Denver's Brown Palace Hotel for the Woolworth Building or the RCA Building; and Pittsburgh's Central Engineering Laboratory for the New York Stock Exchange Annex. This list is composed of a few selected examples, and if the reasons given for using a single set of examples did not apply, the examples could be broken up among every city in the country.

The individual building codes commonly used by different cities before national building codes became popular often reflected different values for allowable stress in a given material or for the required live load for a given occupancy. The values specified were a reflection of conditions in the city in question, including local politics within the various building departments that are beyond the immediate concerns of preservation professionals. New York's early development in tall buildings is probably responsible, in part, for the relatively high values for allowable stress in iron and steel before the American Institute of Steel Construction (AISC) code standardization. Theoretical values that were known nationally have been given here, but in investigating New York buildings, preference should be given to the values in the New York City Building Code. For buildings in other cities, the local codes, if available, should be used. The same local preference should be considered when investigating material types or construction methods. National and international developments affected local events strongly, but here they will only be mentioned in connection with their direct impact on local construction.

Just as structural engineering is relatively new, so is the progressive advance of building technology. Regular change did not occur until the industrial revolution—and even then, not immediately. The development of cast-iron construction in the 1840s did not make wood and masonry construction unusable. The development of cast and wrought iron, coupled with practical elevator technology, set off the development of tall buildings, for which iron construction is inherently better suited than the older materials. This type of interdependent development dates from the mid-nineteenth century and marked the beginning of the technological obsolescence of buildings. Societal recognition of the increasingly technical and nontraditional nature of construction came in New York with the introduction of a comprehensive building law in 1860. This law gave rules on the use of new materials, and new building types, that the "common sense" standard of the courts then in use could not address. Previous laws in New York and elsewhere addressed narrow zoning and safety issues.

There is a pronounced emphasis on iron and steel, with some attention paid to concrete and masonry, in the information provided here. This is a reflection of the current circumstances: in historic preservation or rehabilitation work we are most often dealing with buildings that are between 50 and 120 years old. The vast majority of these buildings have steel or iron framing, ranging from floor beams only to complete skeleton frames.

By the early nineteenth century a concern for "fireproof" construction was evident. This concern accelerated in New York after the fires of 1835 and 1845, which destroyed large portions of the commercial downtown, and in other cities following their largest fires. As more legal restrictions were placed on where wood construction was allowed, its use was greatly reduced in commercial, industrial, and public construction. Wood was and still is used in residential construction, for frame houses where permitted, and for joists in apartment and tenement house construction. Other than changes in the species of wood used and the introduction of some detailing changes such as fire-cutting joists, there is little difference between this type of construction in 1880 or in 1990. The other form of wood construction that survives is the use of heavy timber trusses in church roofs.

This is a specialty form of construction well outside the mainstream of structural technology.

Foundations are also not covered in detail. They are important for the simple reason that the stability of a building's total superstructure depends on the strength of its base, and any renovation that adds load to existing foundations must include a check of the effect. The omission here is because foundations have rarely been a focus of advanced design in New York due to the city's relatively high and sound bedrock. The stone and concrete footings of one hundred years ago bear far more resemblance to their modern counterparts than do the superstructures of the same buildings.

In attempting to describe the development of construction technology, it is important to distinguish between ordinary techniques—used by large numbers of builders, architects, and structural engineers in typical buildings of the city—and techniques used in monumental or civil engineering construction. Bridges, roads, tunnels, and dams are mentioned incidentally, if at all, because of the distinction between their construction and the rest of the city's structures. Any mention of a structural landmark such as the Brooklyn Bridge, which involved a series of technological breakthroughs, is to illustrate a "typical" material or analysis method used elsewhere.

Landmarks have been traditionally overrepresented in the records. A resident of Manhattan in 1930 could conceivably never have used the Brooklyn Bridge, but he or she would almost certainly have lived in a building type that is described here. The sources of information on ordinary structures are scattered. The source most closely linked to the actual buildings is observation by the author during various building renovations and restorations. In the course of the working day, many architects and other engineers have shared their "war stories," some of which appear here. Written sources include direct influences such as the New York building laws and codes and national codes incorporated within the New York codes, such as the AISC *Steel Design Manual*. Less direct sources are descriptions in newspapers and the engineering press of ordinary buildings that became noticeable through failure, by either fire or collapse. Finally, there are sources tangentially related to the built environment, such as obsolete textbooks and design manuals, manufacturers' literature, and architectural histories.

This guide consists of two major parts: a narrative description of the development of construction in roughly chronological order, with connections between the various trades, typical details, and methods of construction highlighted; and appendices containing a summary of New York City building codes, ASTM specifications, AISC specifications, and American Concrete Institute (ACI) specifications that have governed design, and a list of buildings that exemplify the developments described, based on both records and personal observation.

Two trends extend from beginning to end of both the narrative and the appendices: the growth of scientific design at the expense of traditional rules, and the replacement of labor-intensive techniques with technology-intensive materials. The nineteenth century saw the creation of civil engineering as a discipline distinct from architecture, and then the separation of mechanical and electrical engineering from structural engineering. Increasing specialization and modern research methods caused explosive growth in scientific analysis, design, and investigation. The trends described here are therefore part of a significant shift within engineering towards more scientific practice.

If traditional design is represented by the performance-based rules in late nineteenth-century building codes for masonry wall thicknesses, the transition to analytic design is represented by the experimentally based strength formulas for cast-iron columns, and modern design is represented by the experimental analysis and theory-based strength formulas for steel columns. Traditional design is literally centuries old and has a history of successful construction based on trial and error—although in general it has been limited to fairly low buildings and short spans. Early analytic design fueled the first skyscraper boom in the 1870s, but was also advanced by trial and error, errors that were sometimes horrendous failures because of the increasing size and structural complexity of the individual buildings. Modern analytic design is dependent on the quality of the research, but in general it has eliminated structural

failures due to lack of knowledge. Advances in this type of design have been refinements in the models used to represent different materials and systems and the study of misapplications of techniques that resulted in failures.

The most obvious consequence of the economics of construction is the replacement of labor-intensive techniques with technology-intensive materials. When no alternatives to hand construction existed, the amount of labor required to build a thick brick wall was not an issue. Once iron columns could be used in the place of that wall, building designers began to examine the trade-off of the more expensive materials of more modern technology against the more expensive labor of traditional methods. Typically, technology won. It was possible to expect that continued development would reduce the price and increase the quality of new methods—as, for example, crucible steel was replaced by Bessemer steel, which was replaced in turn by open-hearth steel. Advances in labor efficiency were relatively slow and most closely linked to tool use. Almost every new development discussed here can be seen as a forward step in technology at the expense of simplicity and labor use. A modern development such as composite metal deck is an expression of advanced metal fabrication and stress-analysis techniques used largely as a way of eliminating the labor needed to build wooden forms for concrete floors.

## Note on Geography

The layout of Manhattan was fixed, in broad terms, in the 1811 Commissioner's Report that created the gridiron north of Houston Street. There were, and still are, an enormous number of peculiarities to the streets south of Houston and east of Broadway, and those south of 14th Street west of Broadway. During the period of interest here, the grid was still being built further north each year. As any social history of New York points out, the older neighborhoods in the southern portion of the island were rebuilt repeatedly as the entire city expanded. Similar, if less dramatic, city design and construction took place in the other counties that now make up Greater New York, most intensely in Brooklyn and least intensely on Staten Island.

The concern of the text is to describe the physical condition of typical individual buildings, not the entire cityscape. In the interest of clarity, modern names for neighborhoods and streets are used whenever possible, leading to anachronisms such as references to SoHo in the 1860s. If older names are used, the modern are given parenthetically. In addition, direction is given in the modern sense: uptown or up refers relatively to any location north of the one in question and specifically to Manhattan above 59th Street. Downtown refers specifically to the commercial area south of Chambers Street as well as relatively to any location south of that under discussion. Any building or address mentioned is in New York, usually in Manhattan, unless noted otherwise.

# 2. TRADITIONAL CONSTRUCTION
## 1840–PRESENT

The traditional forms and especially the traditional materials used before the development of structural technology after 1840 continued in use in many cases past 1900. The modified heavy timber framing of houses in use before 1830 is similar to the platform frame house construction still used today; the balloon framing developed during the 1840s and 1850s is the basic construction of the majority of wood houses extant; and small buildings are still built with masonry bearing walls following code provisions that are the remnants of the complex rules for height-to-thickness ratios once used.

An analysis of the structure of traditional construction is limited to wood and masonry. Metals were used in construction, but for architectural purposes such as roofing. Vertical supports consisted of masonry walls and piers, wood columns, and primitive wood-stud walls. Horizontal floor support was provided in ordinary construction by wood joists and beams and in monumental construction by masonry vaults and arches. The average building in New York before the 1840s was quite small, even by the standards of the late nineteenth century. Most buildings were a maximum of three stories, and narrow enough that no floor supports were required between the exterior walls. The 20-foot-wide lot became a permanent feature of the city with the 1811 gridiron street plan, and the vast majority of buildings covered only one lot. The exterior walls in houses were either wood or masonry, depending in large part on the finances of the occupant; the exterior walls of commercial buildings were usually masonry.

During the transition from traditional construction to iron framing and fireproof floors, in the 1850s and 1860s, a relatively small group of architects was responsible for many of the more advanced buildings. John Snook, John Kellum, Gamaliel King, Frederick Petersen, and George H. Johnson were part of this group. Their names are now connected with many of the buildings we see as pioneering new structural technique. They were great users of cast iron, for both facades and interior columns, as well as fireproof floor systems using wrought iron and masonry.[1] There is no similarly prominent group of architects from the period known for using traditional construction—partly because succeeding generations have not seen any need to memorialize the structural traditionalists, and because many traditional buildings were built by craftsmen or contractors without the aid of architects.

Structural engineering for buildings as a separate discipline did not truly exist in the 1840s. The need for engineers grew in the 1850s with the advent of wrought-iron beams, which had to be mathematically designed because there was no craftsman tradition to provide tried-and-true rules of thumb. The establishment in 1852 of the American Society of Civil Engineers promoted the rapid spread of technical information, such as records of the experiments with cast and wrought iron performed in England by Eaton Hodgkinson and William Fairbairn. In general, engineers tended to work on those buildings with iron construction. This has not changed, as many low-rise buildings of wood construction are still designed

**20 Henry Street, 1931, south and east facades. Heavy timber and bearing-masonry construction. The piers are required to give the proper amount of masonry specified by the building code and allow thinner walls to permit more light to enter.**

and built without engineers. Planned design cannot be assumed in the examination of older wood and masonry buildings.

## Wood

Wood is the original American material of construction. The New York area was heavily wooded before the Dutch settlement was established, and portions of it remained so into the nineteenth century. The majority of buildings in New York before 1800 were wood, and the majority of building structure was wood as late as the 1860s. "Fireproof" or masonry buildings of the early nineteenth century were masonry skins surrounding a wooden interior, just as the later development of "cast-iron" buildings described structures usually of wood with masonry and iron facades.

The traditional form of wood construction was a heavy wooden frame of posts and beams supporting joist floors. Between 1835 and 1860 this form was replaced with "balloon" framing, consisting of full-height stud walls supporting the floors. Later in the nineteenth century, "platform" framing of one-story studs developed from balloon framing. Nineteenth-century balloon- and platform-framed houses are very close in form to houses built today and require no special analysis.[2]

Fire-prevention provisions were included in the major versions of New York's building laws starting in the 1880s. As a result, few large buildings could be built with wood frames of any type. Wood was used in the interiors, leaving most or all of the vertical support and all of the lateral bracing to the exterior masonry walls. The few existing wood frame buildings in Manhattan were constructed north of the built-up portion of the city, where fire spread was not a problem.[3] More typical of wooden construction was a factory built at 20 Henry Street in the independent city of Brooklyn in 1885, which had plank floors, timber beams, and timber columns within a brick bearing wall.[4]

Heavy timber was used extensively in the construction of churches and synagogues. Before steel framing matured in the 1890s, wood trusses were the only reasonable way for builders to span large auditoriums. Almost every nineteenth-century church still standing has trusses of heavy timber or timber combined with wrought-iron tension rods supporting its roof.[5] These trusses are not necessarily designed for stress any more accurately than the typical joist floors of the same date of construction. In addition, the joints within the trusses very often consist of poorly detailed bird's mouth cuts and iron pins, which produce unacceptably high bearing and shearing stresses.

Most buildings contained wood primarily in the form of joists. Joists were sized by rules of thumb based on previous construction. A given joist size was reused for the same span and building type, with no direct calculation of stress or deflection. Many very old buildings have joist floors that in theory cannot carry the building dead load plus the code live load. Under ordinary conditions this is not a concern, but if the actual live load ever approaches the code level or alterations change the load distribution, the building's safety is questionable. With mathematically based

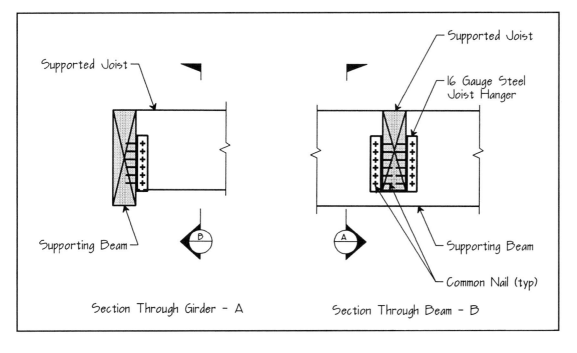

Section Through Girder – A   Section Through Beam – B

engineering analysis of ordinary building components after 1860, joists were sized for allowable
stress, using experimentally determined values of
ultimate stress. Ultimate stress was simplest for
experimenters to determine, since they could simply find the failure loads for different beams of a
given material and then calculate the stress at failure from the known beam geometric properties.[6]

The design of wood joists in the late 1800s
seems to have been concerned solely with stress.
The joists used tended to be wider and shallower
than modern joists, reducing the floor's stiffness
for a given strength. Older floors, such as those
built in hundreds of rowhouses in the 1880s, tend
to be "bouncy" and sag over time in large rooms,
where partitions are not available for intermediate
support.[7] From modern analysis, many joist floors
in apartment houses were not designed for deflection criteria, even after 1900. Before World War I,
the price of the wood used was still relatively high
when compared to the cost of the labor to install
it, leading many designers to mix woods. In the
plans of old buildings, it is not uncommon to find
that short-span joists are spruce or short-leaf yellow pine while long-span joists are made of the
stronger long-leaf yellow pine.

Designers were not ignorant of deflection design of beams, since that had been a part of iron
and steel design since the 1870s. During the
1920s, texts were recommending the use of deeper

joists and deflection limits of the joists' span
divided by 240 or 360.[8] There is no difference
between the recommended treatment of wood-
joist design in 1920 and in 1990.

Even as iron and steel framing and fireproof
floor systems were evolving in the 1880s and
1890s, the details of wood construction remained
much the same. Joists were typically supported by
wood headers with mortise and tenon connections and by masonry walls by simple embedment
of the joist ends. These connections were traditional but far from perfect. The notched end of the
joist in the mortise and tenon is subject to stress
concentration and horizontal splitting. In ordinary
cases, this is not critical, but in long spans and
heavy loading the splitting can be progressive. The
embedment of square-cut joist ends in brick walls
is not problematic most of the time, but during
joist failure in a fire the embedded portions act as
levers to pry the wall apart. Embedded joists are
also prone to destructive biological and chemical
processes such as rot if rainwater easily penetrates.

By the 1890s, an industrial alternative had
been found to mortise and tenon joints. Near the
end of the 1891 version of the building laws was
an ad for the Lincoln Iron Works promoting "Lincoln Timber Hangers," which were light-gauge
steel joist hangers very similar to their modern
counterparts (figure 2.1). The justification given
for abandoning a centuries-old practice was that

**2.2. Bridle Iron Wood Beam Connection Scale: 1" = 1' 0"**

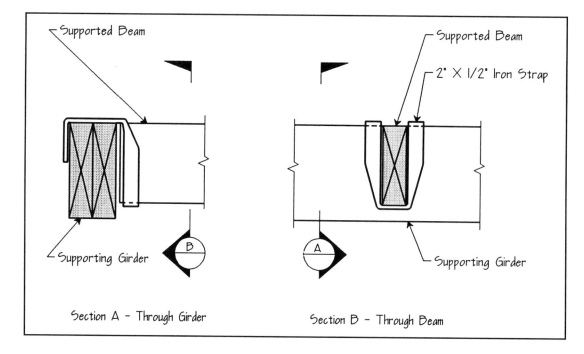

Section A – Through Girder

Section B – Through Beam

the joist hangers were cheaper: the elimination of fitted cuts in two different pieces of wood more than made up for the cost of the hanger itself.[9] While this argument was ultimately true, and connections using hangers were stronger than traditional joints, mortise and tenon joints continued to be used into the twentieth century. The relative increase in carpentry labor cost eventually made the hangers the connection of choice. By 1923, joist hangers virtually indistinguishable from modern designs were available from at least half a dozen national companies.[10]

Joist hangers were a smaller and simpler version of bridle irons, or stirrups, the modern connector used for large timbers. Large beams used as girders in masonry wall buildings and in building forms such as wood-braced frames had traditionally been joined at beam-to-girder and beam-to-post connections with various types of mortise and tenon joints. Unlike joists, which can use floor planking to redistribute loads, the design of wood girders is often controlled by the horizontal shear and bearing at supports. The traditional joints, by cutting away pieces of the wood, weakened the members at exactly the wrong place. Bridle irons are basically large hooks made of iron straps twisted to shape which fit over one beam to support another (figure 2.2). Since the supported end is unworked, it retains its full strength. The elimination of the cut at the top of the supporting

beam strengthened that member as well, although the change was less significant, since that portion of the supporting member is ordinarily in compression, and the projecting end of the supported member can serve to transmit stress across the gap. The only significant issue regarding wood design with bridles is ensuring that both members have sufficient bearing length for the iron. By 1900, bridles had become the most common method of joining large wood beams.[11]

Determining how to embed wood beams in masonry walls without endangering those walls was simpler than connecting wood members to one another. By cutting back the end of each joist at a small angle from vertical—a shape that became known as a fire cut—the builder ensured that, during a fire that destroyed the joists, the stubs embedded in the wall could rotate down and fall without prying up the brick above (figure 2.3). This was one of those lucky inventions that both served its purpose and was inexpensive, and by the 1920s it had become a standard feature. The fire cut was often combined with a small gauge-metal strap embedded in the masonry to supply the portion of the sideways stability of the joist that had been lost through the removal of its top.

The embedded straps provided the first connection for tension in wood-joist floors. In low buildings and rowhouses, the lack of positive con-

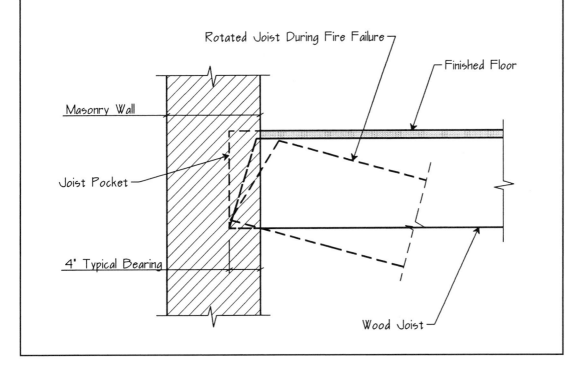

Rotated Joist During Fire Failure

Finished Floor

Masonry Wall

Joist Pocket

4' Typical Bearing

Wood Joist

nection between the floors and walls is usually not a serious problem. Simple friction between the wood joist and the surrounding mortar can provide a reasonable bond between floors and walls. Stability problems in the masonry from lack of bracing provided by the interior floors are most likely to appear in two places: facades parallel to the joists and relatively wide (more than 20 feet across), and facades supporting joists with windows. The lack of stability in the first example is clear: the masonry is only restrained from outward buckling at its intersection with return walls. In wide, single-bay buildings, the span between the side walls may be longer than analysis would recommend for the span of the masonry.

The lack of stability in side walls was not an endemic problem because such a large percentage of those walls were party walls, uninterrupted by openings and braced by floor joists on both sides. At the ends of blocks, where by geometric necessity one building must expose its side wall to the street at each corner, the side walls were often almost as devoid of windows as if they were party walls. Side walls with full rows of windows are less stable than other walls because the floor joists tie the walls in the narrow strip between the window heads below and the window sills above. The

vertical piers between windows are nearly freestanding, concentrating the force required to tie them back at the base past the point where the joist friction is entirely adequate. The solution to both wall stability problems, if any solution was used, was usually wrought-iron building ties. For side walls, these iron rods extended the width of the building and tied the outer faces of the opposite walls to one another. For facades parallel to the joists, the ties were smaller, and usually only extended back several joists. In both cases, the outside face of the tie was kept in place by a large ornamental iron fish plate on the outer face of the brick, often in a stylized "S" or star shape.

Another idea introduced about the same time as fire cuts that did not become popular was the enclosure of each joist end in a sheet-metal box for protection.[12] This increased the cost to the builder without an obvious benefit, and thus it was doomed to disuse. There was even some evidence that such enclosure could be dangerous: in 1909 a six-story factory collapsed from the dry rot of wood column ends enclosed in sealed cast-iron boxes.[13]

The ultimate abandonment of wood for construction other than residential in the less heavily built-up portions of the city was caused by the

**488 Greenwich Street, east facade.** Two-story house of wood joist on brick bearing-wall construction, probably from the first half of the nineteenth century. The pronounced bulge at the second floor is usually associated with a lack of ties between the walls parallel to the joists and the joists behind. The stars are the ends of iron tie rods extending through the building, possibly original, but more likely a retrofit. The building beyond has a cast-iron storefront of construction similar to complete cast-iron facades supporting a typical brick wall above.

fear of fire. While any material is vulnerable to the effect of heat, and wood can be protected from fire's direct effects, wood remains the only construction material that is itself combustible. Any number of horrendous fires could be cited as examples of the dangers of using wood in large or public buildings: that in the Windsor Hotel in 1899 has been chosen because there was nothing out of the ordinary in its construction or destruction.

The Windsor was built in 1873, covering the full block front on the west side of Fifth Avenue between 46th and 47th streets and extending back 140 feet. The exterior was a brick bearing wall; the interior was wood studs and joists. The hotel was built to luxury standards, with straight, wide corridors, open stairwells (with wood stairs), and many passenger elevators and dumbwaiters, all of which served to spread the fire. From the moment fire was first noticed to the time when the interior was a "heap of smoldering ruins," and the exterior walls collapsed inwards, was less than one hour. Sixteen people were killed, but perhaps the reason that the fire made such a large impact on the pub-

lic was that it took place on the afternoon of March 17, during the Saint Patrick's Day Parade. One expert opinion of the disaster was offered by Chief Bonner of the New York Fire Department: "The hotel was constructed under the old building laws of 1873. There was not a fire-proof thing in the place. . . ." The 1892 building code was the first to require masonry cross walls, fire stopping, and fireproof doors in corridors. The general opinion was that fire stops and doors would have greatly slowed the spread of fire even in this wood building.[14]

An editorial in the *Engineering News* carefully explains the concerns that prevented wood construction's continued use:

> The burning of the Windsor Hotel in New York city on the afternoon of March 17 furnishes the greatest object lesson in many years of the great danger to life which exists in every large building constructed in the common manner with wooden beams, studding and lathing and hollow spaces in every floor and partition through which fire can sweep with out hindrance. . . .
>
> This journal is in possession of information which indicates that the fire had been smouldering for hours before it finally broke out. While it was thus smouldering in a confined space, it would naturally generate a considerable amount of combustible gas, which spread through hollow partitions and floors from story to story and which was at once ignited when the flames finally burst out and allowed air to enter. The rapid spread of the flames over the doomed structure is thus fully explained; and it has in fact been paralleled in other fires in buildings of this type of construction where the combustion has begun in the concealed hollows of floors or walls and has gained headway before its discovery.[15]

Even in areas where wood construction was allowed, its use was steadily more restricted. Starting with the 1892 code, New York required that the first and second floors of apartment houses be of fireproof construction. The rationale was that fires were most likely to start in the public areas of the building or in the boiler room, and so these spaces were separated by fireproof floors from the wood-floored residential spaces above.[16] The most interesting exception to this trend was the continued use of wood bearing walls in a form of "fireproof" construction outside the fire zones.

The fireproof wood wall was used only in the construction of low-rise tenements, where there were no fire rating requirements for the interior, only for exterior fire spread. Because of the brief period when such construction was allowed, and the geographic limitations to its use, surviving examples are restricted to the south Bronx, western Queens, and those portions of Brooklyn where tenement construction was active in the 1890s and 1900s, most noticeably Greenpoint and Williamsburg. The exterior walls were constructed of ordinary studs resting at each floor on large girts (up to 4 inches wide by 8 inches deep) in place of the ordinary plates. The spaces between the studs were filled solid with brick, which was the building's fireproofing. The quality of masonry was extremely poor, inadvertently ensuring that the wood carried all of the loads. Even using the fastest slapdash methods possible of installing the brick could not have reduced the cost and required time of this extra step, thus reducing the form's economic advantage over true brick bearing walls.

## Masonry

Masonry wall buildings have existed in New York since its beginning as a Dutch town. The material most commonly used has always been brick, with stone reserved for use in ornament and cornices. A law passed in 1815 required that all buildings in the heavily built-up downtown area be constructed with masonry walls and tile or slate roofs. The line separating this "fire district" from the unregulated area was near Fulton Street; it moved north in 1822 and again in 1829.[17] In the early years of the nineteenth century, when the city was growing rapidly, commercial buildings were commonly built with masonry walls (whether required by law or not) as a symbol of solidity. "Many such buildings were advertised back in the 'twenties as 'fireproof' until the terrific fire in the last weeks of 1835 disproved that hopeful claim by wiping out a large part of the district."[18] After that, the term "fireproof" was not used extensively again until buildings without wood became economically feasible with iron beam and tile arch construction.

Windsor Hotel, 1885, east facade. Built before modern materials allowed "fireproof" construction, the Windsor is of identical construction to the brownstone rowhouses across the street — bearing masonry walls and wood-joist floors, with wood-stud interior partitions.

A more effective means of fire control through the use of masonry construction was beyond the budget of the average building owner. The architect Ithiel Town, to protect his large architectural library, built himself a fireproof house in New Haven, completed in 1836. The library and other portions of the house were built solely of masonry, including the exclusion of wood furring for the plaster finishes.[19] This system was taken to its logical end in the 1841 Merchants' Exchange at Wall Street and William Street, which "was made fireproof by the simple device of omitting from it anything that would burn." The interior floors were all stone vaults, the partitions all masonry.[20]

At the same time that Town was building his own house, he was proposing a unified system for rebuilding the downtown area destroyed by fire. Many of his proposals, such as widening the streets, would have had an obvious benefit in reducing the risk of devastating fires. He proposed that the party walls for five- and six-story buildings be reinforced with buttresses.[21] This would

**Merchants' Exchange, 55 Wall Street, 1860s, north facade. A view of the almost entirely masonry building before it was extended upward two floors using bearing-wall and steel-beam construction in the 1890s.**

**Merchants' Exchange, north facade. View of the building's general appearance roughly one hundred years after the addition.**

not prevent the wooden interiors from burning, but it would reduce the likelihood that the walls, unbraced after the floors burned, would collapse. This proposal was not enacted, and typically masonry walls have not been buttressed in the manner Town described except when exceptionally long.

Masonry bearing walls were rarely, if ever, designed for the loads they carried. In ordinary construction 8 inches, or two wythes, of brick represented the effective minimum wall thickness. In 1830, this was codified as the minimum thickness for party walls.[22] An 8-inch wall, if analyzed for the loads in a three- or even five-story building, is stressed well below the compressive strength of brick. Builders could count on this construction working as it always had, and so could ignore the actual stresses. Trouble occurred only when someone, usually a speculative builder or contractor, ignored the established rules of thumb. There are reports of buildings partially collapsing or being examined during demolition, and found to have bearing walls consisting of one wythe of brick.[23] This is enough material for the compressive

strength if examined in small pieces, but a 4-inch-thick brick wall 8 or 10 feet high is too slender to sustain the level of compression the rule-of-thumb design assumed, especially when lateral wind load is figured in. In addition, a single-wythe wall, built to the normal standards of straightness and verticality, will experience twice the "accidental" eccentricity, producing out-of-plane bending, as a two-wythe wall. These same considerations exist on a larger scale, as two-wythe walls become inadequate for ordinary loads as the unbraced length passes 12 feet, and so on for thicker walls and larger unbraced lengths.

The end of the use of monolithic masonry bearing walls in large buildings can be traced through the events of the 1870s. The decade began with the construction of the [Old] Equitable Building at 120 Broadway. The original design was for a building with interior masonry bearing walls and vaulted floors. George Post received one of his first major commissions in replacing Arthur Gilman and George Kendall, and used the opportunity to modernize the interior construction to brick jack arches supported on iron beams.[24] The

switch was justified as a cost-cutting measure both in terms of the reduced amount of costly skilled mason's work and the speedier erection. As William Birkmire said in defense of the dominance of skeleton frames twenty-five years later: "This question of speed in erection is most important. There being a large amount of capital invested, there should be but one season lost [to construction] before a return is effected upon this investment."[25]

Post and Richard Morris Hunt were the architects in 1875 of New York's tallest buildings with true bearing walls, the Western Union Building and the Tribune Building, respectively. Both had towers well over 200 feet high embedded in a large office block over 100 feet high.[26] Both buildings used wrought-iron floor beams supporting tile arch floors, but were traditional in their treatment of the masonry walls. The Tribune Building was even praised for its pure bearing wall design, for "upright iron supports being cautiously avoided."[27] There were no further structural advances possible in masonry bearing wall construction after this. The type, as then defined, was

[Old] 120 Broadway, 1912: interior courtyard. The wrought-iron beams and cast-iron columns are visible after the fire that destroyed the building. The line of columns at the corner on the right shows the progression in column size down the building. The thin pieces of iron on the near left and at the roof on the right are the mansard framing. The building next door in the center is American Surety (see page 53); the steeple of Trinity Church is in the upper right.

used for tall buildings for another twenty years. As steel frame construction developed, fewer office buildings were built with bearing walls; many of the later buildings are residential and ten to twelve stories high, such as the Chelsea Hotel of 1884.[28] The last major building in New York with pure bearing walls is the United States Archive Building on Washington Street. The full-block building was completed in 1899, and has ten-story-high brick bearing walls designed to carry warehouse loading.[29]

The loss of masonry's most useful niche as monolithic fireproof construction was emphasized in 1877, when Post designed the [Old] New York Hospital on 15th Street as a totally noncombustible building, but in modern materials. The new building replaced a previous nonfireproof building on lower Broadway. The hospital, which thirty years earlier would have had vaulted floors supported on bearing walls, had tile arch floors supported by iron girders and columns.[30]

Masonry was used extensively, and still is, in residential construction for facades and for fireproof party walls. In the early 1880s, at the same time that bearing walls were losing ground in large buildings, there was a rush to build rowhouses on the Upper East Side and in Harlem.[31] These buildings, usually called "brownstones" after the sandstone used for the street facades, were 18 to 25 feet wide, with wood-joist interiors, brick party walls between the houses in a row, plain brick rear facades, and ornate stone veneer over brick front facades. Only rarely taller than four stories above their half-story basements, these buildings suffer from two defects that can be explained best by the builders' rush to put them up and by the beginning of the decline in masonry construction standards. The sandstone is delicate and more permeable than other common building stone, and it has been deteriorating from the day these buildings were built. Sandstone's deficiencies explain the common use of limestone and granite in New York's relatively harsh climate.

The other defect often found in brownstones is the occurrence of continuous vertical joints in the masonry, most often at the corners. Theoretically, if the brick at the joints is not properly toothed together, the wall is tied to the rest of the structure only by the embedded joists. In reality, the continuous rows of houses meant that no one building ever had to resist wind by itself, but this safety cannot be assumed. Over the years, development has often isolated one or two houses, which may be vulnerable to partial wall collapse during construction. These bad joints may be a sign that the most skilled laborers were no longer masons at the end of the nineteenth century. Similar flaws can be found in other masonry construction, including bearing wall churches.[32]

The masonry wall section of the building laws, first written in 1830, grew steadily longer and more complicated with each revision. The legal description of masonry construction was by 1892 the most complex part of the code, with several pages of rules giving the required minimum wall thickness for each possible condition. The complexity was the result of the code's approach to design: the rules are an attempt to give a consistent thickness of brick for similar load levels, whether caused by the height of the building

# CASE STUDY: **BROWNSTONE ROW HOUSE**

The owners of a brownstone on Manhattan's East Side hired a contractor to stop several recurring leaks through the front and rear facades. After examining the building, the contractor recommended an investigation and recommendations by an engineer. The following observations were made on site during the initial investigation and during construction.

The building is a typical New York rowhouse, built to the full width of its 20-foot lot, and roughly 70 feet deep. The house is five and one-half stories high, with a usable basement 4 feet below sidewalk grade. The house is the end of a row, with one of the normally hidden side walls exposed next to an apartment house service alley. The front facade is brownstone (a New Jersey Triassic sandstone), previously painted and patched. The side and rear walls are common brick. Extensions have been built onto the rear of the basement and first-floor levels, creating terraces for the first and second floors. The rear yard is at the basement floor elevation.

Leaks were observed at both rear corners at the third, fourth, and fifth floors, as well as over the first-floor rear window. The front facade was spalling in patches, and much of the mortar on the rear wall and the exposed side wall was loose and lacking cohesion.

Once the roof gutter drainpipes were removed from the rear corners of the building, it was clearly visible that there was no mechanical link between the rear facade and the side walls. On both sides there was a gap approximately ½ inch wide between the brick of the rear wall and the side walls. This discovery accounted for the upstairs leaks.

The rear window at the first floor was a large picture window installed in the relatively new masonry of the extension. There was no lintel used, with the half-dozen courses of brick making up the low parapet at the second-floor terrace simply laid over the window frame. Motion of the brick relative to the frame and

the deformation of the aluminum frame had caused the leak at this location.

The front facade spalling is a well-documented problem with any sandstone but especially common with the brownstone used in New York. The stone is extremely weak parallel to the bedding planes from its sedimentary formation. Even when unstressed, small amounts of weathering will cause failure of the stone along these planes. In the construction of the majority of brownstone rowhouses, the stone was set in the facade with the bedding planes vertical, causing the building to "shed" layers of stone from the stresses of ordinary freeze/thaw cycles.

Finally, mortar used in the nineteenth century often had lime as the primary binding material. This mortar is flexible and weathers well but when exposed to water over long periods of time, the lime leaches out, leaving behind tightly packed sand. This was the observed condition of the exposed mortar.

Having identified the causes of the problems, repair was simple in concept, if somewhat laborious in execution. Brick was removed from both upper rear corners, leaving the adjacent brick "toothed." This work was performed one side at a time, since the building was occupied and shoring could not be used to strengthen the rear of the building during construction. New brick was then laid up and toothed into both the side and the rear walls. The parapet of the second-floor terrace was removed, a steel angle lintel placed over and caulked to the window head, and the brick rebuilt. Since the plan was to repaint the front facade in the same off-white shade it was already painted, the stone was repaired using acrylic patches and painted. Finally, all of the deteriorated mortar was cut out, backer rod placed inside the opened joints, and new mortar with a lime and Portland cement binder forced into the joints and struck for a proper profile.

**Western Union Building, 1885, south and east facades. A good example of the way that the early bearing-wall skyscrapers broke the traditional scale of urban construction.**

**365 West 46th Street, south facade. A smaller-than-average rowhouse, with severe spalling of the brownstone facade.**

ing builders to provide thicker walls in taller buildings.

The 1892 rules provided, generally speaking, for an increase of 4 inches, or one wythe of brick, for each 15 or so feet down from a building's top.[33] Given that the code also required maximum beam or joist spans of 26 feet, the increase in wall thickness was more than required for the tributary floor loads.[34] The additional capacity allowed the walls to act as shear walls in resisting lateral wind loading, although there is no evidence that this was ever consciously designed. Since the rules for wall thicknesses were the same regardless of the width and depth of the building (within certain extreme limits, such as 100 feet width of wall unbraced by cross walls), the lateral load capacity and stiffness of similar-height buildings could vary wildly.[35] A minimum level of stiffness for walls was maintained through provisions increasing the thickness of walls if they were not braced often enough in any one direction.

Several other provisions of the code are of interest. Party walls for dwellings were restricted to a minimum thickness of one foot unless the

above or by wind pressure on the long unbraced rear walls of theaters. By 1890, the range of buildings in New York included tall office towers, large theaters without interior columns, and apartment houses with complicated layouts. These building types are examples of modern architectural types that have been successfully built with modern framing: structural steel or reinforced concrete. The building code's attempt to provide design rules applicable to each type without expressing them in terms of allowable stress within the masonry was doomed to failure in rationalizing the design process, although it succeeded in forc-

A brick rowhouse in midtown Manhattan was purchased by a couple with the intention of using it as a private residence. The building had been built as a single-family residence, but had served for at least forty years as an apartment house with one apartment per floor. The architectural design of the renovation involved very little intentional alteration of the building structure, but the architect wanted an engineer available to study the condition of the building structure as exposed during construction. The following observations were made on site during construction progress meetings.

The house fills the width of an 18-foot 6-inch lot, with approximately one-third of the joists in the center of the house framed to a header at the main stair and the joists in the front and rear spanning the full width. The typical joists are 3x10s, the stair header is a double 3x10, and the stair trimmers supporting the stair header are 4x10s. All of the joist-to-header connections were mortise and tenon with the joist-end tenon approximately one-half the joist depth.

The undersides of the second and third floors were exposed by plaster demolition. Roughly half of the joists showed extensive splitting, typically at the mid-height of the joists near the ends and near the bottom at the center. The most heavily damaged joists were in the center portion of the house, adjacent to the

stair, where the bathrooms and kitchens had been located. Many, but not all, of the splits were accompanied by water damage to the wood. The rear stair trimmer was cut at some time to allow passage of a drain pipe: the remaining section at the cut was roughly 3 inches high. The front stair trimmer was near collapse, with a split extending from the bottom to within ½ inch of the top.

The type of damage strongly suggested overloading, but no ordinary loading in an apartment would overstress joists of this length so badly. Close examination of the wood showed numerous, small, and naturally occurring flaws and checks. It appears that the wood used for the joists may have been of low quality, with corresponding lower allowable stresses.

Given the poor condition of the wood, repair of the joists would be more expensive than replacement. The stair header, which was in good condition, was kept. The damaged joists framing into it were replaced, with the new joists using the existing pockets in the brick wall on one end and attached to the header with light-gauge steel joist hangers on the other. Existing connections were reinforced with joist hangers. After the joists were replaced, the center portion of the floor was shored and the trimmers replaced one at a time. After the new flooring and ceiling are placed, the repairs will be invisible.

---

buildings were 12½ feet wide or narrower, when the party walls could be 8 inches thick.[36] With almost no exceptions, this provision meant that party walls are one foot thick. The code required that "[a]ll buildings not [wood] frame or wood [stud] shall have walls of stone, brick, iron or other hard incombustible material."[37] During the last half of the nineteenth century there was a legal geographical limit, known as the fire line, south of which wood construction was prohibited. Given the relatively high price of stone or iron for ordinary construction, the code forced most common buildings to be constructed of brick.

Finally, the code made provisions for construction other than simple walls. The innermost 4 inches of the required thickness could be hollow tile, to permit easier plastering. The same volume of material required by the provisions on thickness could be collected into independent piers, to permit larger window openings. "Hollow" walls were allowed as long as the amount of material used was the same as required for solid walls and the two portions of the wall were tied together at a maximum of 2 feet on center. From the existing buildings, it appears that no architects took advantage of this provision to build cavity walls; the

The Long Island Historical Society, 128 Pierrepont Street, Brooklyn. An early use of terra-cotta ornament. Unlike some later terra cotta, which was cast in unique forms, the material here is used to mimic carved stone.

early, before the technology was economically viable.[39]

George Post was one of the earliest heavy users of terra-cotta ornament. Between 1877 and 1880 he specified it for the exterior ornament of a small building on 36th Street, the Long Island Historical Society in Brooklyn Heights, and the Produce Exchange downtown. Starting around 1900, McKim, Mead & White favored terra cotta, again for economic reasons.[40] The other argument used for terra cotta was its resistance to fire. Obviously, terra-cotta ornament does not make a building any more resistant to fire than stone ornament, but in cases such as the Potter Building, erected on Park Row in 1883 on the site of a building destroyed the previous year by fire, the terra cotta was almost a symbol of the interior fire protection. The Potter Building was the first in the city literally covered in terra cotta, with hardly any brick showing.[41]

## Present Considerations

Wood, as the only biological material used in construction, is subject to gross destruction from microorganisms. The mechanisms of insect infestation and fungal attack such as rot and dry rot are well documented in sources on architectural preservation and material conservation. Structural wood elements can be reevaluated as reduced sections if portions are removed due to the causes listed above, but this is rarely worthwhile. Under ordinary circumstances, joists, beams, studs, and posts that have been degraded are replaced, in kind for historical preservation purposes or, occasionally, with modern equivalents such as light steel beams or metal studs.

Old wood connections are always suspect. The mortise and tenon joints so often used for joist support have a tendency to split the supported joists along the line of the notch. The supporting girder is often greatly reduced in section at the connections, and inadequate in shear, bending, or both. Finally, the length of the tenon is often too small for acceptable bearing stresses. All of these potential defects should be of concern during the renovation or investigation of wood floors. Older bridle irons are sometimes recessed into notches cut into the wood, creating the same sort of stress concentrations bridles are supposed to prevent.

only hollows normally found in the walls of older buildings are pipe chases. Curtain walls were allowed, but were restricted to a minimum thickness of 4 inches thinner than a bearing wall's thickness in the same location. This last provision was the product of a Department of Buildings familiar with masonry construction examining the new technology of steel framing, and not trusting it.[38]

While the details of the outer face of the masonry walls are better left to architectural history, one nineteenth-century development had implications still felt today. Architectural terra cotta was introduced for the first time by James Renwick in the Tontine Building of 1853 for the ornamental belt courses and cornice. Renwick was hoping that the reuse of molds would make terra-cotta ornament cheaper than "freestone," the ordinary carved limestone or granite ornament. The idea was correct, although Renwick tried it too

One characteristic of old wood structures that often appears to be a problem, but rarely is, is movement over time. Assuming that the building's foundations are adequate, a large amount of apparent settlement and movement will be caused by the shrinkage of the various wood members and the sagging of joists and beams. This is ordinary behavior and, unless accompanied by other symptoms such as splitting or analyzed overstress, can be ignored. It is usually safer to live with sagging joists than to attempt to straighten them, since straightening is difficult to accomplish and is liable to cause splitting and crushing of the wood.

All masonry can be eroded by the action of water, wind, and flying debris, cracked by thermal movement and freeze/thaw cycles of trapped water, and chemically attacked by acid rain and other pollution. As with common wood problems, these are discussed in sources on architectural preservation and material conservation. The structural significance of these conditions varies. Even small defects in the surface of thick walls can be counted as a structural problem if pieces of the masonry can fall. In New York, the death of a pedestrian from injuries caused by falling masonry in 1979 led to the passage of Local Law 10/80, which requires continuous facade maintenance programs or regular facade inspections by a licensed architect or engineer followed by any "ameliorative work" specified by the inspector.[42] Whether or not the facade is due for inspection by law, anyone involved in work on a building should be aware of the possible problems.

In extreme cases, the original masonry construction may have hidden flaws, such as the con-

**McAlpin House, 34th Street and Broadway, east facade. Exposed beams carry wind framing across the light court at lower floors; repairs to the brick show where spandrel beams required repair.**

tinuous vertical joints mentioned above. Any time that such design defects are uncovered during renovation, their effect on building stability must be taken into account. Often, the demolition of adjacent buildings weakens bearing walls, and the now-uncovered party walls bulge or crack. Such walls are also usually more permeable to water than walls meant to be exposed. Through this mechanism, a sound, one-hundred-year-old building can suddenly develop water infiltration problems through open joints.

# 3. CAST-IRON FACADES
## 1847–1875

In the first third of the nineteenth century, iron construction of various types was in use in Europe but had not yet made inroads in the United States. Its first use in New York was in minor architectural elements such as ornamental balconies and railings and store window frames. The use of cast iron in storefront elements suggested the next logical step: the iron structural storefront.

### Invention of the Cast-Iron Front

The first structural iron facades were simple post-and-lintel frames that supported the lower floors of commercial buildings. At the Lyceum of Natural History on Broadway and the Lorillard Building on Gold Street, both constructed during the 1830s, brick front walls rested on iron girders at the facade base.[1] There was essentially little difference between the form of these columns and lintels and those used later in full fronts. The members were cast in one piece and bolted together through flanges in the rear.[2]

The man often referred to as the inventor of the cast-iron front was James Bogardus. After a trip to London where he saw the structural use of iron, Bogardus patented in 1849 a complete iron building composed of individual columns, beams, and infill panels. The columns and facade were all cast iron. The floor beams were wrought iron, and the girders were cast iron with wrought-iron tie rods. The floor plates may have been cast or wrought. All were to be fastened by bolting flanges projecting from the rear or open interiors of the members.

Bogardus never ran a production foundry, but he needed a factory for casting prototypes. His factory at Centre and Duane streets and a warehouse building, the Edgar Laing Stores, at Washington and Murray streets, were the first buildings he designed to be built, in 1848. The two buildings were very similar in appearance. Four stories high, two facades and a corner used the narrow, repeating window module that became cast iron's trademark. The buildings did not follow the patent in several significant ways: The interior floors were supported on wood joists, and masonry walls were used both in the interior of the buildings and for two exteriors. These exceptions left the buildings closer to traditional construction than to the iron ideal described in the patent.[3] In general, the vast majority of "cast-iron buildings" built had masonry walls and wood-joist or iron-beam and masonry-arch floors with one or two, and occasionally three or four, cast-iron facades. Cast iron was thus reduced from a new type of building to a new skin for a typical building.

Because Bogardus contracted out his production work, several large iron works around the city became known as building foundries. These included the Janes, Beebe Company, James Jackson, the Cornell Brothers, and Daniel Badger's Architectural Iron Works.[4] All of these companies also produced iron fronts of their own design. Badger used iron in the most creative ways, and his work is the best known today.

Badger had started his business in Boston and moved to New York in the mid-1840s. Before his association with Bogardus, his business consisted largely of making storefront elements, specifically iron shutters that rolled in tracks contained within decorative columns surrounding the windows. These shutters were similar to modern rolling gates, except that they were better integrated within the storefront design. One of Badger's first commissions in New York was the [Old] A. T. Stewart Department Store in 1846 at Broadway and Reade Street. Badger was responsible for the ground-floor storefronts and some interior structure.[5]

Badger built his New York foundry the following year.[6] The Architectural Iron Works continued in operation past the end of the century, gradually adapting by concentrating less on cast iron and more on wrought iron and steel. The company's catalog, published in 1865, showed cast iron for all uses, including beams of inverted "T" shape and unequal "I" shape (with a larger bottom flange), columns for facades (including those with side cutouts for rolling shutters), and "tension-rod girders" for long spans consisting of a cast-iron beam or truss reinforced in its bottom flange with wrought-iron rods.[7] The last item, a specialty of Badger's, was a tacit admission of cast iron's unsuitability for flexural tension.

Cast-iron construction became popular between 1850 and 1870 through Bogardus, Badger, and other backers emphasizing three points: safety during fires, additional light entry into the building interior, and speed of erection.[8] Because entirely iron buildings were not built in New York, cast iron's advocates could not boast of eliminating superstructure masonry work and thus increasing the speed of on-site construction to the speed of bolting together the iron. Bogardus's second factory, also on Centre Street and completed about 1854, had four iron facades, cast-iron girders, and cast-iron columns, with no masonry.[9] This is a fascinating exception to the general form, mostly because of the question of lateral stability. The bolted connections in cast-iron facades are comparatively very weak; without masonry shear walls, they would be carrying all of the wind load on the building. It is likely that wrought-iron cross-bracing was built into the walls, as Bogardus used it for special structures such as his skeletal fire watch towers. The one surviving fire tower, in Marcus Garvey Park, is of cast-iron construction except for the presence of wrought-iron cross-bracing.

The use of a cast-iron front could gain a small amount of usable space inside a building. The total thickness of the nonornamental portions of the metal facades rarely exceeded one foot back from the building line. Cornices and ornamental trim typically extended out past the building line, and so can be neglected in computing the usable floor space. Masonry exterior walls, even facades parallel to the joist floors that could be considered lightly loaded walls, were often built 16 inches to 2 feet thick. The floor space added might be small, but it was measurable, especially in terms of the well-lit space provided.

One of the true innovations of cast iron was the introduction of large windows throughout commercial facades. Since the designers were relying on direct bolted connections of the individually cast elements, stability did not have to be achieved through mass, as in masonry facades. The reasoned analysis of stresses in the iron allowed relatively slender columns and shallow spandrel beams. The infill within the iron grid could be any material, and usually ended up being glass. The 5-foot-wide, 6- or 8-foot-high windows made possible by the iron fronts dramatically increased the amount of light within the buildings, an effect made more noticeable by the relative thinness of the facade. Windows on the lower floors of old masonry buildings, in addition to being smaller, were set into a thick wall, thus giving a "light at the end of the tunnel" effect. Given the relatively poor illumination provided by gas light during the iron

**Hugh O'Neill Store, 655 Sixth Avenue, east facade. Cast-iron facade on a commercial building far larger than the typical 20- or 40-foot-wide cast-iron warehouse.**

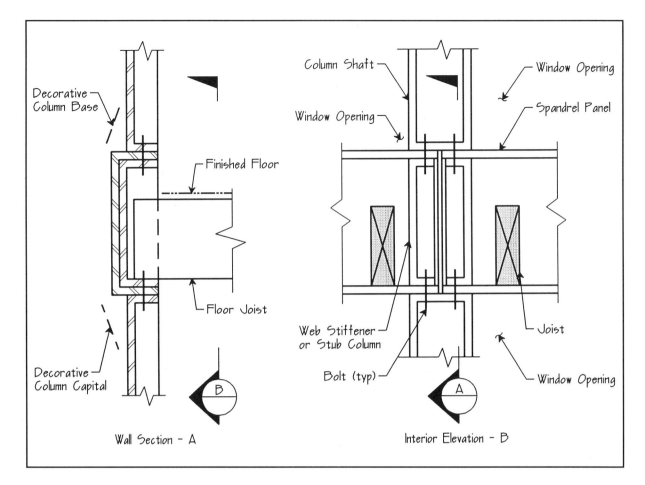

Column Shaft

Window Opening

Window Opening

Spandrel Panel

Decorative
Column Base

Finished Floor

Floor Joist

Decorative
Column Capital

Web Stiffener
or Stub Column

Joist

Bolt (typ)

Window Opening

Wall Section – A

Interior Elevation – B

**3.1. Cast-Iron Bearing Facade Connections Scale: 1" = 1' 0**

**Haughwout Building, 490 Broadway, south facade. One of the best surviving cast-iron-facade buildings, with two complete iron fronts.**

construction period, it is not surprising that cast iron's proponents always mentioned superior daylighting as one of the material's advantages.[10]

Architects and engineers argued over the relative merits of iron even as its use spread. In 1864, the argument was made that cast-iron and wrought-iron beams and columns were preferable to wood because they could be shaped to best resist the forces applied to them, by providing flanges on beams and hollow box shapes for columns.[11] The unspoken portion of this argument is more interesting to us: that it was still necessary so late to convince building design professionals that iron should be used for structure.

The application of iron technology to building interiors developed in a surprisingly different way from the development of cast-iron fronts—although they began concurrently. By 1860, cast-iron fronts had reached their final form. Two remaining examples of early cast-iron fronts are 254–260 Canal Street, at Lafayette Street, both in 1857 and attributed to Bogardus, and the Haughwout Building by Badger at Broadway and Broome Street. Both are five stories tall, with two iron facades, two masonry exterior walls, and wood interiors. The facades use the under-6-foot-wide repetitive column and window module, typical at the time.[12]

Around the same time, the Trenton Iron Works

in New Trenton, New Jersey, belonging to Peter Cooper and Abraham Hewitt, began rolling wrought-iron shapes large enough to act as floor beams. The first batch, 7 inches deep with a round top flange and an overall profile similar to a rail, was used in the Harper & Brothers Building at Franklin Square, the Cooper Union Foundation Building at Cooper Square, and the New York Assay Office on Wall Street. The Cooper Union and the Assay Office were built with masonry exteriors; the Harper printing plant was built with a cast-iron facade designed by Bogardus and cast by the Jackson foundry. The wrought-iron floor beams at Harper & Brothers were carried on cast and wrought composite trussed girders, and carried a brick jack arch floor.[13]

The Harper & Brothers plant was the last building to represent development in both cast-iron and rolled-iron technology. Wrought iron, and later steel, evolved continuously into the twentieth century; cast-iron facades had reached their final form. The difference is clearly shown by the landmark change of 1860: the Trenton mill and the Phoenix Iron Company of Pennsylvania

both began rolling true I-beams.[14] Cast iron had nothing comparable to show.

## Use of the Cast-Iron Front

The structural form of cast-iron fronts changed very little from its beginning to its end. Innovation took place in the use of cast iron and especially in the use of wrought iron between 1850 and 1880, but the cast-iron building stayed the same. The existing rows of cast-iron fronts in SoHo can be dated through their architectural styles, as that aspect of the buildings continuously evolved, but the different styles were all applied to five-story buildings usually on 40-foot-wide lots with interiors of wood joists supported on wood beams supported by the side brick bearing walls and a row or two of interior cast-iron columns. That description can be applied so monotonously to cast-iron buildings that descriptions of large numbers of buildings usually discuss only the decorative influences seen in the facade, the rest of the description being assumed.[15]

Examined closely, the general form of the cast-

**The Bennett Building, 93 Nassau Street, between 1875 and 1880, north and east facades. The Bennett Building is the large building with a mansard roof on the right side of the picture. At this time it was a fairly typical cast-iron facade office building. Later additions brought its height to ten stories. Almost all the surrounding buildings shown in this picture have since been replaced.**

iron building is quite traditional. If described as a relatively non-slender building of wood joists and bearing walls, it is not noticeably different from the form of buildings built in the first half of the century. During cast iron's heyday in the 1860s, masonry buildings were also regularly being built to five- and six-story heights, showing that cast iron was not the driving influence toward height the way that steel would be during the 1890s.[16]

One interesting exception to this general rule is the Bennett Building of 1872 at 93 Nassau Street. The building was out of the ordinary when built: because of the relatively small blocks downtown, three sides of the building were street facades and built of cast iron. In addition, the building was six stories high, as opposed to the standard five. Later, the entire building was made four stories taller, with cast-iron facades matching the originals.[17] 93 Nassau Street is between the concentration of newspaper offices at Printing House Square at Nassau and Park Row and the banks and brokerages at Nassau and Wall Street. The land was (and is) far more valuable than land uptown in the shopping district or in semi-industrial SoHo. The demand for office space on a valuable piece of land seems to have provided the impetus for inno-

vation that did not otherwise exist.

Nothing dramatic happened at the Bennett Building, either in terms of structural nonperformance or design innovation. The iron facades of cast-iron buildings are less well suited to carrying lateral loads than masonry side walls. A cast-iron building with three iron facades would be significantly less resilient than a similar building with one or no iron facades. The wind force applied is approximately proportional to the exposed wall's height, but the overturning moment induced at the building's base is proportional to the square of the height. (The wind force is actually proportional to the building wall's *area*, but for approximately rectangular buildings without setbacks, the difference is negligible. The vast majority of buildings under discussion here, and the vast majority of buildings built in New York prior to 1870, have approximately rectangular elevations and footprints.) Theoretically, the wind stresses at the base of the Bennett Building would be approximately four times those of a typical five-story building. The fact that there were no problems reported can be attributed to the inherent conservatism of the live loads used and the stress levels allowed.

The conservatism of cast-iron design can be seen in the safety factors employed. In 1887, the recommended safety factor from one authority for cast-iron "columns made of the best iron, perfectly molded and with both ends turned" was 6. The recommended practical safety factor of 10 "has been adopted to allow for imperfections in casting: such as air-holes, unequal thickness of metal, etc., deviation of pressure from axis of columns, and the effect of lateral forces accidentally applied."[18]

Despite the metal's limitations, almost any part of a building in the 1860s or 1870s could be made from cast iron. The following list, from 1876, was based on those elements actually being cast and sold. The list is instructive in how it approaches the use of the different members, especially when the typical interior construction of the period is considered, in which wrought-iron beams and girders rested on cast-iron columns.

Castings for building purposes consist usually of foot plates for columns, sill courses, with or without panels, for illuminated tile [cast-iron floor plates with small prisms of glass inset to allow light into the interior space below], the plates for the tile of the various forms used, columns for the interior and exterior lintels, box, tee, or skewback cornices, window sills and caps, spandrel fillings, and the various pieces for ornamentation which are often cast separately and attached afterwards, crest railings, stair ways, both straight and circular, and girders both straight and arched.

Foot [base] plates are used to support other work on the top of masonry, and for columns are usually rectangular or square, from 1 to 4 inches thick in the center, bevelled off from ½ to 1½ inches at the edge. A boss or cross is sometimes raised in the center, and the plate faced down in a lathe outside of this projection for a bearing for the column. If there is no projection, the upper surface may be planed instead of faced in a lathe. These plates are usually set in cement, but if the top of the stone work is sufficiently well dressed (bush-hammered) a sheet of lead is a better, though more costly bearing. Great care should be exercised in setting the plates for columns truly level, and a very common practice of trueing up columns by the interposition of thin pieces of sheet iron between the plate and the column known as "shimming" should not be tolerated. . . .

Iron columns are made either cylindrical or rectangular, and sometimes are left with an opening along one side to be filled with brick work . . . commonly square, and if the panelling is to be thin it will be advisable to make the column in vertical sections and bolt them together by flanges, as the shrink age very often renders them unsafe. . . .

Box lintels are used to carry walls over openings in lower stories. They usually consist of a broad bottom flange with two or more upright tiles, and their strength is computed as for a Tee beam.

Skew back lintels are Tee-shaped girders with one or more vertical webs, with end plates, between which a brick arch is turned, and the idea has been that they only acted as tension rods to carry the thrust of the arch, but in reality any settling of the arch leaves the weight resting on the girder and their strength should be found as for a Tee beam. Hodgkinson's double tee beams and arch and tension rod girders are used, and in our opinion have all the bad features which Sir Wm. Fairbairn finds in trussed cast iron beams. . . . When a Tee or Hodgkinson beam is not strong enough, wrought iron should be used.[19]

The Bennett Building, north and east facades. View after restoration of both the original building and additions.

One of the largest iron facades ever built was the A. T. Stewart Department Store (later Wanamaker's) at Broadway and 9th Street. The building covered the entire block through to Lafayette and 10th streets to the height of five stories, for a total of over 300,000 square feet. There was no true wind bracing system, the building being braced by a few masonry walls, but it was relatively stable due to its squat profile. The iron facades were the typical pattern of horizontal spandrel beams and vertical columns with ornament overlaid; the interior floors were wood joists supported on wrought-iron beams.[20]

## Cast Iron and Fire

One of the properties of cast iron heavily promoted by its early popularizers was its nonflammability. Fire was as great a concern in cities during the 1850s and 1860s as it had been earlier—possibly greater, considering the increasing density of the city. The hope that iron could provide the key to fireproof construction was expressed before the Royal Institute of British Architects in 1864:

As a material for the construction of houses, iron offers the inestimable advantage of being incombustible[;] could we then eliminate the woodwork, we should be saved not only the annual expense of insurance against fire, but relieve our minds from the unpleasant thought of being burned to death.[21]

Bogardus's introduction in New York of the full cast-iron front came in 1848, three years after a fire had destroyed 300 commercial buildings downtown, and thirteen years after a fire destroyed almost 700 below Wall Street and east of Broad Street.[22] The buildings destroyed typically had wood interiors and masonry exterior walls:

> A building three to five stories high was the usual habitat of the business concern. Ordinarily it was of brick, though a few of the more pretentious were of granite from Quincy or the Maine coast. Many such buildings were advertised back in the 'twenties as "fireproof" until the terrific fire in the last weeks of 1835 disproved that hopeful claim by wiping out a large part of the district.[23]

In light of these events, Bogardus's claims that cast-iron construction was fireproof were bound to strike a nerve.[24] Cast iron's use and the consideration given to the material elsewhere were apparently secondary considerations in New York to the ability of the local iron industry to promote itself and the memory of the two fires.

Various fires, including the huge Chicago fire of 1871 and individual building fires in Manhattan, gradually highlighted the deficiencies of exposed cast iron. The beginnings of scientific study of fireproofing and fire prevention quickly called attention to the dangers of exposed cast-iron columns and nonfireproof facades (even if noncombustible).

The degree of safety against damage by fire given by cast-iron construction was questioned even before the Chicago fire. Bogardus's style of construction was described by the press in Europe, specifically in Britain, but the London building laws prohibited similar construction. The specific concern that prohibited cast-iron use in London was the possibility of collapse of iron facades near fires from thermally induced stress.[25]

As cast iron became more widely used, it was inevitable that fires would take place in cast-iron buildings. Early on, there was no conclusive evidence for or against the material. If fire destroyed cast-iron buildings in the 1860s, it also destroyed traditional masonry buildings. The reason is simple. Fires in buildings typically destroyed the interior furnishings, flammable finishes such as varnished wood flooring, and the wood joists used in most buildings before the introduction of terracotta floor arches.

Cast-iron buildings downtown typically were used as offices; those in SoHo and on Broadway or Sixth Avenue were often stores or warehouses full of fabric. The building type that had been described as "fireproof" proved in its built form to be so fire-prone that the area now called SoHo, while officially known then as the Lower West Side, was referred to as "Hell's Hundred Acres." Badger and Bogardus, in their descriptions of cast-iron construction as nonflammable, meant buildings built entirely of cast iron. There was no fire protection gained in replacing one or two exterior masonry walls of a wooden building with cast iron.

The claims made for cast iron's performance in fire can be examined only on their merits in the case of totally inflammable buildings. Bogardus's patented cast-iron building was never built; the closest approach was the Watervliet Arsenal of 1860, ten miles north of Albany, New York. The Arsenal was built by Badger's Architectural Iron Works, chosen for their ability to erect a building meeting the United States Army's requirements quickly and cheaply. The building is basically a shed, 100 feet wide and 196 feet long, with walls and a roof of plate iron. The gable roof is supported on wrought-iron trusses with adjustable tie rods supported on cast-iron columns. The interior girders are "composite" beams, cast-iron beams reinforced at their tension (lower) flange with wrought-iron tie rods.[26] This type of beam was a Badger specialty known as a "tension-rod girder."[27] The Arsenal has had a relatively uneventful life as a cannon assembly plant, which of course neither proves nor disproves cast iron's merits. Ordinary building-use programs do not allow the freedom that existed in Watervliet: if fireproof construction required bare iron walls and roof trusses, it was not going to become popular for use in offices or residences. None of the features of the Armory which render it unique in American construction were new, or unique if examined individually. Almost all of the technical innovations in cast iron construction were based on earlier British and French work.

In the absence of organized testing of materials, the professional perception of cast iron, or any other material, was based largely on anecdotal data. The advantage to basing one's opinions on individual examples proving a point is that almost any point can be proven. The fact that the stories of cast iron's excellent fire performance that can be found before 1880 were gradually replaced by stories condemning the metal is not an indication of a deterioration in its properties but rather a shift in its image. When the wooden interiors of 1–5 Bond Street burned in March of 1877, the cast-iron facades collapsed from the fourth floor to the roof while they remained intact below. Given the gutting of the entire interior and the fact that hot air rises, this is not a remarkable outcome. The four floors of standing iron surrounding empty space can be seen as either vindication of the fireproof properties of iron or, as the *New York Times* stated in an editorial shortly after the fire, "the utter uselessness of the prevailing system of 'fireproof' architecture."[28] To a modern engineer or architect, this fire raises questions about the need for fireproofing building interiors, or about the value of a "fireproof" exterior surrounding an empty burned shell. The Bond Street fire and others with similar outcomes were used as evidence of the fire resistance of cast iron as late as 1956, when the vast Wanamaker Department Store at 10th Street and Broadway burned. The full-block store, built as A. T. Stewart's in the 1860s, had a typical interior of wood plank floors on wood joists supported by wrought-iron beams, cast-iron columns, and exterior cast-iron facades. The fire lasted almost one day, destroying the entire store except the iron facades. The iron showed greater than expected resistance to cracking when, heated by the fire, it was suddenly cooled by water from firemen's hoses. Despite this resistance, the question of how Wanamaker's was in any way made fireproof by its iron construction must be asked.[29]

Some of the most regular users of cast iron recognized the fire risk early in the iron-front boom.

**625 Sixth Avenue, south facade. Close-up view of cast iron meeting masonry one bay from the corner of a cast-iron facade on a large commercial building. The substitution of cast iron for masonry clearly shows that "cast iron construction" is typically just a facade that was applied to a building otherwise identical to any other bearing-wall construction.**

In 1860, John B. Cornell patented a double cast-iron column with the annular interior space filled with "fire-resistant clay."[30] The concept of a decorative iron shell protecting the load-bearing column was later enshrined in the New York City Building Laws as a method of providing an exposed fire-resistant column. Cornell was an owner of the Cornell Iron Works, one of the major suppliers of cast-iron fronts often subcontracted by Bogardus. While his company continued for years to sell cast iron for building construction, fire resistance must have been a great concern to prompt this type of innovation.

As long as iron was considered a fireproof material the columns were generally made of cast-iron, unprotected, with ornamental capitals and bases. The first attempt made to protect columns against the effect of fire was through the use of a double shell, one within the other, the intervening space being filled with plaster. This system was patented, and was required by the New York building laws in an effort to lessen the danger liable to occur from the use of unprotected columns carrying interior brick

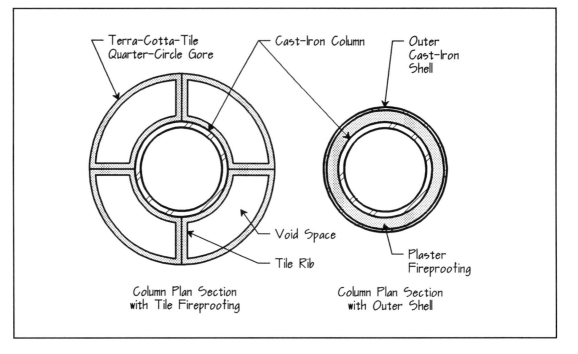

Terra-Cotta-Tile Quarter-Circle Gore — Cast-Iron Column — Outer Cast-Iron Shell

Void Space

Tile Rib

Plaster Fireproofing

Column Plan Section with Tile Fireproofing

Column Plan Section with Outer Shell

walls. The freezing up of the plaster filling, however, in some cases, caused the abandonment of this detail, and the air-space between the two cases was then relied on. Such provision may still be found in the New York building laws.[31]

The double columns later in general use in interior construction could not be easily shaped to meet the geometric requirements of facade members. The building laws as late as 1893 allowed an exception to the requirement of double columns for wall support for "a wall fronting on the street, and columns located below the level of the sidewalk which are used to support exterior walls or arches over vaults . . ."[32] Cast-iron fronts thus escaped the only iron fire-resistance provision in the code.

The turning point in the public's perception of iron's safety was the great Chicago fire of 1871. Obviously, any fire that leveled the major portion of a city built in many different materials cannot be said to have specifically shown flaws in cast iron; and recently claims have been made that Chicagoans, in rebuilding, had no prejudice against cast iron.[33] The fire set in motion the thinking that eventually led to modern fireproofing techniques, specifically that mere use of noncombustible materials was not enough. The definition of a fireproof building that grew out of

building-material research following 1871, and was firmly in place by 1920, was the capability of withstanding the full effects of fire, including intense heat and thermal shock from cold water for enough time to protect the building's inhabitants during their escape, and possibly enough time for the building's flammable contents to be extinguished. Past experience did not recommend cast iron, and the full effect of fire proved to show the metal's greatest weakness.

Metal exposed to the heat of a major fire expands far more than masonry or wood. The subsequent distortion can be ignored for decorative building elements and is one of several concerns in architectural elements, such as window frames, that affect the building's fire performance. But it is deadly for load-bearing structure. During the Chicago fire,

[c]ast-iron, for the fronts of buildings, in streets less than a hundred feet in width, was found to be most treacherous. In several instances these fronts expanded and buckled and fell into the street from the effects of the intense heat radiating from burning buildings on the opposite side of the street, long before their combustible interiors had taken fire.[34]

This is exactly the scenario that had prevented the use of exposed cast iron in London. Individual

## CASE STUDY: **CAST-IRON COLUMN CAPACITY**

In every project that involves altering the loads or modifying the connections in cast-iron columns, determining the capacity of the columns and their connections becomes a chore for the designing engineer. The following method description is based on three retail renovations in New York City.

Without drawings, only the outside diameter of the columns can be easily determined. This information effectively sets a maximum and minimum allowable load, based on the common maximum and minimum wall thicknesses, but these limits leave too much room for error in design. Of the various high-tech methods of nondestructive testing, ultrasound is well adapted to finding the thickness of a piece of metal for which there is only access on one side. Ultrasound is, however, an expensive procedure that is only available from specialty testing labs.

A low-tech testing method that is destructive in the narrowest sense of the word, but nondestructive in a broad view of the possibilities, is to drill small holes in the column and insert a ¼-inch-wide tape measure. The loss of the material at the holes is not significant, especially if the precaution is taken of drilling the holes near a floor level. Since all cast-iron columns are pin-ended by the nature of their connections, any bending stress from nonconcentric loading will be at a minimum at the ends.

Two holes, located at the ends of a column diameter, are required because the inner and outer circumferences of the column may not be perfectly concentric. If the two holes show identical measurements of thickness, that value can be used; if different thicknesses are shown, the average should be used.

Drilling a material as brittle as cast iron is a relatively safe procedure as long as certain precautions are taken. Impact drills should not be used, even though they are effective at converting the piston action to a rotary drill motion. Some of the impact is transmitted to the drill bit, and then to the material being worked. In addition, drill bits should be changed frequently or the bit allowed to cool to ensure that the material is not subject to severe thermal stress created by the bit friction.

The stiffened seat and web lug connections of beams to the columns can usually be exposed to view to allow measurement of all the various pieces. Without a guide to the original design of the connections, if such a design actually existed, a safe upper limit is the shear capacity of the cast seat stiffener plate.

---

building fires demonstrated that city-wide infernos were not necessary to cause enough heat to distort a cast-iron front beyond useful limits. In November 1895, a fire in the Keep Building in lower Manhattan destroyed the Manhattan Savings Bank across the street. Chief Bonner of the Fire Department of New York explained the collapse:

> The heat from the Keep Building on the opposite side of the street acted directly upon the exposed ironwork of the Manhattan Building. The iron resisted the heat, that is, it did not blaze, but so far as the safety of the building was concerned it did something infinitely worse. It expanded under the heat and forced the ends of the beams and girders from their resting places on the supporting piers. The result was inevitable; the floors came down and brought down with them the mass of fire brick used as flooring, the floor-columns, and finally the roof.[35]

It must be noted that the bank was far more fireproof than an ordinary cast-iron-front building: the interior was composed of noncombustible floors. The bank, "while far from representing the best fireproof construction, would ordinarily be called fireproof by builders, landlords, and the public generally. . . . [Bonner] was quoted as saying that there was not then a thoroughly fireproof building in New York City."[36] By this time, cast iron had fallen out of favor for new construction—although fireproofing is obviously an issue that exists during a building's entire life, and should therefore still be a concern in remaining cast-iron buildings.

As late as 1893, William Birkmire of the Architectural Iron Works could still claim that "[c]ast-iron of goodly thickness offers a far better resistance to fire, or fire and water combined, than wrought-iron or steel."[37] This is true only if the metals are directly exposed to the fire's heat, which would usually only be true of cast iron.

The different responses of cast iron to fire, standing in the Wanamaker building with fire only a few feet away and collapsing at the Manhattan Savings Bank with a street separating the metal from the flames, must be attributed to the condition of the buildings. Cast-iron fronts "relax" with time, from either small cracks in the bolted flanges or the loosening of bolts. An older cast-iron front is capable of more movement without the creation of secondary stresses because the individual castings are not held so tightly together as when newly bolted. This reduces the capacity of an old cast-iron front to resist ordinary lateral loads, because the play must be taken up by movement before the facade acts as a unit, but it allows some expansion to take up the play before secondary compressive stresses are formed. The Chicago fire would also have been acting on relatively stiff cast-iron fronts, with an amount of heat far greater than that caused by single building fires.

## Obsolescence of Cast-Iron Facades

By the late 1870s, developments in building technology were leaving cast iron behind. From their beginnings at the Cooper Union and Harper & Brothers, rolled floor beams had quickly become popular. They represented the only long-span technology for buildings at that time, were capable of supporting the relatively heavy fireproof floors becoming available, and were steadily dropping in price as competition and the mill technology both developed. Wrought iron continued in use through the 1880s, with the use of steel starting in the late 1870s.[38] The number of sections available grew steadily, as did the depth of the heaviest sections. Wrought-iron beams were being rolled as deep as 15 inches by 1875; so they were the only practical method for carrying brick arch floors for spans over 20 feet.[39] Cast iron could not develop in a similar fashion because of its nature. Even if

methods of casting could be developed that eliminated the flaws in castings (and the shortcuts of unscrupulous foundries), the metal would remain brittle, and so would be limited in its safe use.[40]

Stone, brick, and terra cotta were the preferred materials for the facades of residential and office buildings. They represented solidity in architecture in a way that iron could not, although they were increasingly supported by metal frames. The development of tall and slender office buildings after 1875 forced examination of the load-carrying frames of buildings, which sped development of the steel frame.

Despite the continuing architectural character and interest of the cast-iron front, the building type had reached a dead end in structural development. The disappearance of cast iron by the end of the century was the result of lessons learned about fire protection, the economic competition from steadily improving steel frame and curtain-wall design, and the increased need for building heights impractical in cast iron.

Cast-iron columns continued in use, especially in relatively small buildings, past the turn of the century—about twenty-five to thirty years after the boom in cast-iron facades passed its peak. There were several reasons for the different patterns of use. Interior columns could be fireproofed as easily as columns of steel; the first interior column fire protection of terra cotta was designed for round cast-iron columns. In bearing-wall buildings, where masonry walls provided the lateral load resistance, the interior columns supported only floor loads. In those cases, where columns would not be subjected to bending from sidesway, cast-iron columns could be considered as stressed only in compression, to their strength. In small masonry buildings with interior iron columns, there were no examples of inherent weakness in design, as shown in the Keep Building fire.

The 1891 building laws had rules defining where fireproof construction was required, either by location within the fire district or by overall size. The old cast-iron districts were by this time well below the fire line, and "fireproof" included noncombustible interiors as well as exteriors. The option still existed for exposed iron exterior walls in fireproof construction.[41]

Cast-iron fronts could still be constructed,

although certain restrictions in the laws limited the freedom that the material had provided earlier. The laws required that the "dimension of each piece or combination of materials required shall be ascertained by computation . . . [based on the theory of] the strength of materials." The required minimum factors of safety for all materials were 3 for bending, 6 for compression, and 6 for tension tie rods.[42] The generally accepted factor of safety for cast iron was higher, varying from 6 to 8. It is likely that some of the existing iron bearing walls supporting floor framing do not meet this standard for factors of safety, or for rigorous structural analysis, especially at their connections. In addition, there was a prohibition on cast-iron beams longer than 12 feet. Above that cutoff, wrought-iron—and by extension, steel—beams were acceptable, as were arched cast-iron beams with wrought-iron tie rods.[43] In practice after 1885, steel beams were used for spans of this size and cast iron used only for relatively small wall lintels of inverted "T" or sideways "E" configuration. These shapes give a lower tension flange much larger in area than the compression flange, and thus reduce the risk of brittle flexural failure.[44]

Cast iron has never been eliminated from the building codes, only from use as a structural material. Modern curtain-wall construction employs sheet glass, unprotected aluminum, and other noncombustible but heat-sensitive materials. Cast-iron facades are still practical, if desired, as the sheathing of a curtain wall.

## Present Considerations

The flaws in cast iron described here are inherent in the material. No method of approaching renovations for use or modern building code requirements will make a cast-iron facade any safer than it is now. The reduced number of lethal and destructive fires in SoHo is due to the increased efficiency of modern fire-fighting and the change from commercial and industrial use to residential. The building code now prohibits exposed structural ironwork and exposed wood plank and joist floors within the fire district, except for certain occupancies, and when the building is protected by sprinklers and other active fire systems. In the continued use of these structures, consideration should be given to architectural and mechanical

45 Greene Street, east facade. Typical one-lot (20-foot) wide cast-iron facade. The first-floor cornice is missing, leaving the second-floor spandrel panels hanging, and exposing the tops of the first-floor columns.

fire safety provisions such as sprinklers and rated enclosures around exit stairs.

Iron facades, when examined at close range, often exhibit numerous minor failures. The small pieces making up applied ornament, such as the acanthus leaves on Corinthian capitals or small pediments over windows, are often loose, cracked, or broken. This disturbs the architectural integrity of the building but is not of structural concern. More serious for the continued use of the building without facade replacement are the common cracks in the rear flanges used to bolt together the individual spandrel columns and beams. If the connection cannot be rebolted, it may be necessary to replace otherwise useful members. Often, significant amounts of interior finish must be removed to expose these connections. Finally, new bolts put into cast-iron members cannot be pre-tensioned, as is common in modern bolted connections. A tensioned, high-strength bolt would almost certainly break the cast iron it was fastening.

Generally, the quality control during the original construction seems to have been at a fairly high level. Since cast-iron facades and columns were the first Industrial Revolution–based technological forms of construction, more attention and care were given to their use. The structural fabric of these buildings, while often designed in an incorrect or obsolescent manner, was usually well built. Most damage visible today consists of individually overstressed members and related mechanical failure. Damage to typical facade elements tends to be concentrated on the lower portions of the facade for two reasons: the increase in applied load compared with the upper stories, and the exposure to human contact. There are two types of destructive behavior that are prevalent: disregard of the building and improper care. The first has been common in SoHo, with its industrial past. Many cast-iron fronts have suffered the loss of minor decorative elements and damage to structural elements through impact from delivery trucks and hand carts. Improper care can range from lack of maintenance on paint to the use of cement sealants and other maintenance materials that are not compatible with the iron.

# 4. THE EMERGENCE OF THE STEEL SKELETON FRAME
## 1870–1904

During the last years of the nineteenth century and the first years of the twentieth, the steel skeleton frame became the dominant structural form in large building construction. During the transition period, from 1870 to 1905, several simultaneous developments combined to create the steel skeleton: the change from bearing walls and vaulted or joist floors to frames of independent columns and beams supporting floor slabs, the change from cast iron as the dominant structural metal to the more adaptable steel, and the change in structural theory from masonry-based compression forms such as bearing walls and arches to flexure-based design such as knee-braced portal frames. Each change was significant by itself and is in wide use today. Combined, they created the structural form that made the skyscraper possible.

## Bearing Walls, Cages, and Skeletons

The primary impetus for the development of frame-type structures was not engineering theory but economics. Even with the introduction of the passenger elevator as an integral part of commercial construction, space on lower floors continued for some time to command higher rents than space higher up. The early elevators were noisy, unreliable nuisances that did not immediately make penthouses attractive. Gradually developers and builders realized that bearing-wall buildings lost large amounts of rentable space to the increased thickness of internal and exterior walls when compared with frame buildings.[1]

Buildings such as the Tribune Building, erected in 1875 at Park Row and Nassau Street as the tallest building in the city, with 260-foot-high bearing walls in its central tower and bearing walls well over 100 feet high all around, represented the first wave of tall building construction in the United States but were the end of development for their type.[2] Skeleton and cage frames became popular so rapidly after 1880 that the United States Archive Building at 641 Washington Street was, in 1899, one of the last major buildings built with bearing walls in New York.[3]

The language of designers did not entirely keep pace with the development of structural types. During the period of transition, there were three major types of large buildings: those with iron or steel floor beams supported by brick bearing walls and cast-iron columns, those with more or less complete iron or steel frames but with self-supporting exterior walls, and those with complete iron or (usually) steel frames supporting curtain walls. These three types are listed in the order of their appearance in construction, although all were in use simultaneously between 1880 and 1900. The first type was generally called "bearing-wall" construction after its distinguishing feature. The second and third types were both called "cage" or "skeleton" construction depending on the source consulted. For the purpose of clarity, the third type, which is still in general use, will be called "skeleton" construction, because the term is also still in general use. The second type, transitional between bearing-wall buildings and true skeleton frames, will be referred to as "cage" construction, a term which has long been archaic.[4]

The cage form of construction, because of its general lack of resistance against lateral load, was particularly vulnerable to failure if not properly built. The definition of cage construction in general use during the 1890s was that of a frame

building surrounded by self-supporting walls. This does not automatically imply that the frame was not designed for wind loads, but the evolution of the cage type resulted in buildings with substandard wind bracing or none at all. In a discussion of wind load design after a tornado hit central St. Louis in 1897, Julius Baier argued for skeleton over cage construction because a local wall collapse would not affect overall building stability.[5] Baier recognized, at least unconsciously, that the exterior walls provide stability to cage-

type buildings. In bearing-wall construction, which had been the standard through the 1880s, and was still extremely common in 1897, the lateral loads were resisted by the masonry walls, both interior and exterior. This masonry shear wall action was not specifically designed but was built into the height-to-thickness ratios prescribed by the building code.

The growing use of terra cotta in internal construction began the degradation of the structural design of the bearing-wall system. Terra-cotta partitions for interior use were cheaper and easier to erect than brick walls. The lightness of terra cotta allowed partitions to sit on individual floor beams or girders. This allowed the architect much more freedom in planning internal spaces than brick walls, which by their nature were continuous in a vertical plane from their tops down to a dedicated foundation. The structural implication seems not to have been noticed: terra-cotta partitions are too brittle to carry the substantial shear forces caused by lateral loads, although they can serve to stiffen a building. One source of lateral resistance had thus been removed. Terra-cotta floor arches had also driven out brick arches for most uses, but this was a structurally less significant change. Floor diaphragms were subjected to far lower stress levels from wind loads than vertical walls, and the stresses were largely compressive, which terra cotta could resist nearly as well as brick.[6]

The change to cage construction was inherently dangerous when coupled with the use of cast-iron columns. In tall buildings without diagonal bracing, the flexural stress in columns caused by wind loads can exceed the compressive stress caused by the gravity loads. Cast iron in tension fails in a brittle and unpredictable manner. This type of failure could result from unbalanced load moments or wind moments on cast-iron columns or connections. The potential danger was recognized at the very beginning of cage use. William Birkmire, a lifetime employee of the Architectural Iron Company of New York and an advocate of cast-iron use, admitted in 1893:

There is, however, considerable distrust of cast-iron in *high and narrow building*, especially in relation to the connections with the floor and wall girders. Brackets and lugs are apt to break suddenly and completely, but with wrought-iron and steel will bend a great deal without breaking, and that rivets are stronger than bolts. To this objection it can be said that the brackets and lugs, instead of being cast with the columns, can be put on with angle-knee connections, drilled holes in the columns and with any number of bolts, which in a great many of our high buildings has proven entirely satisfactory, *where lateral bracing is not required.* . . . Where high and narrow buildings are concerned, much attention is given to the bracing against wind forces—that is, in the stiffness of the joints and the stability of the structure upon the foundation, and when the bracing is a portion of the frame construction, the difficulty of doing it properly with cast-iron columns is very great, but with wrought-iron or steel these difficulties are largely removed. . . . But we are entering on an age of steel. Rolling mills produce it quicker and cheaper than any other metal. . . . One after another the advocates of cast-iron have fallen in line in favor of wrought-iron and steel in high and narrow buildings, but for buildings with a large base cast-iron will continue to be popular.[7]

The gist of Birkmire's logic is made clear by the phrase "when the bracing is a portion of the frame construction." The only available means for a structural designer in 1893 to brace a building other than with its frame was through the use of masonry shear walls. Thus Birkmire emphasizes that the arguments against cast iron are unimportant in the design of bearing-wall or cage type buildings when the exterior walls act as shear walls.

The exception of lateral bracing meant that even with the improved form of connections described, cast iron could not be used as part of a wind-resisting frame. The type of connections Birkmire promotes may be superior to the standard, but they were rarely if ever used. Beam-to-cast-iron column connections typically were made through integrally cast web lugs and seat brackets for as long as the material was used. Obviously, drilling holes and bolting on separately cast connection pieces would eliminate a large portion of cast iron's economic advantage over wrought iron and steel. Having sold cast iron to builders as not requiring expensive mill work to create connections, its proponents could not convincingly argue

33 West 13th Street, interior. Typical 1880s loft building, with wood joists supported on brick bearing walls at the lot lines and beams and columns between. The distant wall is a bearing wall; the columns are plain circular cast iron with add-on decorative capitals. Directly over the capital the cast-iron "pintle" is visible, with bolted lug connections to the double wrought-iron girders. The original ceiling, in the background at the left, is pressed tin.

for more complicated connections with more individual pieces.[8]

It should be noted that steel's resilience in bending was a major factor in its widespread adoption. In the ordinary range of working stresses, in both tension and compression, steel exhibits nearly perfectly elastic behavior, with a linear relationship between the stress caused by applied loads and the resulting amount of deformation. When the stress is raised beyond the "yield point," the elastic behavior changes to extremely plastic behavior, where deformation increases disproportionately to the increase in stress. The allowable stress in steel design, until the development of plastic design and load-and-resistance factor design, was defined as the yield-point stress value reduced by a safety factor. In the AISC codes, this property later appeared in the allowance of "non-elastic, but self-limiting, deformation" in the description of frame action. Steel connections can bend to find an equilibrium position without failure because of the large amount of nonelastic deformation that can take place between the allowable stress limit and ultimate failure. In cast iron there is no distinction between nonelastic behavior and brittle fracture, and the stress at ultimate failure in bending is much lower than for steel or wrought iron. In order to provide a cast-iron seat connection with as large a margin of safety as a steel seat angle, the cast iron must be substantially thicker. The increased thickness increases the seat stiffness, reducing to negligible amounts the deflection before failure.

Darlington Apartments, 59 West 46th Street, 1904, south end of lot looking north. This photograph was taken by a neighbor the day after the collapse, and shows the complete loss of structural integrity. Almost all of the members visible are steel beams from the upper floors, distinguishable by their "I" shape and their twisted and bent form. The rectangular cast-iron columns shattered into small pieces.

In recognition of the cast-iron problem, designers typically attempted to balance loads on cast-iron columns. Where possible, beams were framed in on opposing sides to reduce gravity-induced moments. This was not always possible in a given architectural floor layout, and it did nothing to reduce the effect of wind-induced moments on the columns.

The conversion to cage construction reduced the efficacy of the exterior walls as lateral load shear walls. During construction, the erection of the frame could now proceed ahead of the walls, a sequence not possible in a bearing-wall building. If the frame was built with cast-iron columns, it was vulnerable during construction to damage from high winds or lateral loads induced by equipment such as cranes. Another problem was potentially more serious. The stability of the building would depend on the ties between the masonry shell and the iron interior. The two elements were tied together by brick piers built integrally with the walls that served as column fireproofing, by metal straps from the facades back

to the frame, or by both. If these ties were insufficient to transfer stress between the frame and the shell walls, the building was unstable. If the cage-type building was built with riveted wrought-iron or steel columns as part of a true wind frame, the ties were needed only to brace the wall locally to the frame.

Particularly in apartment house design, cage-type buildings became increasingly tall and slender during the 1890s and early 1900s. The logical conclusion of the trend came with the collapse of a cage building during construction. The collapse of the Darlington Apartments on March 3, 1904, killed twenty-five men and attracted national attention to such a degree that it seems to have been the end of the use of cast-iron columns in buildings taller than one story. The cause of the collapse was closely investigated by prominent structural engineers and reported in the engineering press. As is often true in spectacular failures, there was more than one cause, but all of the various causes were related to the use of cast-iron columns in what was a de facto moment-resisting frame, without any form of cross-bracing to resist lateral loads.

The Darlington was to have been thirteen stories high, 148 feet tall, 55 feet wide, and 90 feet deep. At the time of the collapse, the frame had been built up to the eleventh floor and concrete floor slabs had been poured up to the ninth floor. The floors were built of steel beams, supported by cast-iron columns that were surrounded by brick piers at the building's perimeter. The building was a typical cage with a self-supporting exterior wall, which by this date was considered an almost obsolete form of construction not used for large office buildings but common in apartment houses. There were no internal walls in the building, only terra-cotta partitions that rested on floor beams.

The most dangerous details of the design were the column connections. The columns were one story high and spliced by integral, unstiffened end plates that were attached to one another by pairs of bolts (figure 4.1). Typically, the columns were loaded on only one side by girders, with all of the floor beams framing into the girders. The framing layout was a rectangular double ring, with columns at the perimeter and surrounding the central core (figure 4.2). The columns were thus loaded through seated connections with an eccen-

tricity as large as 12 inches, and without balancing moments. The bolt holes were oversized to allow for alignment, but without any means of stiffening the connections. Not surprisingly, the frame swayed noticeably during construction. Several weeks before the collapse, observers noticed that the frame was leaning 18 inches out of plumb. It

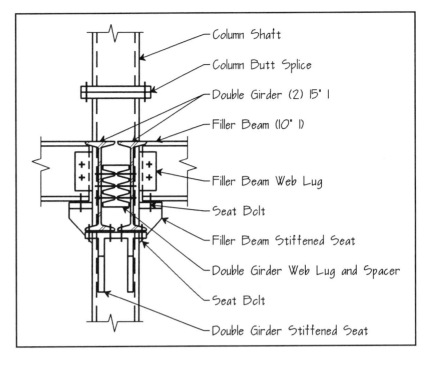

**4.1. Cast-Iron Column Connections**
Scale: ¾" = 1' 0"

**4.2. Darlington Apartments— Schematic Floor Plan**
Scale: ¹⁄₁₆" = 1' 0"

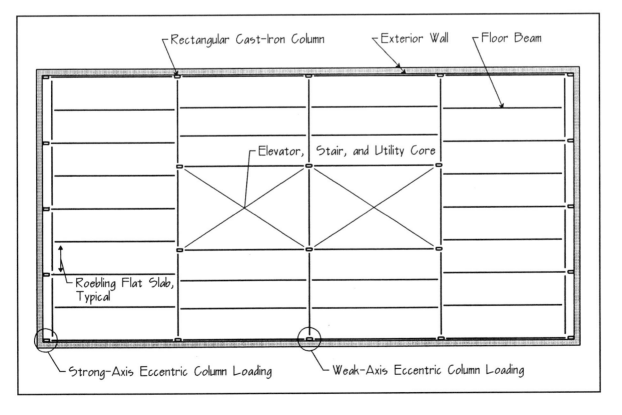

**The Ansonia, south and east facades. General view, 1904.**

was pulled back into place through the use of block and tackle, and temporary bracing was installed.

The building had been designed by its architects, Neville & Bagge, with the construction company of Pole & Schwandtner acting as the engineers of record. In a throwback to bearing-wall buildings, the beams had been designed as simply supported, and the columns for gravity loads, without consideration being given to frame action. The immediate cause of the collapse was excess play in the bolted connections: the columns were effectively unbraced for the full height of the building because of the sideways motion the connections allowed. After one column buckled, the frame failed progressively, with most beams and girders bending as they were pulled down, and nearly all of the columns breaking at their midpoints or near the splice flanges.[9]

The existence of buildings such as the Darlington was certainly no surprise to the majority of engineers and architects involved in large building construction, and the general reaction was to dismiss the type as unworkable.

Steel skeleton or cage construction in its complete development gives us a rigid steel framework connected and braced like a bridge in all directions, carrying walls, floors, and finish merely as so much clothing, and able to safely bear all loads which come on either the completed building or the framework alone. Structures of the "office-building" type are practically all of this kind. But very many other buildings, from five or six-story warehouses and stores to twelve (and in at least one case seventeen-) story apartment buildings, are radically different in structural respects, though masquerading under the general name of "skeleton" or "steel-frame" construction. They have floor beams and girders of steel, columns of either steel or cast-iron designed for vertical, central loading only; they are provided with no bracing, and the connections between the beams and columns are unfitted to resist any calculable bending or twisting moments. In such buildings the walls and floors give lateral strength and stiffness to the structure, while vertical loads are carried by the frame. Evidently this is quite different from the condition in an office-building structure, both after and during erection. Of course, it is known that when the walls are in place and firmly set, they supply sufficient lateral resistance, though the amount of this resistance cannot be calculated, especially when the wall is pierced by windows in every panel.[10]

The seventeen-story building referred to is the Ansonia Hotel, built in 1901 on upper Broadway. The Ansonia is not a true cage, however, because it contains internal bearing masonry walls of extreme thickness. The building became so famous for its sound-deadening acoustics that many musicians chose to live there, not realizing that the soundproofing was an accidental side effect of the structural design.[11]

The collapse of the Darlington not only discouraged further use of cast-iron columns and the cage type but also called their previous use into question.

The use of cast-iron columns for buildings of considerable height is a practice which was retained from the practice in low buildings with large natural stability. Good engineering has tended away from cast-iron columns during some years past, even for low buildings, and their use for buildings higher than five or six stories has been regarded by nearly all competent structural engineers to be doubtful practice at best. Yet they have been very generally used in light tier-building work for much greater heights, and in apartment houses their use has been almost universal.[12]

The Darlington was still being referred to in the 1930s as the end of cast-iron construction and the beginning of near-universal use of steel columns and riveted connection in metal buildings. The general consensus was that cast iron was suspect in the minds of too many people to be used again, whether or not it was truly to blame for the collapse.[13] The switch to all-steel construction removed the final support from the cage type, a development that was eventually reflected in the removal of the building code provisions requiring curtain-wall thicknesses similar to those used in bearing walls. The decline in cast-iron use, specifically its disappearance from large office buildings before 1900, was inevitable for reasons beyond frame design. By 1898, it was recognized that the use of cast iron left the engineer open to

the liability of defects occurring in the process of manufacturing cast-iron columns, the shifting of the cores which entail variations in thickness of opposite sides, concealed cavities, blow holes, cinder, imperfect union of two currents of metal in the mould, and initial strains due to unequal cooling, added to the opportunities for intentional departure from specified thicknesses, and the use of inferior qualities of pig-iron on the part of unscrupulous founders. . . . The connections between riveted steel columns are generally liked better than the connections of cast-iron columns, being more rigid and stable . . . [and] especially desirable where unequal or eccentric loads are placed on columns.[14]

None of the criticism of cast iron was based on new information or recent developments. All of the flaws in cage construction and cast-iron columns that were highlighted by the Darlington

collapse had been illustrated by previous failures. For whatever reason—the midtown location, the residential use, the high death toll—the collapse of the Darlington attracted far more national attention than the problems in the erection of the Diamond Exchange Building, or the partial collapses of the Ireland and Brown Company buildings.

The Diamond Exchange was under construction on Maiden Lane in 1893, and had its cage frame erected up to the tenth story of the thirteen planned, when construction was stopped because of the visible tilt of the frame. The cause of the out-of-plumb erection was one of the primary causes of the Darlington collapse: the use of oversized holes in splices of the cast-iron columns. Simple bolts in these loose connections allowed out-of-plumb erection. The play in each connection built up over the height of the building as each tier of columns was slightly further from vertical than the one below. The solution was simple: the unbraced cage frame was converted to a braced frame through the addition of diagonal rod bracing with adjustable tension fittings.[15] The rods were gradually tightened to straighten the frame to its correct position and were left in place to serve as lateral bracing—thus preventing bending moments or tension in the cast-iron columns from lateral loading.

The Ansonia, 2101 Broadway, south and east facades. Close-up shows the extreme density of masonry construction at the base, including around the storefronts, an indicator of a bearing-wall or cage-construction building. The masonry wall is far thicker than a skeleton frame's curtain wall, but is not the four or five feet that would be expected of a bearing-wall building this height. The Ansonia is a cage frame.

Approximately half of the Ireland Building on 3rd Street collapsed during the final stages of construction in 1895. The examination of the ruined building by the Manhattan coroner's office, the Superintendent of Buildings, and various professional engineers revealed so many design flaws, construction errors, and accidental circumstances that no single cause for the failure could ever be pinned down. The building was a simple and outdated design of exterior masonry bearing walls and one center line of cast-iron columns supporting steel floor beams and tile arch floors. Two of the four columns failed for their full height, pulling the beams cleanly out of their pockets in the walls. Sixteen laborers were killed, setting off a long series of legal actions, starting with the arrest of the plaster contractor on site for overloading the upper floors with plaster materials and ending with a finding in a civil suit that John Ireland, the building's owner, was personally responsible for the death of one of the men. The contractor was excused from liability because he was following the instructions of the architect, Charles Behrens, who was Ireland's direct agent.[16]

The best evidence indicates several contributing factors. The entire iron and steel frame was designed only for gravity loads, and the lightest gravity loads possible were used. The foundation material was suspect, with several reports of quicksand and footing settlement coming to light during the coroner's inquest. The footings were extremely light for the loads, consisting of a 4-foot square and 1-foot thick granite slab resting on a 5-foot square and 1-foot thick unreinforced concrete slab. The plasterer had almost certainly overloaded the sixth floor with gypsum, driving the columns through the footings in a punch-through failure. The column shafts above all shattered from the secondary stresses induced by the sudden settlement.[17]

While most of the causes listed are the result of obviously slipshod design and construction supervision, the condition of the rectangular cast-iron columns is of interest. The coroner's inquest was useful in that it examined each piece of the failed structure and recorded the various conditions. "The broken sections of the iron pillar were carefully examined. . . . and the metal was found to be of poor quality. It was so full of blowholes that

some pieces looked like worm-eaten wood." This evidence is significant even to those without a great deal of structural knowledge, although it was apparently no surprise to James Cornell of the Cornell Iron Works, the steel and iron subcontractor, who said that such condition was normal and does not affect the column's strength. The engineering press was less blithe, with an editorial in *Engineering News* stating that the "predicted failure of a building erected with cast iron columns has taken place in New York city. . . . There are other buildings with cast iron columns now in the course of erection in New York city, larger and taller than the one which has just failed."[18] The Darlington proved to be one of these larger buildings without trustworthy columns.

The partial collapse of a just-constructed soap factory on Twelfth Avenue in 1897 was eventually traced to the same root cause as the Darlington collapse seven years later. Rectangular cast-iron columns in the four-and-one-half-story building were "simply abutted near the floor levels and bolted through the flanges with four loosely fitting bolts at each joint. This is the crudely inefficient old way that has so many times in the past given trouble and which no competent civil engineer in this country would peril his reputation by using. . . ."[19] The result was

a vertical series or tier of cast-iron column sections four stories high, essentially unbraced laterally, with joints practically without stiffness, with the [live] load mainly at the top [floor, where numerous tanks were located] and actually all load applied with great eccentricity entirely on one side of the tier from top to bottom. Failure was thus actually designed into the structure. . . . Beam and girder connections were loosely bolted, whereas they should be riveted. . . . With steel columns and ordinarily good connections the "accident" would have been impossible. Cast-iron columns are a source of grave danger wherever they are used, as we have frequently shown, and no safety factor can remove that danger.[20]

The failure of the engineers to acknowledge that such practices continued despite the profession's disavowal of cast-iron and unbraced cage frames, and the failure of the Building Department to deal with the Ireland and Brown collapses, led

New owners took over a restaurant space in an office building in TriBeCa, and decided to renovate in order to clearly distinguish their new restaurant from that of the previous tenant. Their architect needed engineering assistance to design modifications to a mezzanine and to fill in openings in the main floor adjacent to the basement stair.

Unlike some buildings where the original structure can require an extensive probing course to determine, the various elements of this cage frame were obvious. The restaurant space had once been a showroom for the company that had built the building as its headquarters, and had a ceiling roughly 25 feet above the first floor. Within this large open space, occupying roughly 40 percent of the first two floors of the building, were four exposed cast-iron columns 16 inches in diameter. In line with the columns at the exterior walls are masonry piers surrounding 12-inch by 16-inch rectangular cast-iron columns. The columns form a regular 20-foot grid, despite the irregular shape of the building, which is a curved "L" at an oblique street intersection. The building is eight stories high, and the exterior wall at the base is more than 2 feet thick. The girders at the ceiling level of the restaurant space (the underside of the third floor) were architecturally treated as they projected below the surrounding floor, connecting the columns and end piers.

The extreme thickness of the exterior walls indicated either a bearing wall or a cage frame. The visible presence of the columns within the exterior wall piers was enough evidence to assume that the building was, in fact, a cage frame. The previous tenant had created a mezzanine to expand the amount of seating available for the restaurant. This mezzanine was in part supported on new steel pipe columns that passed through the first-floor structure to new footings, and in part on beams simply pocketed into the exterior walls. It is not good practice to use the exterior walls of a cage for beam bearing, but examination of the bearing conditions showed no sign of overstress of the brick.

Portions of the existing mezzanine were scheduled to be reused. The new mezzanine area was designed to be supported only on new and existing pipe columns. Bearing within the exterior wall was avoided in order to prevent the formation of "hard points" in the wall, which lead to cracking and general masonry deterioration.

The portion of the first floor to be filled in consisted of two open wells on either side of the "T"-shaped stair to the basement bar area. Since this had originally been solid floor, there was no question that the columns could resist the additional load. The opening edge beams were measured and their capacity checked: as expected, they could resist the additional load without reinforcing. The new edge beams for the infill slabs were attached with welded double angles to the existing beams.

to the soap factory failure. Since no conclusions were drawn from the Darlington that were not obvious from the Brown factory, it must be assumed that twenty-five deaths on Fifth Avenue were a more effective persuasion than two on Twelfth Avenue.

## Cast Iron, Wrought Iron, and Steel

The transition from bearing-wall buildings to skeleton frames was made over a forty-year period, with lengthy sidetracks into cage frames and other intermediate forms. The transition was further blurred by the concurrent use of three structural metals. The advocates of steel and wrought iron naturally preferred the true skeleton frame over other forms because it minimized the use of materials other than metal. Brick and stone masonry would be used only for decorative curtain walls, terra cotta only for floor arches and partitions, and concrete only for foundations and later as a replacement for terra cotta in floor construction. Builders who preferred cast iron

**Produce Exchange, Two Broadway, 1890, south facade. A cage building with extensive wrought-iron interior framing.**

needed the stability of masonry shear walls provided by bearing-wall and cage construction. At the same time, steel and wrought iron briefly were the subject of debate over the future of ductile metals.

As late as 1903, the Carnegie Company felt threatened enough economically by cast iron to include the following propaganda in the company handbook. Although slanted for commercial purposes, the passage does express well the decline of cast iron before the Darlington collapse.

> The subject of fireproof construction is steadily growing in importance. The need of fireproof buildings in the business centers of our great cities has been well demonstrated, and their superiority has become so generally recognized that at present but few structures of any size or importance are designed which are not more or less of this type. . . . The substitution of steel for iron in beams may be cited as a radical improvement in this direction, and, simultaneously, the introduction by this firm of new patterns for its steel beams. . . . Another change which is gradually taking place is the substitution of steel for cast iron in the composition of columns. Cast iron is a material so uncertain in character that its use has long since been abandoned in bridge

construction. In buildings the loads are generally more quiescent and the liability to sudden shocks is more remote than in bridges; yet, on the other hand, the columns seldom receive their loads as favorably as in bridges; in most cases there exists considerable eccentricity, that is, the loads on one side of the column are heavier than those on the other side, and the bending strains arising therefrom increase the strains from direct compression materially. . . . As a protection against these contingencies resort must be had either to the crude and uncertain expedient of a high safety factor, not less than 8 or 10, or a material such as rolled steel must be adopted of a more uniform and reliable character than cast iron.[21]

## Skyscrapers

A great deal of debate has taken place since the boom of the 1920s about the origin of the skyscraper form. The arguments break down into the regional camps of New York versus Chicago, and the academic camps of varying definitions of the term "skyscraper." The most commonly accepted formula is that a skyscraper is a high, relatively narrow building with elevator service and, most importantly, a skeleton frame supporting a curtain wall. This standard leads to Chicago's Home Insurance Building of 1880 as the country's first skyscraper. What is most interesting about this conclusion is that it would not have made sense to a large number of builders during the years in question, 1880 to 1900.

George Post had the opportunity and ability to build the first full skeleton frame in 1880, when his design for the New York Produce Exchange was submitted, but declined to leave the cage form for practical reasons. Post, as both an architect and an engineer, understood the implications of iron and steel framing but obviously believed in the cage type as the optimal structural form. The Produce Exchange had all of its interior wrought-iron floor framing supported on cast-iron columns. The only exception was the support of the spandrel beams on brick piers surrounding the perimeter columns. The wall panels between columns were carried on the spandrel beams; thus they were also supported on the brick piers. The frame was a hybrid which has been described as "coming literally within inches of being a complete iron skeleton without masonry supplements."[22]

Another example of Post's cage design is the Havemeyer Building of 1892. A frame of wrought iron with moment connections was set within a self-supporting exterior wall. The frame was similar to that which would be used in a modern-style supported-wall building, but did not extend into the exterior masonry to support it. The building was 193 feet high on a lot 60 feet wide—proportions that would have been difficult or impossible to use productively in bearing masonry, and probably ambitious for New York's steel-framing builders at that time, only three years after their first skeleton frame.[23]

To an engineer and architect of Post's ability, the cage form offered two advantages over the skeleton that would exist until architectural fashion simplified building facades and waterproofing improved. First, the complex facades of the era, with numerous cornices, pilasters, and corbels were difficult to support from the frame's spandrel beams, requiring hung lintels, outriggers, and braces to support the eccentric masonry weight. This problem has never been eliminated for skeleton frames, only reduced by the simpler facades of the twentieth century. In cage frames, the connections between the steel and masonry are identical regardless of the facade design, since the facade is entirely self-supporting.

Second, cage construction provided thorough protection of the frame of the building from weather. In a cage building, there was always a minimum of one foot of masonry between the outer air and the metal; in a skeleton building the protection was often as little as 4 inches at the spandrel-beam flange tips. Brick and terra cotta are permeable and will admit water even when solidly built. In many early skeleton-frame buildings, the lack of adequate flashing over the beams and cavities within the walls to shunt rainwater back towards the exterior served to trap water against the steel. In 1894, Post "thought there was reason to doubt the long life of the steel structure imbedded in masonry. In his own practice he had already found occasion to remove beams on account of rust, though these beams had been encased in solid brickwork."[24]

The peculiarities of the New York Building Code added to the delay in the use of skeleton construction. The 1892 code allowed a reduction of only 4 inches for a non-bearing wall as opposed to a bearing wall.[25] This was an impediment to cage construction, since the cage-wall height was figured the same way as a bearing wall, so that the walls would be relatively the same thickness. Curtain walls in skeleton frames, in contrast, are only as tall as the floor-to-floor height, and so would be substantially thinner in the lower floors of a building than cage or bearing walls.[26]

**The Tower Building, 50 Broadway, circa 1915, west facade. The first building in New York with a true (but partial) skeleton frame, squeezed between two "modern" office building before it was thirty years old.**

**Printing House Square from the west, 1890s, Tribune and Pulitzer buildings, west facades. The Tribune Building (right, with spire) had already been extended upward from its original height through the simple addition of more masonry on the bearing walls. The Pulitzer Building was a prominent example of a commercial building of cage construction.**

The structure inside the Statue of Liberty, built on Bedloe's Island in New York Harbor in 1883 through 1886, "represented the most extensive system of wind bracing so far employed for any American structure other than a bridge," or in other words, a skeleton frame.[27] The sculpture was the work of Frederic Auguste Bartholdi, but the interior structure was the design of Gustave Eiffel, an engineer who had already designed some of the world's most advanced structures in Europe in his "crescent" railroad bridges, and was working on his Parisian tower at the same time. This example, unlike those in the 1850s, could not be ignored, especially given Eiffel's prominence, and probably helped to encourage the experiments in frame design in New York.

The first attempt at a skeleton frame in a New York building was the Tower Building at 50 Broadway, completed in 1889. The lot for this building was less than 22 feet wide on the Broadway end, and thus uneconomical for bearing-wall construction. The New York building code fixed the thickness of bearing walls for buildings, and would have removed approximately ten feet from use at the ground floor for the eleven stories and 129 feet

planned.[28] The architect, Bradford Gilbert, received special permission from the Building Department for the final scheme, which could best be described as a four-story bearing-wall building sitting on top of a seven-story skeleton frame.[29]

The building was narrow enough to be designed as one structural bay wide. The cast-iron columns supporting the lower floor beams and the upper bearing walls were contained within the walls and sat on the foundation lot line walls. The floors and lower curtain walls were carried by wrought-iron girders, as were the bottoms of the bearing walls. Lateral stability was provided by diagonal cross-bracing at the ends in the plane of the walls.[30] The diagonals reduced the columns' role in wind bracing to that of compression struts, for which they were reasonably well suited.

One year later, the new headquarters for the *New York World* newspaper, the Pulitzer Building at Park Row and Frankfort Street, was completed. The Pulitzer Building was the tallest building in the city when built, at nineteen stories including a tower, and probably the tallest cage-type building ever built.[31] Despite the examples of the Tower Building and a few examples in Chicago (including the Home Insurance Building), this building was built as a cage, with self-supporting exterior walls, because "its architect, Mr. George B. Post, strenuously insists that the cage principle—the outer walls built self-sustaining and independent of the frame work of iron or steel columns and girders which supports the floors and roof—is better than the skeleton principle. ..."[32]

The first building in New York with a complete skeleton frame was the American Surety Building at Broadway and Pine Street. The architect was Bruce Price, and the building, completed in 1895, was briefly among the tallest in the city at 303 feet and twenty stories.[33] The distinction of having skeleton form means less than meets the eye because of the theorizing and construction of the preceding years. Just as New York architecture was less "pure" in theory than the Chicago School, New York structural engineering was less centered on pure structural forms, often mixing bearing walls and cages or cages and skeletons.

The phrase "skeleton construction" was first officially used in the 1892 building code in an attempt to describe the new building type. If this is taken as the final acceptance of self-supporting metal frames as a legitimate form of building, then the development of the steel-frame building was a slow process. Before the development of the balloon frame in the middle of the nineteenth century, wood houses in North America had been constructed of basically independent frames covered with a "curtain wall" of sheathing. Isolated examples of metal frames with curtain walls were built in New York as early as the 1850s. James Bogardus had built in 1856 a shot tower for the McCullough Shot and Lead Company in the form of a cast-iron octagonal frame 175 feet high that completely supported the brick infill panels; at nearly the same time he built a similar tower for the Colwell Lead Company. These towers, built for the manufacture of round lead shot, needed to be high enough for the molten lead to harden through air cooling before it hit bottom. The towers obviously could not be built of wood, and Bogardus must have been able to build his towers cheaply enough to compete with the traditional masonry form.[34]

The fire towers erected during the same years as the shot towers were more interesting from the structural viewpoint. The shot towers used iron to carry some gravity load and to ease construction, but did not contain any form of iron bracing; instead, the tight masonry panels acted as shear walls to brace against wind. The fire towers were built for observers to watch the cityscape for signs of fire and, being constructed by the city government, had to be as inexpensive as possible. Julius Kroehl was the engineer for the still existing tower in Mount Morris Park (now Marcus Garvey Park) on upper Fifth Avenue, Bogardus for those since demolished at 33rd Street and Ninth Avenue and MacDougal and Spring streets. All three were built primarily of cast iron, with wrought-iron-rod cross-braces. The braced form was familiar to the builders from bridge construction, if nowhere else. The 33rd Street tower had its legs anchored into the bedrock with split castings and wedges, suggesting that Bogardus was familiar enough with the effect of wind to recognize the uplift forces that would exist at the bottom of a skeletal structure approximately 75 feet high and only 20 feet wide.[35]

The other two examples of complete metal frames in this early period were even more isolated in their effect on building design. The New York Crystal Palace, built in 1853 in imitation of the 1851 London Crystal Palace to house a fair, was a complete frame of cast-iron columns and wrought-iron "ribs," although it was proportionally very broad for its height. The Crystal Palace's contents burned and destroyed the building in under fifteen minutes five years after completion, effectively removing it from consideration as a

**The American Surety Building, 96–100 Broadway, 1929, west facade. The first complete steel skeleton-frame building in New York, ten years after the first in the United States, the Home Insurance Building in Chicago. The new 120 Broadway is to the north, dwarfing it.**

Fire tower, Mount Morris Park, Fifth Avenue near 120th Street, 1880s. Cast-iron columns and beams and wrought-iron tie rods.

nomic base for tall buildings, the conceptual newness of skeleton frames slowed their spread. Despite the examples listed above, to support heavy brick and masonry walls on seemingly flimsy iron or steel beams demanded a reversal in how architects, builders, and engineers looked at construction. The idea of supporting floor beams on walls had, in 1890, forty years of successful "fireproof" building construction, and twenty years of tall building construction behind it. This success can be blamed for some of the resistance to the idea of supporting walls on floor beams.

## Present Considerations

Through an economic coincidence, relatively few cage office buildings survive in New York. Most were built as part of the first skyscraper boom in the 1870s and 1880s downtown. This area remained the first choice in location for many years and is still seeing new office construction. Most of the early skyscrapers have long since been demolished, replaced by larger buildings. The use of cage frames died out fairly quickly in commercial construction, so there were few examples from the beginning.

Cage frames, and specifically cage frames with cast-iron columns, were continuously used in residential construction at least until the Darlington collapse in 1904. There are many apartment houses and hotels in Manhattan, some in midtown but most on the Upper East Side and Upper West Side, that date from the 1890s and 1900s and may well have this type of structure. The fact that they are standing after more than ninety years of use is proof that they are not inherently inadequate, but any modifications or renovations must proceed with care.

A cage building with cast-iron columns is dependent on its exterior masonry walls—and possibly any interior masonry walls as well—for lateral stability. Unlike a curtain-wall building, where window openings of any size and configuration can be cut into the exterior without concern for any structure beyond the new lintels required, a cage building must be treated for the purpose of renovation as if it had bearing walls. This consideration grows geometrically with the wind-induced stress as one travels further down the structure, making the replacement of a solid wall

type model.[36] The last effort was an 1860 grain elevator in Brooklyn designed by George Johnson for Daniel Badger, and similar in form to the shot towers.[37] Architects and engineers were apparently even less likely to go to Brooklyn to look at a grain elevator than they were to look at the shot and fire towers in Manhattan, since no mention was made of any of these forerunners when complete frames returned twenty-five years later.

The interval of nearly forty years from the shot and fire towers to the use of the term "skeleton construction" can be attributed to two primary factors: the relatively late development of the elevator and the resistance of designers and builders. Until the elevator developed to the point where the upper floors of office buildings became desirable space, there was no general desire for tall buildings. It is notable that the [Old] Equitable Building at 120 Broadway was both one of the tallest structures in the city and the first commercial building designed with a passenger elevator when finished in 1870.

Even after the elevator had established the eco-

Crystal Palace, before 1858, north(?) facade. This structure is entirely cast- and wrought iron and glass, and compares favorably for window area with modern curtain walls. The building was destroyed by the heat of its contents burning: the glass shattered and the iron weakened and deformed.

at grade with a modern storefront one of the most problematic renovations.

The capacity of cast-iron columns must always be questioned. During demolition of cast-iron column buildings, flaws in the iron such as porosity or shifted cores resulting in thinner-than-expected column walls often come to light. Tables are available in old design manuals, and formulas are given in Appendix E, for theoretical allowable column loads. At any time the design load approaches the theoretical load, the concentricity of the applied load should be examined to see that gravity load moments are not being introduced. The inability of cast iron to predictably carry ten-

sile stress makes eccentric loading extremely dangerous.

Finally, the chemical composition and sectional integrity of wrought iron or steel in old buildings should be examined before modifications are built. Old metal may not be capable of producing acceptable welds, and welding is the accepted method of attaching new steel to old members because of the difficulty of creating adequate and regular bolt holes on site. The less-efficient furnaces and rolling mills of the past sometimes produced rolled sections with variable metallurgical quality or geometric discontinuities with reduced load carrying capacity.

# 5. THE FIREPROOF BUILDING (I)
## 1872–1910s

New York did not experience major fires of the type that destroyed most of Chicago in 1871 or a large section of Boston in 1872 during the period of rapid development in building technology, but many of the changes that took place after 1872 were driven by attempts at fire damage prevention. Fires in New York in 1835 and 1845 destroyed large portions of the commercial downtown area below Chambers Street. They had sparked demand for fireproof buildings, and of course there was the occasional spectacular building fire that highlighted the problem.[1]

One reaction to the increased interest in buildings able to resist fire long enough to allow firemen to work before they faced a blocks-wide firestorm, as in 1835, was the establishment of the "fire line," a legal division of the city into a fireproof construction district and an unrestricted district. The fire line moved steadily north up Manhattan Island, following the fringe of heavy development. In 1866, the fire line had been moved to the midpoint of the block between 86th and 87th streets, extending from the East River to the Hudson River; by 1887, to 155th Street.[2] This ensured that fireproof construction was required by law for all of the large buildings downtown and midtown during the development of the first wave of skyscrapers. By the time the southern portion of Westchester County became "annexed territory," the future Bronx, the fire line had moved far enough north to cross the Harlem River. Any large apartment houses in northern Manhattan were included in this area. As the fire line moved north, the definition of buildings required to be fireproof broadened. Nonfireproof buildings were restricted first in width, then in height, then in usage.[3]

During the nineteenth century the definition of fireproof construction changed several times. Before the 1835 fire proved otherwise, commercial buildings with masonry exterior walls were referred to as fireproof.[4] The definition that was created out of the new technology and became standard in usage after the Civil War was all-masonry construction of walls and partitions, iron beams and columns, and arch floors. The individual components varied over the years, as steel replaced iron and terra cotta and then concrete floors replaced brick arches, but the basic concept remained unchanged from the late 1860s until at least 1900.[5]

The cast-iron front was one result of the search for better building technology. The two great proponents of cast-iron construction, James Bogardus and Daniel Badger, became successful constructing buildings that had wood-joist floors and exterior walls (other than street facades) of masonry.[6]

The selling points of cast iron—its perceived advantages over traditional building materials— were its fire resistance, its slender proportions and light appearance, and the speed with which it could be erected.[7] The last two factors were never controversial. The high strength of cast iron in compression relative to masonry meant that the between-window piers in a facade could be reduced in width, greatly increasing the possible window size. The individual cast-iron pieces making up a facade merely had to be bolted to one another to be set in place. This work did not require the type of skilled labor that laying up a brick or stone wall did.

The other cast-iron advantage, fire resistance, was a less clear-cut issue. Individual cast-iron buildings had suffered damage in fires from the time they were first built. Currently this is not attributable to any single cause, largely because of the possibility that any given building may have been built improperly. It was generally recognized that proper detailing in cast iron was essential because of the brittleness of the metal. Overly restrained bolting flanges or panels could crack from ordinary thermal movement, or could contain within them partial cracks from post-casting shrinkage.[8]

The first indication that there might be problems on a large scale with cast iron was the Chicago fire of 1871. Although it has been suggested by some that cast-iron columns actually withstood the firestorm better than stone and brick, that was not the perception at the time.[9] The opinion was stated that

[c]ast-iron, for the fronts of buildings, in streets less than a hundred feet in width, was found to be most treacherous. In several instances these fronts expanded and buckled and fell into the street from the effects of the intense heat radiating from burning buildings on the opposite side of the street, long before their combustible interiors had taken fire.[10]

Over twenty years after the Chicago fire, a history of construction in New York was even more specific. The Chicago and Boston fires "for the time, drove iron in all forms and all places out of favor."[11] The advantages of wrought iron were so great that nothing would delay its use for long, but the link is clear: iron "in all forms" had been called fireproof. The obvious failure of iron columns and iron beams under the extreme conditions of the Chicago fire were an indication that this might not be so.

Traditional buildings did not survive the Chicago fire very much, if any, better than iron-front and iron-column buildings. The destruction of the extensive wood-frame-house residential neighborhoods is no surprise now, and was not then, but the destruction of the central portions of the city was a direct repudiation of the then-cur-

rent definition of "fireproof." A building with masonry or cast-iron exterior walls was usually referred to as fireproof, despite the fact that most of the commercial buildings had wood-joist floors. The wood-joist roofs proved especially vulnerable, as burning debris spread by high winds set fire to roofs of buildings far removed from the general fire. After 1871, no building with wood floors would ever again be called fireproof. Photographs of Chicago immediately after the fire show the same pattern of damage for all buildings, regardless of construction: the ground-floor facade mostly complete, and less wall remaining at each level, until nothing remained above the fourth floor. The interiors were all entirely gutted.[12]

The *New York Tribune* headquarters designed by Richard Morris Hunt shows the effect of the Chicago fire on the logic of building construction. The Tribune Building was built in 1875 at Park Row and Nassau Street and was the first "tower"-type building in downtown Manhattan: a large, nine-story block with a tower extending up 260 feet. All of the floor beams were supported on the exterior and internal masonry walls.[13] As a contemporary description said, "upright iron supports [have been] cautiously avoided; experience having demonstrated that they cannot be relied upon when exposed to great heat."[14]

There were many arguments made in favor of cast iron as a fireproof material, including examples of buildings in the Chicago fire that had collapsed entirely except for portions of cast-iron facades or internal cast-iron columns. There were similar examples made of individual building fires in New York. Tests were performed with isolated, loaded columns in laboratories, where the metal was heated and then water-cooled to simulate the effect of a fire. Under these conditions, the wrought-iron and steel columns buckled and collapsed, but the cast-iron columns, although cracked, stood.[15]

The problem with these arguments is that they were based on false assumptions about the isolation of building elements' reaction to heat. The truth was gradually learned and became generally known, so that by 1895, when the Keep Building burned, the expansion of the Manhattan Bank Building's iron facade could be singled out as the cause of that building's collapse.[16] The boom in cast-iron facade construction ended during the 1870s, partially because of the inherent impossibility of fireproofing an exposed iron facade.[17]

Wrought-iron floor beams were developed concurrently with the cast-iron facade. From their first use at the Cooper Union, the Harper & Brothers plant, and the Assay Office, wrought-iron beams became the standard floor supports for commercial buildings by 1870, and for high residential buildings by 1890.[18]

## Noncombustible Buildings

The development of the late-nineteenth-century standard fireproof type proceeded quickly from the iron-frame prototypes. The Trenton Iron Works and the Phoenix Iron Company of Pennsylvania were both marketing true "I"-shape wrought-iron beams by 1860.[19] The Phoenix company also began selling in 1862 the patented "Phoenix column," a round iron column composed of four or six arced segments with protruding flanges. The flanges were bolted together to form a complete circle. The advantages to the Phoenix column were the relative ease of connection for beams provided by the protruding flanges, and the increased visibility of any potential defects in the column interior.[20] In 1865, the first wrought-iron beams were rolled in Pittsburgh, marking its beginning as the country's structural iron and steel center.[21]

The [Old] Equitable Building at 120 Broadway was built in 1870. George Post, an architect and engineer, was hired to redesign a building originally to have been built entirely of masonry. Specifically to reduce the cost of construction of the five-story, 130-foot-high building, Post "kept the exterior granite piers but introduced a far from novel construction of brick partitions, brick floor arches, and wrought-iron beams in the interior."[22] The Equitable Building's height was made economical by the use of a passenger elevator; it is considered to be one of the first true tall buildings. The standard fireproof type was directly tied to tall buildings: the law required fireproof construction for such large buildings, and the newer technology developed simultaneously with the architectural and structural design.

During this period, wrought-iron I-beams were available in more sizes and weight from more manufacturers each year, up to 15 inches deep by 1875. It was also in 1875 that the Thomson Steel Works of the Carnegie Company began operation in the Monongahela Valley as the first large commercial Bessemer steel plant in the United States.[23] Bessemer steel was the first commercially viable method of mass-producing steel. The Thomson plant was originally designed to roll only rails, but the use of steel floor beams in the deck of the Brooklyn Bridge in 1877 marked the first beams of that type in New York.[24] The first steel I-beams commercially available were from Carnegie in 1884 and were so popular that by 1893 Carnegie had stopped production of wrought-iron beams.[25]

## Fire Protection

The one innovation that, by modern standards, substantially increased the fire resistance of the buildings called fireproof was the introduction of terra-cotta column and beam enclosures. Such enclosures were a direct outgrowth of the use of terra-cotta jack arches and flat arches for floor construction. Before terra-cotta arches were used, floor arches for fireproof construction had been built of common brick. This technique was expensive, because it required skilled masons, and by its nature it left the bottom flanges of the beams exposed. The bricks did not interlock; thus they had to rest on the top surface of the beam bottom flanges to resist vertical loads, and bear against the beam webs to resist the arch thrust.

In 1871, George Johnson and Balthasar Kreischer patented a system of terra-cotta floor flat arches (figure 5.1). Johnson was an engineer who had worked for Daniel Badger. He was living in Chicago, where the great fire inspired his invention. Kreischer was an established manufacturer of brick and clay products on Staten Island. The patent was for what would later be called side construction: each terra-cotta block was a prism, rectangular in plan and in section perpendicular to the arch span, trapezoidal in section parallel to the span. There were trapezoidal "cells" running perpendicular to the span. These were voids that constituted up to 80 percent of the sectional area, to lighten the floor dead load and allow the arch to work through the ribs remaining. The slope of

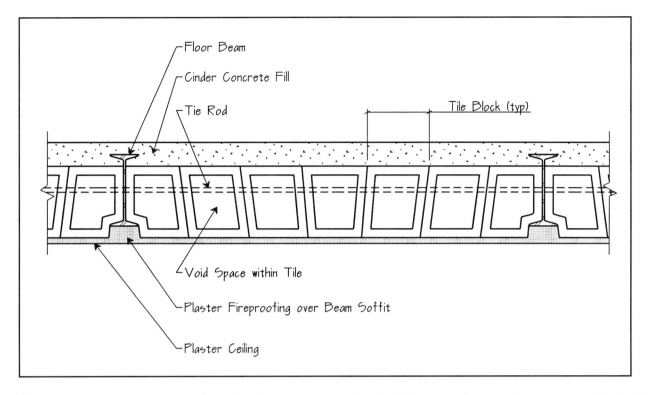

Floor Beam

Cinder Concrete Fill

Tie Rod

Tile Block (typ)

Void Space within Tile

Plaster Fireproofing over Beam Soffit

Plaster Ceiling

**5.1. Typical Floor Section—1871 (Original) Flat Tile Arch Scale: ¼" = 1' 0"**

the trapezoids was opposite on either side of the arch centerline, running in both cases from the bottom of the arch nearer the centerline toward the top of the arch nearer the beam. The center piece was a traditional keystone shape, narrow on the bottom and wide on top.[26]

The original patent did not include soffit tiles, small trapezoidal pieces later used under the beam bottom flanges. The arch blocks were designed to be the same depth as the beams, but to sit several inches lower. The first designs simply had this upward discontinuity in the flat arch underside filled with plaster before the general ceiling coat of plaster was applied. Fill, usually loose cinders or cinder concrete, was placed on top of the arches and beams to create a level floor. The beams were thus entirely encased in noncombustible material: the fill on top, the first arch blocks on either side, and the plaster below.[27]

The first known use of the patented terra-cotta arch system was in the [Old] United States General Post Office, at the southern tip of City Hall Park, built in 1872–73. "At about the same time that Mr. Johnson introduced his method, a similar but heavier construction was used in New York City by Mr. Leonard F. Beckwith . . . ," the architect of the Post Office.[28] The patent was voided several years later due to competing claims for invention,

thus opening the field to manufacturers. The open competition, coupled with the use of semi-skilled labor in assembling the off-site manufactured arches, led to the extremely widespread use of terra-cotta floor arches during the 1880s and 1890s.[29]

The use of plaster on the underside of the beams was known to be an imperfect form of fireproofing. The plaster was far more vulnerable to shoddy construction and degradation with time than the off-site-manufactured and well-protected terra-cotta blocks. The Mutual Life Insurance Building on Nassau Street in 1883 introduced two important innovations: the soffit tile supported by the skewback blocks (the end blocks of the arches that rested on the beam bottom flanges) and the subdivided interior cell. All prior designs had been in cross section simple closed shapes, usually trapezoids. The subdivided blocks had interior ribs that provided greater stiffness and impact resistance to the individual tiles (figure 5.2).[30] Until this time, the bottom flanges of beams had been exposed or covered with a coat of plaster directly applied to the metal, and so were still vulnerable to heat weakening during a fire. While the failure of an individual floor beam is less important than the failure of a column, because of the smaller tributary floor area supported, the beams

[Old] General Post Office, 1880, Park Row and Broadway, south facade. The first use of patented tile-arch floors. Every visible surrounding building has been replaced.

were far more vulnerable because of rising heat being trapped directly below the floor slabs.

Flat side-construction tile arches, in the form described above, were used for as long as tile construction remained economical. It was not the most advanced floor system available, but it was fireproof and simple to build, and did not seriously decline in popularity until after 1900. The first serious competition for the fireproof floor field was the variation called end-construction tile arches. In this floor system, the internal ribs in the terra cotta were perpendicular to the beams within blocks that had parallel formed sides and sloped cut ends (figure 5.3). This form did not directly increase the fire resistance of the floor, but it was stronger and so had extra capacity in extreme loading conditions. The additional strength came from the solid line of horizontal and vertical ribs that stretched continuously from beam to beam. The vertical ribs, which in side construction are not continuous, provide an additional load path and thus more capacity for the same given amount of material. End arches were also easier and cheaper to create, since all blocks had the same

basic cross section. One long piece of terra cotta could be mechanically extruded, and then cut into slope-faced blocks after firing. The lower cost and additional strength eventually resulted in the almost-exclusive use of end construction at the expense of side construction.[31]

It should be noted that end construction is more modern in that it is not theoretically similar to traditional masonry arches. Side arches are built and act in the same way as a stone flat arch translated to the new material. End arch action is based on the conscious design decision to align the ribs for load carrying, which has no direct analog in solid masonry. The more efficient form is gained at the expense of traditional structural understanding. The only change from the basic end construction form was the adoption of a side "keystone" block within an otherwise end arch. This form, the combination flat arch, provided the end construction advantages with the additional material of side construction at the most delicate location (figure 5.4).[32]

To complete the fire protection of the floor systems, the girders parallel to the arch spans also

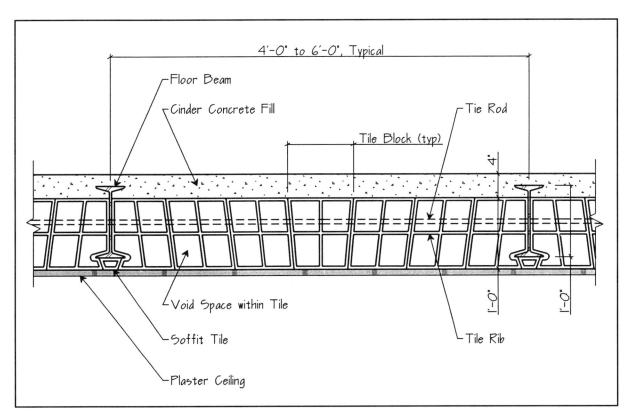

**5.2. Typical Floor Section—Side-Construction Flat Tile Arch.**
**Scale: ¾" = 1' 0"**


**5.3. Typical Floor Section—End-Construction Flat Tile Arch.**
**Scale: ¾" = 1' 0"**

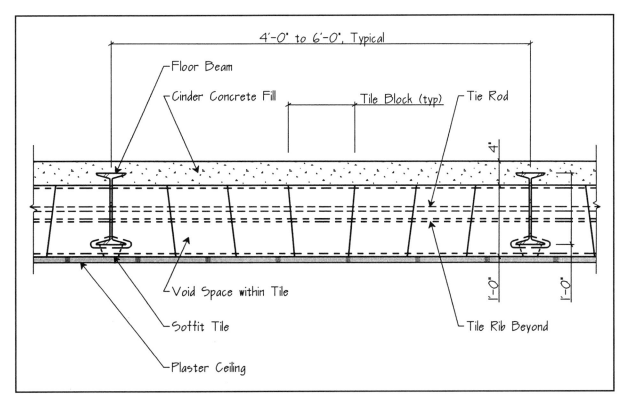

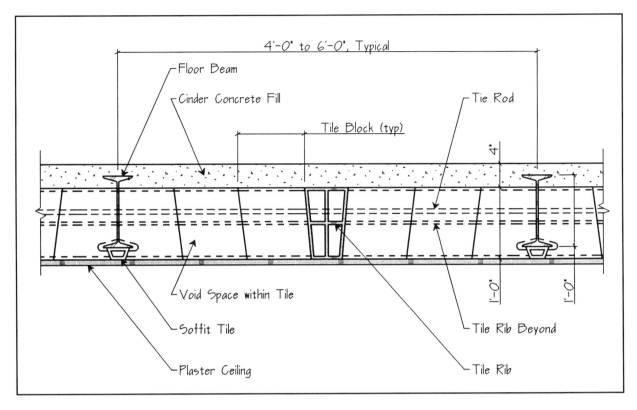

4'-0" to 6'-0", Typical

Floor Beam

Cinder Concrete Fill

Tile Block (typ)

Tie Rod

4'

Void Space within Tile

Soffit Tile

Plaster Ceiling

Tile Rib Beyond

Tile Rib

1'-0"  1'-0"

**5.4. Typical Floor
Section—Combination
Flat Tile Arch
Scale: ¼" = 1' 0"**

had to be covered. This was done with simple rectangular prism blocks of hollow terra cotta. The next logical step was taken by Johnson, in the form of terra-cotta block fireproofing for metal columns.[33] The first external fireproofing for columns, around 1880, consisted of quarter-circle "gores" approximately one foot long that were designed to fit between the protruding flanges of Phoenix columns. The idea was extended to circular arc blocks designed to be laid up around the standard round cast-iron columns, and then finally to rectangular blocks that could be built up around square cast-iron columns and flanged rolled or built-up wrought-iron and steel columns (see figure 3.2).[34]

The first application of column fireproofing was, not surprisingly, inspired by a horrendous fire. An office building on Park Row, between Beekman and Ann streets, formerly used by the *New York World* newspaper, burned in January 1882. The total destruction of a ten-story building, filling half a city block and located in the heart of "Newspaper Row," naturally received attention. The replacement, the Potter Building of 1886, was the first building in the city with terra-cotta column fireproofing.[35]

The construction of noncombustible partitions

around columns was one of the recognized methods of fireproofing interior columns. The New York building laws encouraged a system of double columns in cast iron, the interior load bearing, the exterior ornamental, and the intervening space filled with plaster—later left empty—as insulation against heat. The detail was used for columns supporting sidewalk vaults and was one of the alternatives offered for columns supporting curtain walls, the other option being a brick pier around the column built integrally with the curtain wall (figure 5.5).[36]

## Fire Protection and Water Damage

The use of integral brick piers as fireproofing for spandrel columns may have begun as a fire-protection measure, but it took on a bizarre life of its own. This detail, which has been revealed through facade inspections of recent years to trap water and lead it directly to the vulnerable metal, was commonly used through the 1930s. This defect was known as early as the 1890s, but rather than being used as an argument against brick piers, was used as an argument for cast-iron columns. Finally, the vast majority of buildings that use this detail have columns at all external corners. Building piers at these columns meant

# CASE STUDY: **COLUMN AND PIER DETERIORATION**

The facade of a commercial high-rise was first examined because of extensive cracking at the exterior corners and window heads in the mid-block courtyard. The courtyard facades are unornamented common brick, with numerous corners created by the irregular plan used to provide windows to the interior of the building. A waterproofing company had been retained by the owner to repair the facade, and an engineer was retained at the same time to provide general specifications, monitor the work, and observe the building structure for any repairs needed other than brick replacement. The following observations are a summary of the numerous investigation and construction site visits.

The building was built as speculative office space in 1917, bought in the 1960s, and used since exclusively by a large corporation. The building has received a moderate amount of exterior maintenance, but its back-office use and location next door to the corporation's headquarters building made it a low priority for thirty years. The structure was known to be a steel frame, twenty stories high, with a brick curtain wall.

When the exterior corner in question was first examined from a hanging scaffold, two cracks ran from the roof down five stories, one on each facade roughly 18 inches from the corner. The cracks had opened as much as one inch at the roof parapet. The contractor was instructed to remove the face brick over the spandrel beam-to-column connections at the roof level and the topmost floor. From work elsewhere on the building, it was known that only one wythe of brick was used as covering over these beams. When the openings were prepared, the immediate danger to the building's stability that was present was obvious: the column had rusted away literally to nothing. The brick pier in the corner, which was built as fire and water protection, was supporting the ends of the spandrel beams, which were in

acceptable condition. There was a hollow space in the center of the pier where the column should have been located, but at the roof level the column consisted of a few traces of rust on the brick. One floor down, there was some steel left of the column, but less than half of the original section, irregularly shaped. The scaffold was supported on a roof-mounted frame, rather than being parapet-hung, and so was as safe a work platform as the roof itself. Shoring was needed to support the building structure for the safety of the occupants and to secure the scaffold.

The evidence seen at this location, once the entire corner was stripped of brick from the roof until the limit of damage five floors below, showed that the damage was a cycle of water entry causing the steel to rust, the rust's expansion pushing out the brick, and the displaced brick admitting more water. The cracks aligned with the edge of the steel column, roughly 4 inches from the edge of the pier.

Given that the beams were intact, including most of the plates and angles that made up the riveted top and bottom flange angles and the web double-angle connection, the decision was made to replace the column in its original configuration. Structural steel was used as shoring, attached to the column that was below the limit of damage, and supporting the floor beams on each side at the framing levels to the roof top. As the shoring was erected, stage by stage to support the steel framing of the twentieth floor, the twenty-first floor, the attic level, and the roof, the danger of collapse from workers disturbing the masonry was eliminated. The pier and column remnants were then removed and a new column (in the original built-up box form of two channels and two plates) bolted to the bottom column splice and the "hanging" beam connections. Deteriorated plates and angles in the connections were replaced in kind, and all of the ¾-inch rivets replaced with high-strength bolts.

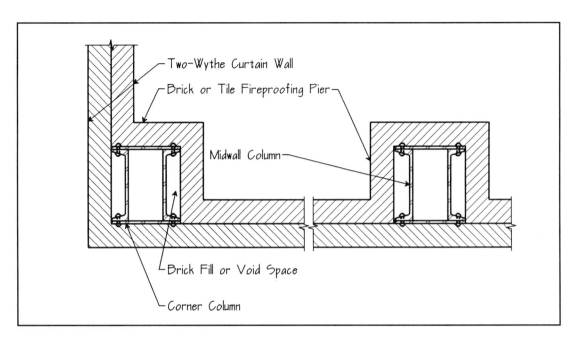

that the locations that in a modern curtain wall would receive expansion joints were instead stiffened, and thus subject to cracking, exacerbating the corrosion of the columns (figure 5.5).

The following, from 1893, is typical of the sort of partisan argument, based on a known design defect, that was made for one of the three "competing" metals:

The constructors and producers of cast-iron advocate its use as the only material for the columns inclosed in the walls. They claim also that the oxide of iron paint so commonly used for coating iron soon dries out, leaving a coating of dry, broken scale or powder. Between the columns and the outer air are only a few inches of brick or some fire-proof material, through which dampness soon finds its way. In wrought-iron, they claim that rust honeycombs and eats entirely through the metal. Mild steel rusts faster than wrought-iron at first, then slower. Cast-iron, on the contrary, slowly oxidizes in damp situations; rust does not scale from it, and the oxidation when formed is of a much less dangerous kind, extending only a little way into the metal to about the thickness of a knife-blade, and then stops for good. Cast-iron of goodly thickness offers a far better resistance to fire, or fire and water combined, than wrought-iron or steel.[37]

In cases where the columns were waterproofed, by the application of asphaltic paint or other materials that bind to the metal surface, the water damage over the life of the building has been reduced. Recognition of this fact existed before 1900:

Advocates of riveted steel columns insist that such columns, when properly encased in fireproof and waterproof materials, as the intent always is that they shall be, are protected permanently from injurious influences. High buildings are erected for permanency, to last for centuries. Years from now the question will be practically determined whether skeleton structures are a wise or foolish method of building, whether they are stable and lasting, or secure and reliable for only a comparatively short number of years.[38]

## Nonstructural Fire Dangers

With hindsight, it is obvious that factors other than the fireproofing system used for the building structure greatly affected building performance during fires. Some of these are long known, others became clear only after well-publicized fires. Iron shutters on lot-line windows other than those on street facades have been a part of the building laws since the 1860s. They were required for the same reason as noncombustible exterior walls: to prevent the spread of fire from one building to those adjacent. The theory was correct, and still has its place in building codes in

A new retail tenant on the first floor of a large residential building on Manhattan's West Side hired an architect to strip the space to the bare structure and design new finishes. The following description is based on the review by the building owner's engineer.

The apartment house, built in 1908, covers a full block. The basic structure is a steel skeleton frame, although the building is old enough so that some secondary structure, such as the basement framing supporting a courtyard slab, uses cast-iron columns and bearing walls. The floors are end-construction tile arches covered with loose cinder fill. Wood sleepers are embedded in the fill and provide nail locations for wood plank flooring.

The new construction required hangers for sprinklers; air ducts; electrical conduit; and a new, complicated, gypsum-board ceiling. The original design showed separate hangers being installed for each mechanical system and the black iron ceiling supports. The architectural detail called for the ceiling hangers to be a steel strap detail described in the New York City building code as a carry-over from the days when tile arches were still being built. This detail is not unacceptable, but it is based on loading the bottom flange of the tile and letting that piece of terra cotta span in bending from one end of the tile to the other. This form of loading is not desirable and certainly should not be used in any great concentration. The mechanical drawings did not specify the connection for the hanger rods, leaving the definition of the word "insert" up to the contractor. The combined effect of the ceiling and mechanical hangers could have been as many as half a dozen hangers in a 4-foot-by-4-foot area. This could not be tolerated by the tile arches.

After the initial review, the reviewing and designing engineers met on site to investigate the existing hung ceiling. This ceiling was scheduled for removal, but the hangers were of immediate interest. The hangers were found to be inserts of the type described above. There was no sign of distress where the hangers met the tile, despite loading by the acoustical tile ceiling, various pipes, and air ducts. Since the new loading was roughly equivalent to the old, the final decision was to reuse the existing hangers wherever possible, and to make all new connections required directly to the steel floor beams. In this manner, new loading of the tile was avoided.

the requirements for fire-rated lot-line windows, but the execution was imperfect. Tenants in the spaces using the windows would want the shutters open, to maximize the amount of light available. Even if the shutters did not actually rust in the open position, their fire-protection qualities depended on the deliberate effort of someone within the building, which could not be counted on during a fire panic or late at night.

The danger posed by flammable interior finishes was also recognized quite early, but to no great effect. Varnished wood flooring was used in fireproof buildings, often nailed to furring strips embedded within cinder concrete fill or concrete slab top surfaces. This was enough, under the proper circumstances, to destroy the contents of the building regardless of its structural integrity.

The danger posed by mechanical shafts and chases was not recognized until the end of the century. Such vertical openings act as flues during a fire, providing escape for smoke and creating a draft to pull in more oxygen. The use of open stairwells, wood-stud wall-enclosed plumbing and elevator shafts, and unstopped, open sleeves through slabs all contributed to the speed with which fires spread through buildings.

## Present Considerations

The major features of fireproof buildings, established by 1900, are still in use today: the skeleton frame supporting a noncombustible curtain wall, the fireproofing of all steel members by covering them with an insulating material, and the use of truly fireproof materials for floors. The changes

**Four Park Avenue, south and east facades. Facade restoration is in progress. Close-up of south wing shows a corner stripped of brick and the steel columns and hung lintels behind.**

**1501 Broadway, east facade. Close-up showing a steel column encased in brick fireproofing exposed during storefront renovation. Vertical "stripes" are the column flange tips.**

since then have been in detail and material—for example, the substitution of gypsum plaster on wire lath and then spray-on cementitious materials and gypsum board enclosures for terra-cotta block as column fireproofing.

Certain details described here may require special attention during renovation or assessment of fire safety. Older forms of tile arches may not have soffit tiles, leaving the bottom flanges exposed or protected by plaster only. The plaster may be capable of providing a modern fire rating, but this cannot be assumed without investigation of the existing detail. Hung ceilings added after the original building construction may not be adequate fire protection (and may, in fact, be unsafe under normal conditions) if they are hung from fasteners attached to the tile instead of to the beams. This detail is not uncommon and, depending on the fasteners used, can be dangerous if small lips of tile flange are supporting the weight of substantial areas of ceiling. Tile arch floors built before 1900 may lack protection for girders that are deeper than the filler beams and so project downwards from the underside of the continuous tile surface.

The use of masonry piers as fire protection can cause confusion in determining the form of the structure. If the curtain walls are more than one foot thick, as was common before the 1910s, the spandrel columns can be "lost" inside the masonry envelope. In the same way, the pier required for fireproofing may not be noticeably thicker than the complete curtain wall, including furring tile and plaster. This can lead to the incorrect conclusion that an old curtain wall is actually a bearing wall. Under normal circumstances the fireproofing pier can inadvertently carry load meant for the columns, a condition that is exacerbated if the columns or beam-to-column connections have corroded substantially. In extreme cases, the pier can be found to be carrying the full column load because the column itself has been reduced to nothing but delaminated rust. Obviously, the condition of the steel must be examined before the masonry piers can be removed to be replaced with less bulky forms of fireproofing.

The architectural problems described above are common in older buildings, and the solution of them is, with the installation of sprinklers, the most common fire-protection renovation performed. Changes in finish rarely affect the structure, but the replacement of stairs or enclosure of open stairwells may. In wood-joist interiors, such as those used in rowhouses and tenements, the framing around the stair is often the weakest in the building. The addition of partition loads in this area can be impossible without structural support such as using the new partitions as stud walls.

# 6. STANDARDIZATION OF STEEL FRAMING
## 1870s–1921

Standardization of building materials and fabrications, in the sense of the standardization in machinery production that took place in the United States during the first half of the nineteenth century, was slow in arriving. It was as bizarre a concept as any that builders and designers had to deal with. Metal skeleton construction could be related to timber-framed buildings without a tremendous leap of faith, but since every building had been one of a kind and hand built, the rationale for standardization was difficult to find. It seems to have come primarily from the wrought-iron and steel fabricators, the people in construction closest to the industrial revolution.

In timber and masonry buildings, there was a certain level of standardization: joists were cut 3 inches by 8 inches or a similar size; bricks were manufactured in a limited number of sizes that still exist, such as common and roman; and ashlar masonry was often cut to common sizes. These standards were the minimum necessary for building to take place, and were as much imposed by the industries that cut trees into lumber and produced brick as by any perceived need for "interchangeability" in the industrial sense or for industrialization of the building process. Cutting timber and masonry into regular sizes is common to construction of any historical era, and as such is not an industrial standardization.

Cast iron, in general, falls into the same category as wood. It was produced in certain regular sizes, but not as part of a common system of standard shapes. Because cast iron is produced from handmade molds that were difficult to make, the iron laborers were among the most skilled workers involved in the building trades, ranking with skilled masons. As long as cast iron was used in an architectural manner, the molds for each facade element were different. A particular column form might be reused in a number of buildings, but there was no relation between that column's exterior and the exterior of any other column. The greatest degree of standardization in cast iron was in the interior connections, either those hidden within the floor for interior columns or those hidden in the rear of an iron facade. These connections, consisting of interlocking flanges, were similar in many buildings. Most often, the similarities follow the individual iron works' pattern, but late in the cast-iron era, after 1880, certain connections became relatively common. The most frequently repeated cast-iron connection is an interior beam-to-column connection consisting of a seat plate with vertical stiffening lugs below the beam and a vertical lug bolted to the beam web.[1]

The common iron connections fail one simple test for true standardization: rational design. The seated connections used in cast iron produce relatively large bending moments in the columns, tension in the seat lugs, and possibly load transference through the bolted web lug, none of which were specifically designed. The connection sizes were based on the beam sizes, and most likely work because of the inherent conservatism of the high floor live loads used in the iron era and the lack of significant wind forces in low bearing-wall or cage-type buildings.[2]

Wrought-iron and steel sections, in contrast to

the materials described above, were produced entirely by machine. The rollers for a mill were hand cut to produce the correct beam profile, but those rollers, once production began, would turn out endless repetitions of the same beam. Because of the difficulty in working metals, mechanical tools were used to punch or drill holes for bolts and cut the shapes as required for a specific project. A series of floor beams for a building, even if rolled in different mills and worked in different shops, would be similar enough to be interchangeable within the ±½-inch tolerance so often used in frame construction.

The relative economy of wrought-iron floor beams was one of the primary reasons for their quick adoption as a standard in the 1860s. The displacement of wrought iron by steel twenty years later was also largely economic. Truly standard connections and patterns for creating built-up columns and plate girders created a significant economic advantage by permitting the use of prefabricated pieces and more or less unskilled labor. The large rolling mills were financially interested in the national advance of their industry, as opposed to the mere sale of product, in a way that the smaller, local casting foundries had not been. The Carnegie Company published its first handbook in 1876. This book listed the design properties of the various shapes, showed standard connections and details, and had tables of design aids (e.g., maximum distributed floor loads for various beam sections) intended both to sell steel construction to engineers and to provide for standard usage. During the 1880s the other steel companies published similar handbooks. These books were reprinted frequently until the publication of the first AISC *Steel Manual*, and still exist in reduced format.[3]

## American Standard Beams

The first lot of wrought-iron beams rolled in Trenton were called "rail beams" or "bulb-tees." This section, with a wide bottom flange, short thick web, and bulbous top flange, had several advantages at the time.[4] The shape was familiar in a smaller size from railroad construction, and the wide bottom flange provided a seat for brick floor arches that the narrower top flange did not interfere with. Given the theoretical information available at the time, many builders believed that the shape was inherently more efficient than doubly symmetric sections. This incorrect assumption was based on experience with cast-iron beams.

Most of the column and beam experimental data available before the 1880s was based on experiments that Eaton Hodgkinson and William Fairbairn had performed in England during the 1830s and 1840s. Given the technological edge that Britain had over the United States, those results were probably as accurate as any needed here before the 1870s and the beginning of tall building construction. Steel was not a building material at the time of the experiments, so the tests had concentrated on cast iron for columns and had been split between cast and wrought iron for beams. Cast iron's brittleness, irregular material properties, and weakness in tension led Hodgkinson to concentrate on cast-iron beams with bottom flanges in the range of six times as large as the top flanges. This reduced the tensile stress in the bottom flange to acceptable levels, while taking advantage in the top flange of the high allowable compressive stress. This form became known as a "Hodgkinson beam."[5]

Wrought iron and steel, properly worked and rolled, have almost exactly the same materials properties in tension and compression. A Hodgkinson beam—or its close equivalent, a rail beam—is a waste of metal in either ductile metal. The allowable load on such a section could be increased by equalizing the flange areas. In addition, the heavy web that is required to prevent explosive shear failures in cast iron could be reduced in ductile metals, although this reduction was introduced gradually.

The advantages of wrought iron over cast iron for flexural members had been established by the early experiments. Once Cooper & Hewett solved, for the United States, the technical problems of rolling iron beams, the advance was relatively rapid. By 1856, Cooper & Hewett were rolling true I-beams, 9 inches deep.[6] Other companies followed into the field, which expanded to include, by 1875, I-beams up to 15 inches deep.[7]

The vast majority of I-beams rolled in wrought iron and those rolled in steel before 1900 share certain geometric characteristics. Unlike modern beams, the depth designation is the exact depth in

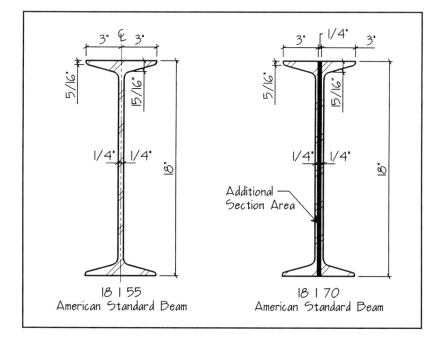

**6.1. American Standard
Section Comparison
Scale: 1½" = 1' 0"**

inches of the beams. Beams of a single depth from a single manufacturer increase in section by increasing web thickness, with an identical increase in flange width. This can be visualized best (although obviously the description is not related to the actual manufacturing process) by imagining a beam sliced vertically through its web centerline with a new "plate" of material inserted between (figure 6.1). As in modern sections, a size was designated by its depth in inches and weight in pounds per running foot (e.g., 12" I 31½#)—although the weight was sometimes given in pounds per yard rather than the more common and usable pounds per foot. The interior flange surfaces (the bottom faces of the top flange and the top faces of the bottom flange) were sloped, almost always more than 8 percent and usually 16⅔ percent from horizontal. These characteristics can be related to the manufacturing process, where slabs of red-hot metal are rolled between contoured rollers. The weight of beam sections of a given depth was increased by moving the rollers fractionally further apart, thickening the web and increasing the flange width by the same amount. The exact shape of the flanges was dependent on the contour cut into the rollers, and thus varied from company to company and, on occasion, from mill to mill within one company. Sloped flanges were produced by triangular indentations in the rollers, which were used to ensure complete

penetration of the hot metal down into the rollers' voids.

Once the idea of "I"-shaped beam sections had truly taken hold, and the manufacturing process had been thoroughly developed, the differences between beam sections produced by different mills and companies were easily reduced. Since everyone used the same manufacturing techniques, and since differences in construction were created during the post-rolling fabrication process, rolled beams were a commodity, not the specialty that cast-iron shapes had been. Nothing presents this more clearly than the fact that no changes were required for builders to switch from wrought-iron beams to steel. Other than the metallurgical properties of the beams, the two were indistinguishable. (An obvious question thus raised is the honesty of builders during the metal transition period: Who would know if wrought-iron beams were substituted for the steel specified? This is one of the reasons for coupon testing in alteration projects.)

The Association of American Steel Manufacturers (AASM) adopted a classification of "American Standard beams" in January 1896. The standards were defined in terms of weight per foot, flange width, interior flange slope, and web thickness. Other sections had previously been called "standard," but the term had no true meaning before the AASM classification.[8] The specifications included material properties and chemical composition for open-hearth and Bessemer steel.[9] This specification, a precursor to ASTM specifications a few years later, provided that an engineer could call for a "12" I 31½" without specifying a company and without wondering if accidental variation would reduce the safety of his design.

Shapes other than I-beams underwent a similar process of simplification around the same time. Many of the rolled shapes available in 1890 became less common in the following decade, although few totally ceased production. Bulb angles, where one leg is teardrop-shaped, and bulb channels, with one teardrop flange, were never common shapes, although they were occasionally used in columns, where the bulbous flange helped increase the radius of gyration without increasing the column's overall dimensions. Zees were used in a similar manner, and also were used in certain types of built-up beams. Zees were more readily mixed with other sections than the bulb shapes, since both flanges and the web could be riveted, and were used in locations where angles or small channels would be used today, such as roof purlins and small posts. Ultimately, the zee is an unstable shape, since it is almost impossible to load in bending without causing torsion that the shape is not capable of withstanding. Angles share this flaw and so are rarely used in flexure except for extremely light loads or in situations where they are restrained against rotation. Ultimately, the decreasing price of I-beams and true channels led to the substitution of them in situations where other shapes and built-up beams had been popular.

## Wide-Flange Beams

Despite code changes, the actual steel members used in modern buildings are very similar to those used one hundred years ago. Although coupon testing is recommended in any case where the load on existing members will be increased or welding to existing members is contemplated, the tests often find that the steel is metallurgically similar to currently used mild steel. The theoretical stresses allowed in those members vary from the old building codes to the first AISC code to the current code, but the actual stresses calculated by any single method are similar. While there have been many refinements in theory, such as plastic design and LRFD design, there have been two refinements in construction worthy of note: the use of specialized steels and the use of wide-flange sections. The first will be addressed in chapter 10, "'Modern Steel' Construction," since it is a post–World War II phenomenon; but the second was a major factor in the increase in flexibility of steel framing during the first half of the twentieth century.

American Standard shapes were rolled in the same manner as plates, angles, channels, and zees: red-hot slabs of steel were passed between a series of horizontal rollers a set distance apart, which squeezed the steel down to the required web thickness and forced it into grooves in the rollers that were of the same cross section as the required flanges. In this method, the flanges were not truly rolled, but instead were formed by "dragging" the

hot metal into the flange "form" within the rollers.[10] This does not necessarily produce members with consistent properties throughout their section, and it is strictly limited in the width of flange it can safely produce. Extremely few American Standard shapes with flanges over 6 inches wide were ever rolled.

Basic physical mechanics theory is responsible for the I-beam shape: In a homogeneous material such as wrought iron or steel, the largest useful moment of inertia, the gauge of a member's resistance to bending, is produced by an "I" with the flanges perpendicular to the direction of load. For a given depth of web, the most efficient use of additional material is to increase the area of the flanges. This is almost opposite the means of increase in size for American Standard shapes. The majority of heavier sections were created by simply setting the rollers further apart. This is efficient in the mill, but it increases the area of section for a given depth by increasing the web area. Because of the limitations on forcing steel into the flange cutouts in the rollers, there was no practical method to roll more efficient sections.

The limitations on the moment of inertia in available sections were not very important for typical floor beams in an era where the beam span was rarely over 20 feet and the beam spacing rarely over 6 feet. Even in early bearing-wall and cage buildings, however, the American Standard shapes were inadequate to serve as girders. The most common solution to this problem was to use two beams side by side as a girder. A "double girder" was a complicated trade-off of material and fabrication costs: each girder was a standard beam section, and therefore cheap and readily available, but the connections to columns were far more complicated than for single girders and the gain in section strength was not great enough to allow for increased spans. There were code provisions for small iron "spacers" to tie the pairs of beams together; but because all beam-to-girder connections in such a situation are inherently unbalanced, uneven loading or improperly designed alterations can create torsional loading of one or both of the beams, and therefore reduce the beams' flexural capacity.

The other common method of creating a section large enough for girder loading or a greater

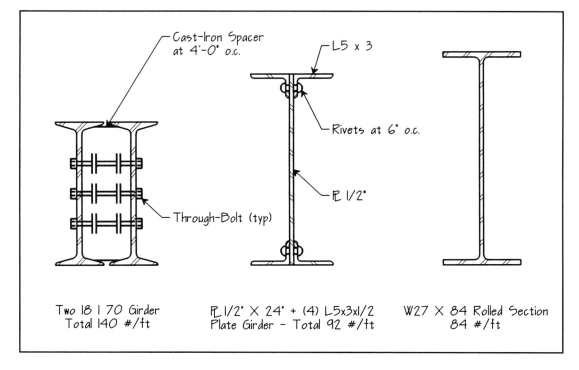

Cast-Iron Spacer
at 4'-0" o.c.

L5 x 3

Rivets at 6" o.c.

℔ 1/2"

Through-Bolt (typ)

Two 18 I 70 Girder
Total 140 #/ft

℔ 1/2" X 24" + (4) L5x3x1/2
Plate Girder - Total 92 #/ft

W27 X 84 Rolled Section
84 #/ft

than usual span was to build up a heavier "I" section from plates and angles or, more rarely, from plates and zees (figure 6.2 shows a girder section comparison). By using large angles in pairs to make up the flanges, efficient heavy sections could be riveted together. This scheme appeared in wrought iron as early as 1859 in the A. T. Stewart Department Store on Broadway and 9th Street.[11] While structurally efficient and flexible in possible use, plate girders required relatively large amounts of expensive, skilled, time-consuming labor in fabrication. They have never been cost-efficient for use as typical girders in buildings, and were only used in that capacity in exceptional circumstances, such as overlarge spans.[12] Riveted plate girders are also inherently less efficient in their use of material than rolled beams with the same section properties (or welded plate girders, which were not a realistic option until well into the twentieth century). The loss is due to the reduction in area of the tension flange caused by the rivet holes; the compression flange is not considered reduced because the loads can be transmitted through the rivets. A typical provision was that provided in 1887 in a foundry handbook, where the safe loads for built-up wrought-iron girders noted this loss of section and reduced the allowable flexural stress from 12,000 psi to 10,000 psi to compensate.[13]

The permanent solution to creating larger-capacity beams was the invention of mills capable of rolling "wide-flange" beams. The definition of wide-flange beams is difficult to pin down because the current selection of "W" sections includes beams with flanges narrower than some American Standard beams. The first distinguishing feature of wide-flanges is the rolling method; all other features are derived from that. The "Universal" or "Grey" rolling mill has four rollers—two major horizontal rollers which create the web and the inside faces of the flanges, and are narrower than the beam is deep, and two vertical rollers which finish the outside faces of the flanges. The first Universal mill was patented in 1887, and was popularized by Henry Grey from then to the turn of the century.[14]

Because each portion of a beam rolled by a Universal mill is directly rolled, the material properties are more consistent than for American Standard, and no inherent limitation is imposed on the proportions of the sections created (figure 6.3). The four-roller mill allowed easy creation of beams with reduced slope or even parallel face flanges. Today, this is the primary means of identification of wide-flange shapes. In addition to the extra strength that wide-flanges provide when used as beams, the wider flanges provide larger moments

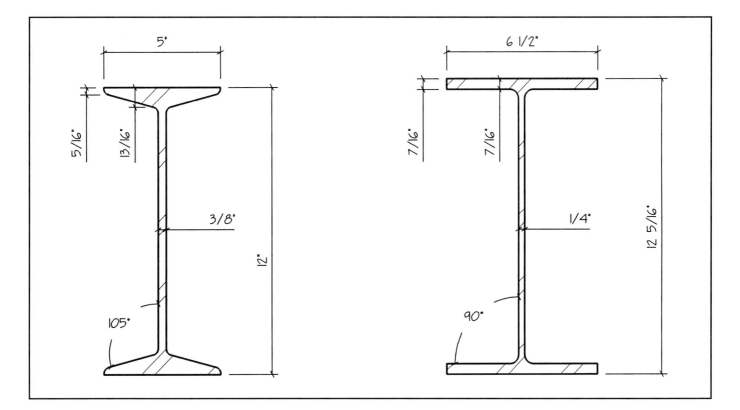

of inertia in the weak axis, parallel to the flanges. This is marginally important in beam design, where it allows greater unbraced lengths, but is extremely useful in column design.

American Standard shapes were not typically used as columns because their relative weakness in one direction greatly reduced their load capacity. The Bethlehem Steel Company introduced wide-flanges into common use as columns, specifically to replace the built-up plate "I" and box shapes previously used. The wide-flange shapes are suited to column use because of the high moments of inertia around both axes. They became so popular in that capacity that the phrase "Bethlehem column" was used for years to indicate wide-flange columns.[15]

By 1907, the Bethlehem Steel Company was advertising three classes of wide-flanges, all with a 5 percent inner flange slope: "special I's," "girder beams," and "H columns." These three classes have gradually blended into one another, but they led to most of the current "W" shapes. The special I's were designed to replace American Standard beams in ordinary use, with similar moments of inertia with a beam weight savings of approximately 10 percent. The girder beams were

designed with moments of inertia equal to two of the lightest American Standard beams of the same depth, with a beam weight savings of approximately 12½ percent. The primary difference between the two series was a proportionally increased flange width for the girder beams. The H columns were wider still, with a flange width equal to the nominal section depth, to provide the maximum possible resistance against weak-axis buckling.[16]

The practical use of the Bethlehem sections were best explained in a 1922 company handbook:

> The Bethlehem I-beams are designed to give greater inertia than the standard I-beams for the same weight, by making the flanges wider and the webs thinner. The Bethlehem girder beams are intended to take the place of built-up girders and are much heavier than the Bethlehem I-beams for the same depth, the increased weight being due to the wider flanges and thicker webs.[17]

Other companies gradually began rolling similar or identical sections, usually with parallel flanges, although the first edition of the AISC

**6.3. Beam Section Comparison**
**Scale: 3" = 1' 0"**

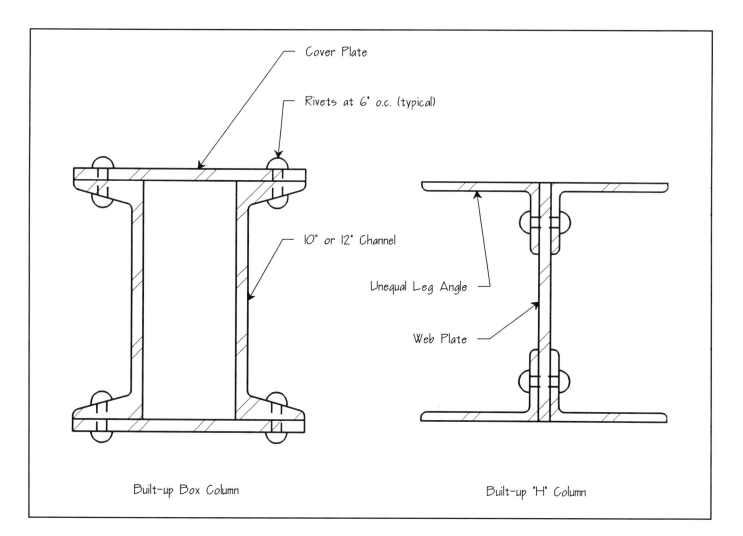

Cover Plate

Rivets at 6" o.c. (typical)

10" or 12" Channel

Unequal Leg Angle

Web Plate

Built-up Box Column

Built-up "H" Column

**6.4. Ordinary Built-up Column Sections**
**Scale: 3" = 1' 0"**

## Column Types

For more than sixty years after the introduction of rolled beams, wrought-iron and steel columns were commonly built up of plates, channels, and angles. The amount of additional labor required to rivet together the individual pieces of each column was enormous, especially when the relatively high number of columns needed in the past for a given floor area is taken into account. The extra expense of rolled-section column assembly was one reason that cast-iron columns were popular for so long after engineers had begun to doubt the material's safety. There was no difference in detailing between wrought-iron and steel columns, and no change in column form was caused by the total surrender of the wrought-iron market to steel during the 1890s.[18] The only difference between the

two metals in column form was the popularity of Z-bar columns in wrought iron, which was not repeated in steel. This distinction was caused by the decline in the popularity of zees in favor of more stable sections simultaneous with the decline of wrought-iron use, not by any connection between the two trends.

The shape of built-up column sections was based on the idea of increasing the moments of inertia about both axes as much as possible for the cross-sectional area provided. Since the technology of driving rivets was basically unchanged after 1870, there was little change in the form of built-up sections from their initial popularity in the 1870s until their decline in the 1910s. The two most popular shapes were boxes built out of two channels and two plates and "H's" built out of a web plate and four flange angles, with two flange angles added for heavier sections (figure 6.4). Among the less-common types, zees could be

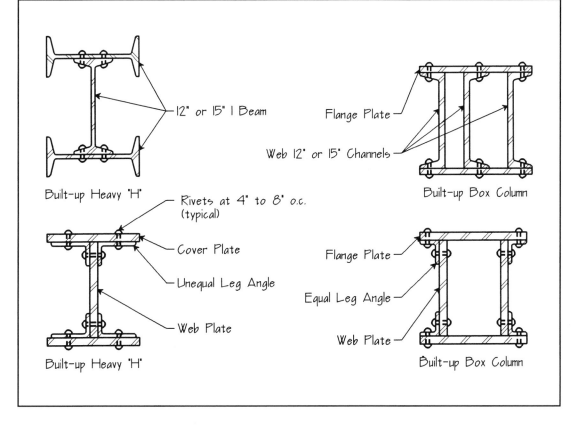

used in groups of four to form an "H" with extra flanges, which could, for lighter columns, provide easier connections to beams.[19] Phoenix columns, originally patented by the Phoenix Iron Works but later produced by several other manufacturers as well, were round wrought-iron columns composed of four, six, or eight circular arc sections with flanges at each side radial to the circle's center. Variations included octagonal and square arc sections. The sections were assembled into full circles and riveted through the flanges. The flanges also provided the location of beam connections. The primary advantage of Phoenix columns was that, as basically circular sections, they provided the highest possible moment of inertia about any central axis for their weight. For heavy use, the columns were often reinforced with plates cutting across the circle's diameter at beam connections, in order to better distribute the imposed load to all of the column area. The decline in their use, started by their proprietary nature to the Phoenix company and the peculiarities required in detailing wind connections to their flanges, was completed by the failure of Phoenix or any other company to roll them in steel.[20] Phoenix columns seem to

have been most popular in the 1880s in cage buildings without true wind systems, where their detailing difficulties could be kept to a minimum.

By the 1920s, when built-up columns were dying out in ordinary use, the basic box and "H" sections described above were still popular. For heavier loads, becoming more common as buildings grew taller and column spacings grew wider during the skyscraper boom, boxes with a third channel in the center, triple-webbed H's and H's of three rolled beams were used (figure 6.5).[21]

Built-up sections were needed for columns as long as American Standard shapes were the only rolled-beam sections available; efficient column design requires sections with fairly similar moments of inertia about both axes. American Standard sections, with their narrow flanges, required much more material to provide a given column capacity than a more efficient wide-flange. An 1893 explanation of the popularity of built-up box columns stated: "It is a well-known fact that the metal near the neutral axis of a column is good for little, and that the capacity of equal areas varies as the metal is removed from the neutral axis."[22]

After the introduction of wide-flanges as Bethlehem columns in the years after 1900, they became the most popular column shapes when their advantages could be used. As late as 1924, built-up columns were still considered to have an economic advantage under certain conditions: when complicated beam-to-column connections were needed and the wide-flange sections had flanges or webs thicker than one inch; when material price was of overriding concern; and when column loads were extremely high.[23] The first condition was expensive because the rivet holes, ordinarily punched, had to be drilled in thick material, which was several times more expensive per hole than punching. Built-up sections typically were made larger by adding more relatively small pieces; wide-flange sections of a given depth series grow by increasing web and flange thickness. The material price was a concern because of fabricators' concern about using shapes primarily produced by only one company, even if that company was the second largest in the country. The last condition was the result of the still small number of sizes of wide-flange columns available. For very large loads, engineers often specified wide-flange columns with flange cover plates. The riveting needed to attach the cover plates was just as onerous as that required to assemble a box column out of channels and plates—possibly worse if the holes in the wide-flange had to be drilled and those in the built-up column could be punched.

## Column Formulas

One of the few areas where theoretical knowledge and practical construction directly intersected was in the design of metal columns. The theoretical basis for beam design is relatively simple once the basic assumptions of material homogeneity, lack of section distortion, and small deflections are made. Each of these assumptions is readily defensible, and in numerous tests made during the nineteenth century, wrought-iron and steel beams followed the predicted elastic behavior. Column theory was not as easily simplified. The complexity of column analysis is caused by the extremely nonlinear behavior of slender elements under compression. At some load well below the material yield stress, a slender element buckles sideways, losing almost all of its ability to resist load. Unlike braced-steel or wrought-iron beams, which

will exhibit continuous linearly increasing deflection under increasing load until the bending stress reaches the yield stress and gradual bending failure results, columns loaded almost to their buckling load appear as strong as they did when unloaded. This appearance is deceiving, because a small percentage increase in load will result in almost instantaneous failure. The situation is still more complicated because of rolling defects that cause initial out-of-straightness, initial internal stress in the members from unequal cooling after rolling, and the impossibility of loading a column in a perfectly concentric manner. Initial curvature, initial stress, and eccentric loading lower the stress at which buckling takes place, but not in a readily apparent manner.

The mathematician Leonhard Euler introduced a theoretical column formula in the 1700s based on initially straight and unstressed, perfectly concentrically loaded columns. The formula is dependent only on the smallest moment of inertia for the column section (or the smallest radius of gyration, which is directly related), the length of the column, and the elastic modulus (the stiffness) of the material. Euler's formula variations are usually stated in the form

$$F_a = \frac{K \cdot E}{\left(\dfrac{l}{r}\right)^2}$$

where $F_a$ is the allowable stress, $K$ is a constant based on the mathematical derivation and including the safety factor, $E$ is the elastic modulus, $l$ is the unbraced length of the column, and $r$ is the smallest radius of gyration. Euler's formula, by ignoring the complicating factors, can relate directly the stress at which a column buckles to its slenderness. This implies that buckling is the only important factor in determining the strength of a column.[24] This assumption, and the stresses predicted by Euler's formula, are increasingly inaccurate as the column under consideration shortens.

A stub "column" only several inches long fails, according to common sense or when tested, at the material yield stress. Long, slender columns fail by buckling at stresses very near those predicted by Euler's buckling formula. The columns in between these two extremes in slenderness are not clearly defined by similarly straightforward theory, and unfortunately are the columns most often used. There is no need for columns so large as to be stocky enough to fail by yielding, and columns of ordinary floor-to-floor height in buildings rarely carry so little load as to make possible the use of sections slender enough to fail by Euler buckling. Faced with the need to define allowable stress in situations where it had not been mathematically defined, engineers fell back on a position that could be defended by logic: they used existing tests of columns, and occasionally performed their own tests, and then fit curves to the test results that could be reduced to usable formulas.

The creation of column formulas required specialized knowledge possessed by few engineers and no laymen. In retrospect, this field fell between beam formula design, which also required above-average mathematical knowledge, and building structural system selection. Once the formulas for various beam loadings and conditions were written, they were fixed; today they are still used in almost exactly their original forms. Building system selection was largely a matter of personal taste for the designers until the 1910s, when pure skeleton frames finally became dominant. The different types of column formula were debated in the engineering press, especially in the *Transactions of the American Society of Civil Engineers*, for years. The result was a form of standardization: everyone involved agreed on the general trend of the experimental data and agreed on a small number of formulas designed to explain the data.

Four families of column formulas exist, resulting from various mathematical models used to fit experimental data. The most popular during the nineteenth century, and the first of the "intermediate" column formulas to be introduced, was the Rankine formula, later slightly modified and presented as the Rankine-Gordon formula. This formula class is of the type

$$F_a = \frac{F_c}{1 + \left(\dfrac{1}{F_b}\right) \cdot \left(\dfrac{l}{r}\right)^2}$$

where $F_c$ is the allowable compression for a short

or braced column and $F_b$ is the allowable bending stress.[25]

The formula can be seen as the maximum possible allowable compressive stress, $F_c$, reduced by a factor representing the increasing likelihood of buckling with increased slenderness. This is a form of analysis common to the intermediate column formulas and noticeably missing from the Euler formula, where the allowable stresses do not enter into the calculation. The intermediate formulas, based on empirical data, are less abstract in their treatment of the material.

Another common formula was the Straight Line or Johnson's formula, introduced in 1886. The New York building laws used Straight Line formulas from 1901, when they replaced Rankine-Gordon formulas, until 1923, when the AISC Rankine-Gordon formulas were substituted. The Straight Line formula is the easiest to understand: from the maximum allowable stress at a continuously braced column, the allowable stress was reduced in direct proportion to the increasing slenderness. The formulas are of the type

$$F_a = F_c - K \cdot \left(\frac{l}{r}\right)$$

They were typically used with the Euler formula, with the designer switching from one to the other, as the column slenderness increases, at the curves' point of tangency.[26] A Straight Line was used for columns by the code of the American Railway Engineering Association (AREA) before 1935.

Two newer forms were the Parabolic and Secant formulas, named after their curves. The Parabolic formulas were of the form

$$F_a = F_c - K \cdot \left(\frac{l}{r}\right)^2$$

the Secant formulas,

$$F_a = \frac{\dfrac{F_y}{S}}{1 + K \cdot \sec\left(\dfrac{l}{r}\sqrt{\dfrac{S \cdot f_a}{E}}\right)}$$

where $F_y$ is the material yield stress, $S$ is the factor

of safety, and $F_a$ is the actual axial stress.[27] As might be expected, the complex form of the Secant formula made few friends, and it was used primarily in the codes of the AREA.[28] The AREA and the American Association of State Highway Officials (AASHO, now the AASHTO) both used Parabolic formulas in the 1940s.

The first New York law to explicitly state a column formula, in 1892, used one of the most common formulas of the late nineteenth century:

> The strength of all columns and posts shall be computed according to [Rankine-]Gordon's formula, [$F_a = F_u /(1 + Kl^2/d^2)$, according to the code's reference, $F_u$ = ultimate compressive stress or "crushing weight," $d$ is the least column dimension, and $K$ is 1/800 for cast iron, 1/3000 for wrought iron or steel, $K$ being experimentally determined and unitless] and the crushing-weights in pounds per square inch of section, for the following-named materials shall be taken as coefficients in said formula, viz.—Cast iron, 80,000 [psi]; wrought or rolled iron, 40,000; rolled steel, 40,000.[29]

The slenderness measured by $l/d$ was limited to 20 for cast iron and 30 for wrought iron or steel. It is interesting to note that no distinction was made between steel and wrought iron, and that the allowable loads are substantially higher for cast iron than for either ductile metal. These results were somewhat leveled by the larger safety factor for cast iron. By 1898, the formula had not changed except for the substitution of 48,000 psi as steel's ultimate compressive stress.[30]

Three years later, New York had switched to a Straight Line formula for all three metals, substituted the more accurate $l/r$ for $l/d$ as the measure of slenderness, and imposed maximum slendernesses of 70 for cast iron and 120 for wrought iron or steel.

For cast-iron columns,

$$F_a = 11300 \text{ psi} - 30 \text{ psi} \cdot \left(\frac{l}{r}\right)$$

for steel columns,

$$F_a = 15200 \text{ psi} - 58 \text{ psi} \cdot \left(\frac{l}{r}\right)$$

and for wrought-iron columns,[31]

$$F_a = 14000 \text{ psi} - 80 \text{ psi} \cdot \left(\frac{l}{r}\right)$$

Here the switch to steel in most construction is apparent, as is a certain loss of confidence in cast iron. This type of formula remained in use until the AISC code was issued. As a recognition of the problems with cast iron, its formula was revised down to

$$F_a = 9000 \text{ psi} - 40 \text{ psi} \cdot \left(\frac{l}{r}\right)$$

while the steel formula was revised to[32]

$$F_a = 16000 \text{ ksi} - 70 \text{ ksi} \cdot \left(\frac{l}{r}\right)$$

The New York code did not instantly recognize the AISC specifications, and as late as 1926, still had the revised Straight Line formulas.[33]

In examining buildings, the AISC code is of primary importance. The first edition of the AISC code, in 1923, used a Rankine-Gordon formula,

$$F_a = \frac{18000 \text{ psi}}{1 + \left(\frac{1}{18000}\right) \cdot \left(\frac{l}{r}\right)^2}$$

with an absolute maximum of 15 ksi. The second edition was identical, except that the previous maximum slenderness of $l/r$ of 120 was raised to 200 for secondary and bracing members. (For perspective, it should be noted that ordinary building columns rarely exceed an $l/r$ of 80.) The third edition in 1937 switched to a Parabolic formula,

$$F_a = 17000 \text{ psi} - .485 \text{ psi} \cdot \left(\frac{l}{r}\right)^2$$

for main members, with a maximum allowable slenderness of 120. Secondary and bracing members continued to use the old formula, with a maximum slenderness of 200. The fourth edition continued with these formulas. The fifth edition, in 1947, still used these formulas, but for the first time allowed main columns to have slenderness as high as 200, using the formula

$$F_a = \left[17000 \text{ psi} - .485 \text{ psi} \cdot \left(\frac{l}{r}\right)^2\right] \cdot \left[1.6 - \left(\frac{1}{200}\right) \cdot \left(\frac{l}{r}\right)\right]$$

where the last term serves as a modification factor

increasing the safety factor as the slenderness approaches 200. The 1963 sixth edition of the AISC code was the first to introduce ASTM A36, which has now become the most commonly used steel, as well as high-strength steels. The column formula introduced can be considered a modified parabolic formula, although it has been modified so heavily as to be almost unrecognizable, and is substituted with Euler's formula at large slendernesses.[34]

An examination of the development of these formulas reveals an interesting fact: the design of columns for pure compression became steadily less conservative over the years. There are a few aberrations, but the change can best be attributed to further research during and after World War II on the behavior of metal structures, triggered by aircraft design.

## Live-Load Reduction

Early live loads used in building codes seem unrealistically high when compared with current standards. The 1892 New York code required a minimum of 75 pounds per square foot for all building interiors, regardless of use, with higher values for heavier loading.[35] Dwellings of any type, including high-rises, are now designed for 40 psf except for public hallways and lobbies. This conservatism is the natural result of a new field of building regulation: the earliest versions of any portion of a code are usually more restrictive than necessary. This perhaps can be attributed to the mindset of legislators who have been persuaded to include new material but have not yet heard enough about its real-life effect to worry about easing it.

Anyone familiar with live-load code provisions knows that the actual load, most of the time, is noticeably less than the design load. The amount of variation in load from day to day depends on use. The actual load in a residence is made up primarily of furniture weight, and is far more consistent than the weight in retail areas, where both furniture and crowd weights are important factors. Open assembly spaces have the largest variation, since all of the live load consists of transient population weights. While there is no way to predict that any given piece of a floor will not be loaded to the full design value (or occasionally beyond), large areas rarely are. For most load types, the larger the area considered, the less likely it is that it will all be fully loaded. The concept is often stated in terms of probability theory; the probability of full loading at any given time is inversely proportional to the area involved. This behavior is limited to spaces used for consistent, daily uses, such as residential or office space. Assembly spaces are filled to capacity on occasion, no matter how large they are.

The first recognition of this probabilistic effect was in the 1900 Chicago and New York building codes. Both allowed reduction of the live load calculated for columns in high-rise buildings, because the probability of all of the floors being simultaneously loaded was small enough to ignore. In reality, even if all floors were loaded simultaneously, a column designed with reduced load would not fail, since the increase in load would most likely be less than the safety margin. The New York provisions were extremely conservative for live-load reduction. Columns with a total height over five stories could have their load reduced. The full code live load was used at the roof and top floor; a 5 percent reduction from the full value was allowed at the next floor down, and an additional 5 percent each floor down, to a minimum live load of 50 percent of the full value at the lower floors.[36]

Live-load reduction has always been optional, and even today, many buildings that could be designed with it are not. Designing for reduced loads produces a lighter frame, which is more subject to sideways drift, local deflection, and vibration. In a "prestige" building, or one that is noticeably slender or has thin floor slabs, and therefore is already subject to serviceability design problems, live-load reduction is often neglected. When the first code provisions for reduction were introduced, these considerations were not as important as they would later become, because buildings were designed in general to be stiffer than modern buildings. The relative lack of old buildings designed with live-load reduction is probably more a function of general designer conservatism and the concentration of building cost in labor instead of in material. Reduction would not be widely used if it were suspect and simply saved a few pounds of steel.

The lack of enthusiasm for live-load reduction, as well as its position as a new and less-restrictive part of the code, kept the provision for live-load reduction the same in the New York code through 1926, a period that saw liberalization of many other portions of the code, including the design live loads.[38]

## Connections and Wind Bracing

The history of engineers' and builders' conversion to steel skeleton frames described earlier is largely a description of the maturity of structural riveting. Riveting is an ancient technique, but one not applied to building in any important way until the use of wrought-iron beams began. One of the primary conditions of successful riveting is that both the rivets and the base metal pieces to be joined must be ductile. This eliminates from consideration all materials other than metal, as well as non-ductile metals such as cast iron. The great cast-iron facades of the 1850s, 1860s, and 1870s were all bolted together. This detail made the style of construction possible, but it robbed those facades of much of their potential rigidity. Bolt holes in cast iron were typically cast with the piece, rather than being drilled later, which was far more expensive. (This does not apply to the smaller connectors needed to hold in place applied architectural ornament, such as the acanthus leaves at Corinthian capitals. They were often attached with screws using undersized pilot holes or no pilot holes.) The bolt holes, being oversized by approximately an eighth of an inch, allowed bolt movement that kept the joints from rigidity.

Early wrought-iron construction was often bolted as well—in part because there was no perceived need for rigid joints, and in part because the technology of riveting had not yet caught up. As long as wrought-iron beams were used simply as floor beams and usually supported on cast-iron columns, the only place that riveted connections could have been used was in beam-to-girder connections, where there was no great advantage of one type of connector over the other. In the modern AISC codes, beam-to-girder connections remain one of the few places that unfinished bolts, which are almost identical to the bolts of 120 years ago, can be used safely. The relatively small advantage to using riveted construction in bear-

**St. Paul Building, at the foot of Park Row (at Broadway), before 1912, facing south, north and east facades visible. One of George Post's few skeleton-frame buildings. The heavy cornices and street facades are carried on outriggers from the spandrel beams; the plain brick curtain wall at the side lot line is more clearly indicative of the construction form. The tall building in the distance is American Surety.**

The first large, prominent building to take advantage of the idea was the *New York Times's* new headquarters at 42nd Street and the intersection of Seventh Avenue and Broadway (renamed Times Square at about the time of the construction). This building, constructed between 1904 and 1909, used the full reduction allowed on its twenty-five-story-high columns.[37] For engineers, the use of reduction in this building is made even more interesting by the building's shape: in plan it is a narrow triangle squeezed between the converging avenues, requiring wind bracing as elaborate as the similarly sited Flatiron Building of 1903.

ing-wall or cage buildings probably delayed its general application. Rigid connections are of the utmost importance in wind bracing, to prevent joint slippage under lateral load. The stiffness of a skeleton frame of a building several hundred feet high can be substantially reduced by a small fraction of an inch slippage in its bracing connections or in the "rigid" joints between wind girders and the columns. The majority of early skeleton frames built in New York had their rivets driven while red-hot by hand, by laborers using mauls. As early as the 1850s steam-driven semi-automatic riveting machinery had been introduced. Because of the size of the machinery involved, this was not particularly useful for field connections.[39] During the twentieth-century skyscraper boom, shop connections were cold-riveted using heavy hydraulic presses and field connections were riveted (usually hot) using smaller, hand-held pneumatic guns.[40]

When examining the evidence on connections around 1890, certain contradictions are found. The building code had fallen behind the advance in technology, and referred to "iron or steel" beams being "strapped, bolted, anchored, and connected together and to the walls in a strong and substantial manner."[41] This sounds good, but it leaves the form of the connections undefined, despite a few additional clauses giving minimum sizes of connection angles. The maximum shear given for bolts and rivets was identical, which was either unsafe for the bolts used or, more likely, overly conservative for the rivets.[42]

Two buildings by George Post, the Havemeyer Building of 1892 and the St. Paul Building of 1899, illustrate well the level of sophistication in connection design at that time. The Havemeyer Building was another of Post's cage buildings and, at 193 feet high on a 60-foot-wide base, extremely slender for the structural form. Post was too good an engineer to make the mistake of relying on the detached bearing wall for resistance to lateral load. The building had a portal-braced wind system of rigidly riveted wind girders and columns, but Post was still concerned about deflection during construction, and so had threaded rod cross-braces built in. The area exposed to the wind is greater after construction, when the building is enclosed, than during construction, when it is a bare skeleton. The inclusion of cross-braces specifically

Flatiron Building, Fifth Avenue and Broadway, at 23rd Street, north facade. Towers as slender as this have been constructed of masonry, but rarely for office use, and never with glass storefronts at the base. In 1902 this building was a dramatic statement of steel framing.

designed for construction is a clear sign that the exterior wall was meant to be used for wind deflection control, if not specifically for wind-load stresses. Another interesting aspect of this building is that the bracing, despite being described as temporary, was left in place permanently, implying that in the final analysis Post had more confidence in the steel than in the masonry. The portal connections consisted of relatively shallow brackets attached at the tops and bottoms of the wind girders.[43]

The St. Paul Building is one of the few true curtain-wall buildings Post designed, with its exterior wall carried at each floor on double spandrel beams. The wind system was a series of portal connections as described above, but enlarged for the taller building height.[44]

A description of common steel framing in 1902 showed that certain forms had become the standard for tall buildings. The assumption was that the steel skeleton would resist all applied forces

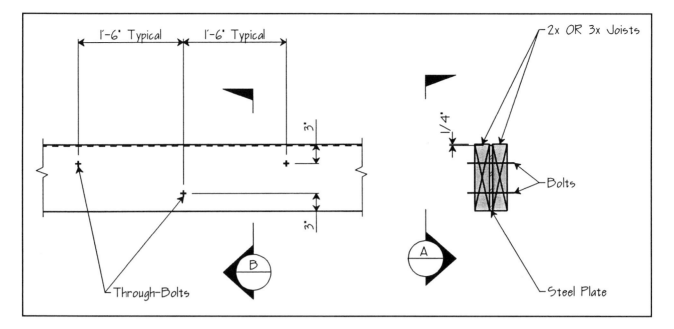

and carry the masonry curtain wall. Beam-to-girder connections described were the same double-angle forms common today, riveted together. Wind connections were riveted rigid connections, occasionally made stiffer by the use of knee braces or portal frames. Diagonal bracing was used where possible, where the blocking of passage through some internal bays was architecturally acceptable.[45] These details had hardly changed twenty years later, except for the elimination of portal frames as commonly used framing. Wind connections ranged from riveted top and bottom flexible angles on small or stocky buildings to triangular gusset-plate stiffened brackets on taller and slender buildings.[46]

The only technology that at the turn of the century held any promise of superseding riveting in the creation of rigid connections was resistance welding. Welded connections have the advantage that, if properly designed, they are as rigid as a connection can possibly be because the separate pieces of metal cannot move relative to one another along the line of weld. Various welding techniques were patented and improved after 1877, including resistance welding, arc welding, and thermite welding. Welding was expensive at that time and difficult to use on heavy structural shapes; it found its niche in railroad-car building and some aspects of ship building.[47]

One of the few applications of welding in building construction was in the roof trusses of

the Electric Welding Company of America in Brooklyn in 1920. The majority of the structure was riveted, but the exposed truss joints were welded. Given the limited application of welding in the building, and the requirement of the Brooklyn Department of Buildings for a full-scale load test before construction, this appears to be more an isolated forerunner of structural welding used for its advertising value than the beginning of modern connection types.[48] Structural welding in buildings did not begin in any substantial way until after World War II.

## Joists and Miscellaneous Steel

The description of connections given above applies to the connections of rolled sections in ordinary steel-framed connections. There are other areas where similar standardization took place, notably miscellaneous steel for framed buildings, particularly lintel details, and the various types of members used to fill the gap in strength between ordinary wood framing and rolled steel beams.

Structural steel has an allowable bending stress in the range of twenty times that of wood, depending on the wood species. This translates fairly directly into the actual difference in strength between wood beams and steel beams, since the floor-plate thickness is more or less the same regardless of the construction materials. Architects could not allow extra depth for the floor structure simply because wood is weaker than steel; often the floor plate is thinner in the residential buildings where wood can be used. There was, and still is, a need for beams that do not have the strict limitations on strength and length that are associated with wood but are less expensive and less massive than structural steel.

One solution to this problem is the oldest form of iron beam: the flitch plate (figure 6.6). As soon as the rolling of plate iron in large quantities became practical in the 1840s, builders began experimenting with strengthening wooden beams by bolting iron plates onto them.[49] This form of construction, while cheaper than iron-beam floors, was not noncombustible, and so was not used in any of Manhattan's building booms to any appreciable extent. Experiments were performed in the 1870s that confirmed the action now counted on in flitch plates: if the different materials were fastened together correctly, then the overall flexural strength of the beam would be equal to the sum of the ordinary strength of the wood section plus the strength of a wood section the dimensions of the plate, increased by the ratio of the moduli of elasticity of the wood and iron.[50] While never common anywhere, and outlawed in most of Manhattan by the fire laws for almost the entire era under discussion, flitch plates were occasionally used. One example is the Wucker Warehouse, at Third Avenue and 125th Street, built in the 1880s before the fire line had reached so far uptown. Its floors are supported on wooden joists, carried on 16-inch-deep flitch plates spanning between wooden and cast-iron columns.

Another old idea that has continued in minor usage is the creation of structural members from light-gauge, cold-rolled iron or steel sheets. The design of such members is complicated because of the extreme thinness of each piece of a section. When flanges and webs are on the order of an eighth of an inch thick, buckling concerns dominate all design in a way that the heavier rolled sections avoid. Light-gauge construction is limited by the capacity of the members to shorter spans and lower buildings than rolled steel, although it has found consistent use since the 1940s as an incombustible replacement for wood-joist and stud framing. Light-gauge steel's most common use now is as the framing for partitions.

The use of light-gauge iron in New York dates back to 1839, when galvanized sheet iron was introduced as a roofing material. In the form of corrugated sheets, this roofing became popular for industrial buildings by the 1850s.[51] In 1855, "the building for the Bank of the State of New York was constructed. That building contained members which were cold formed from $\frac{1}{16}$ in. and $\frac{1}{8}$ in. thick steel sheets."[52] This building is another example of an idea whose time had not yet come: joist construction remained an oddity until after 1900, when its use in industrial roofing and as a replacement for wood became more common.[53] The typical details in the 1920s were similar to those used now: cold-rolled sheets of steel were bent into "C" shapes, with return flanges as stiffeners at right angles to the outer edge of the top and bottom flanges. During the 1920s and 1930s, they were often used with a thin slab of concrete on expanded mesh to provide a totally incombustible floor.[54] In 1946, this form of construction was finally accepted as legitimate with the publication of the American Iron and Steel Institute's "Specification for the Design of Light Gage Steel Structural Members."[55]

Starting in the 1930s, structural-steel prefabricated assemblies, in the form of open-web joists, were used where the loads were smaller than normal or where the depth of the supporting beams was not critical. The joists are nothing more than small Warren trusses, most often with a pair of angles for the top and bottom chords and rods for the web members. Long-span joists use angles for the webs as well. Open-web steel joists were used from the beginning as a substitute for wood joists, placed at close spacing and supporting thin concrete slabs on expanded metal mesh (figure 6.7).[56] The design of open-web joists follows the AISC steel specifications, with certain additional constraints because of the extreme weakness of the joists out of their plane.[57]

Open-web joists can be seen as the descendants of the "tension-rod girders" produced by Daniel Badger in the 1860s. Those girders were combinations of cast iron in the compression flange and wrought iron in the tension flange, and were built expressly to take advantage of the characteristics of the two metals.[58] The tension-rod girders showed, in a spectacular fashion, that the majority of material in the web of an ordinary

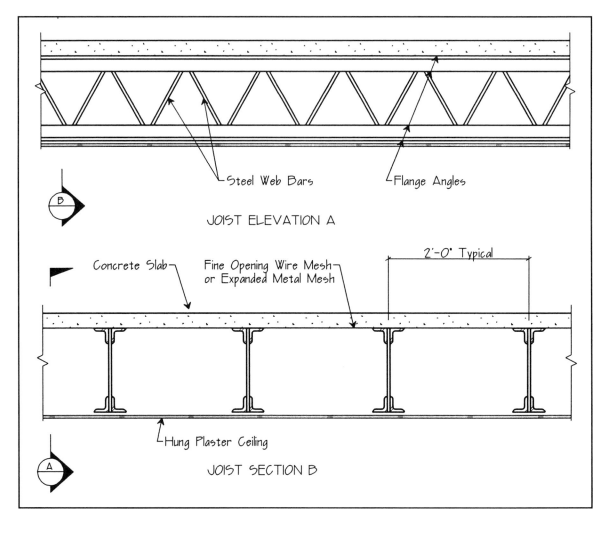

Steel Web Bars    Flange Angles

JOIST ELEVATION A

Concrete Slab    Fine Opening Wire Mesh
or Expanded Metal Mesh

2'-0" Typical

Hung Plaster Ceiling

JOIST SECTION B

**6.7. Typical Floor Section—Open-Web Steel Joists**
**Scale: ¾" = 1' 0"**

beam is stressed well below its capacity, and can be removed safely. This economy is seen today everywhere in open-web joists and, primarily in Europe, in castellated beams.

The switch to skeleton frames and supported curtain walls created a new class of framing: hung lintels. Unlike steel beams, which were conceptually identical to wood beams, or concrete slabs, which were conceptually similar to masonry arches, hung lintels were a structural requirement for supporting masonry walls that extended both above and below each floor level of a building. The only precedent was the use of small beams embedded in masonry walls to serve as door and window lintels.

In traditional masonry construction, wall openings were usually spanned by an arch built integrally with the wall. In the form of low segmental arches, this solution was occasionally still used in the early twentieth century in New York in low-

rise apartment houses with brick bearing walls. The other traditional solution, used only when hidden, was to embed a wood beam in the wall over the opening. Provided that the wall above the opening is solid for a height approximately equal to the opening width, the loads within the masonry will spread sideways into the portions of wall on either side of the opening. This is usually referred to as "arching action," because the distribution of loads resembles that created by a hypothetical semicircular arch over the opening: the vertical load distributed to the sides and a horizontal thrust created at the elevation of the opening top. The only load not supported by this action is the masonry directly over the window, below the "arch." A lintel installed over the window serves to support the wall during construction of the area over the window and to support that area once the wall is completed high enough to permit arching action to begin. These loads are relatively small, and in a high wall with identical openings at each floor, the lintel loads will be identical at each floor.

During the 1850s, as iron construction became popular, cast-iron beams were used as lintels in brick walls. The ordinary objections to cast iron's use in flexure were less important for lintels. The loads on a lintel are typically dominated by the wall dead load, and therefore are subject to fewer and smaller fluctuations of the kind that might cause a sudden failure. Also, since lintels are entirely buried within the supported wall, they can be shaped in whatever form is structurally expedient. For cast iron, this usually took the form of an "E" on its back for a two-wythe wall, or a variation for thicker walls (figure 6.8). This shape provided additional material and lower stress for the tension face of the beam, again reducing the likelihood of tensile failure.[59] This detail was used through the end of the nineteenth century, and lost its last popularity during the general dismissal of cast iron in the years after the Darlington collapse.[60]

Wrought-iron angles and I-beams were first used for lintels several years after cast iron, but became the standard by the 1890s, particularly for longer spans. The 1892 New York building code required that lintels, as well as ordinary beams, longer than 12 feet be constructed of wrought iron or arched cast-iron beams with wrought-iron tie rods. At this late date, there was no longer any question that wrought iron was cheaper and more reliable than tied cast-iron beams.[61]

Once wrought iron (and, by extension, steel) had been established as the lintel standard for long spans, it was a small step to use it for hung lintels. The earliest form of a hung lintel was a beam or several small beams set at the support elevation and supported by outriggers, or small cantilevers, from the main building columns. This is a messy detail that can require an extra set of spandrel beams at every floor, and always requires hung lintels capable of spanning the full bay width. After more efficient lintels had been devised for ordinary floor construction, this detail was still used for heavy cornices.[62]

Wall profiles varied widely from building to building, because of varying dimensional requirements for the different degrees of ornamentation

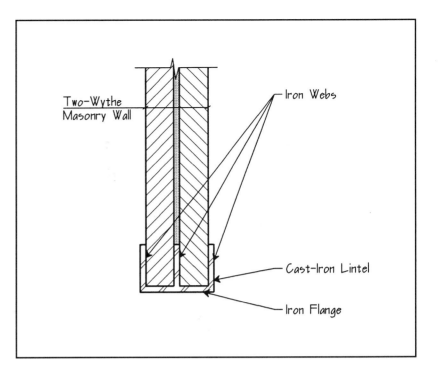

and material, varying practices within different architectural offices, and the evolution of the building code. The code progressed from requiring curtain walls nearly as thick as bearing walls to allowing the minimum thickness of wall that could resist wind loads. The most common forms were hung directly off the spandrel beams (figures 6.9 and 6.10). The spandrel beams were usually multiple small beams, although in the 1910s this was usually reduced to two I-beams or one wide-flange. A flat brick wall would be supported directly on top of the floor slab, on top of the beams, except for the outermost wythe, which ran continuously past the spandrels. This facing was usually supported by a series of angles supported by a zee connection from the bottom flange of the outermost spandrel beam. As the amount of ornamentation protruding from the wall increased, more angles, small channels, or small I-beams were added to this basic detail until all of the pieces were supported.[63]

The typical tall building during the 1910s, 1920s, and 1930s had a two-wythe curtain wall, with the inner wythe supported directly by the framing and the outer by a hung lintel. The wall was built in the "common bond" or "American bond" pattern, which consists of simple running bond with a header course every sixth or seventh course. The headers serve to tie the wall together,

**6.8. Cast-Iron "E" Lintel**
**Scale: 1" = 1' 0"**

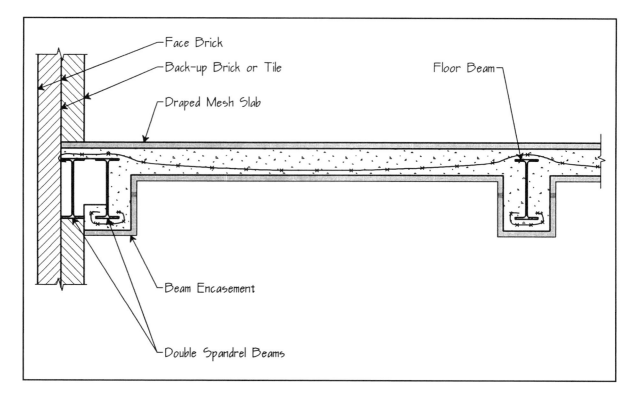

Face Brick
Back-up Brick or Tile
Draped Mesh Slab
Floor Beam
Beam Encasement
Double Spandrel Beams

**6.9. Curtain-Wall Double Spandrel Beam Section**
**Scale: ¼" = 1' 0"**

and often a header course was placed near the top of the spandrel beams, at the lowest point at which the outer wythe could be tied to the inner. Brickwork is less precise than steel, and the recommendation made in 1924 that "[s]upports for brick, terra cotta or stone should be bolted to the spandrels instead of being riveted, if the building codes will permit" was to allow the final adjustments to the steel to even out the masonry.[64]

In 1928, the first building in New York with corner windows was built. This feature, which was popular in Art Deco apartment houses, requires a right-angle hung lintel. This is slightly more complicated than an ordinary lintel in that the lintels cannot be braced to prevent rotation in the ordinary manner. The solutions to this problem were never standardized, partly because corner windows were always in the minority and partly because the architectural details varied so widely. In the same way, cornice support details are different on every building, with outriggers (sometimes referred to as "outlookers") that di-

**6.10. Curtain-Wall and Shelf-Angle Spandrel Section**
**Scale: ¼" = 1' 0"**

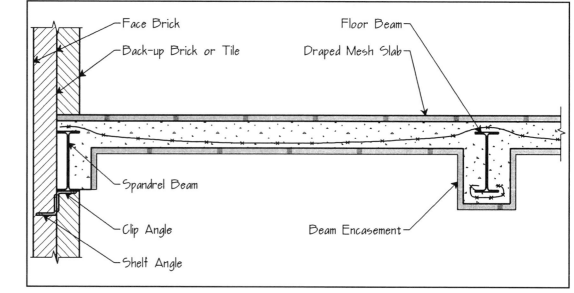

Face Brick
Back-up Brick or Tile
Floor Beam
Draped Mesh Slab
Spandrel Beam
Clip Angle
Shelf Angle
Beam Encasement

# CASE STUDY: OBSOLETE FRAMING

During a commercial redevelopment project in midtown Manhattan, two small commercial buildings with historic street facades were to be incorporated within a new office and retail complex. One building's interior structure was wood joists, which had to be removed to meet the building code fire provisions, but the other building's interior structure was scheduled for reuse.

The building was built in 1907, long after skeleton construction had become the standard for large commercial buildings. It is a midblock bearing-wall building, with brick party walls on the side lot lines, a brick rear facade, and an ornate stone front facade. The five floors and roof were all flat tile arches on steel I-beams, with one girder line running down the center of the building. The girder is a double I-beam, originally running between two round cast-iron columns. The first- and second-floor front windows are storefronts filling nearly the entire facade width. The masonry above is carried on double "I" spandrel beams pocketed into the masonry on the ends and supporting outrigger angles to carry the second- and third-floor cornices.

The building served as a store and offices without incident for seventy years. The planned combination of the building with surrounding properties required the removal of the rear wall and one side wall, along with the column supporting the girder adjacent to the side wall. The steel alterations were simple in concept, requiring the re-support of the beam end previously bearing on the rear wall and the girder end previously bearing on the column at new girders, part of the new steel frame being erected adjacent to the building. In order to get the beam ends to align with the new girder locations, which were fixed by the architect's location for the new columns, extensions were full-penetration welded on the beam flanges and webs, making them roughly one foot longer than they were originally. The increase in moment from the increased length was offset by the decrease in dead load provided by the change in floor system described below.

The necessity of removing portions of the tile arch floors and removing the tie rods in the end spans made it necessary to remove all of the tile arches. The "domino effect" of arch thrust ensured that an uncontrolled movement of the end arch during demolition would allow the adjacent arch to move and then to collapse, and so on. Rather than saving tile arches that would be invisible to the tenants and would constitute no more than 10 percent of the retail floor area, the design allowed for replacement with a concrete slab on composite steel deck. This floor system, in addition to the obvious advantages, was lighter and reduced the total dead load by roughly one-third. The metal deck was placed on small channel bolsters to provide the correct top of slab elevations.

rectly supported the cornices as cantilevered beams, and brackets from the columns supporting a secondary set of spandrel beams as the two most popular types.[65]

## Codes

The trend toward standardization included an increasingly broader scope for material codes. The steel company handbooks from the 1890s and the first quarter of the twentieth century show different values for different cities, especially for the more complex issues such as column formulas. By 1923, when the first national specification from an impartial source, the American Institute of Steel Construction's *Steel Manual,* was issued, officials across the country had all managed to accept certain basics, such as an allowable bending stress of 16,000 psi (16 ksi).

Before national standards existed for the use of "nontraditional" materials such as steel, unqualified legislators were put in the position of codifying structural engineering work in a rational manner. In some areas, such as floor construction systems, they failed, and the results skewed the selection of materials and techniques. Floor systems, despite the difficulties caused by advancing technology, could be compared to the traditional masonry arch floors. Steel construction, on the other hand, was not readily accessible to any attempts by laymen, and the legislature wisely let recognized authorities serve as technical references. The 1892 New York building laws cited "Trautwine's Handbook" for computing material actual and allowable stresses, revised in 1898 to "Haswell's Pocketbook."[66] By 1901, the actual numbers were being quoted in the code, but came for the most part directly from the Carnegie Company manuals.

In comparing the New York building codes between 1900 and the publication of the first edition of the AISC code with the steel company handbooks, few variations from the pattern can be found. The allowable bending stress for braced beams was fixed at 16 ksi in all of the codes until the AISC raised it to 18 ksi. The AISC, as a professional organization encouraging research (and, of course, the increased use of structural steel), was in a better position to examine the experimental and usage-based empirical data on stress than any other group. The increased allowable bending stress was a recognition of the overly conservative nature of the existing codes for ASTM A9 steel, and of the consistent quality of steel being produced.[67]

For comparison, the allowable shearing stresses in rolled sections, rivets, and bolts gradually rose, from a range of 7 to 10 ksi in the 1901 New York building laws to a range of 10 to 13.5 ksi in the AISC code. All other allowable stresses, with the previously noted exception of column compressive strengths, showed similar behavior: a gradual national uniformity to the values, coupled with a slow increase, usually ending with another increase in the 1923 AISC specification. The subsequent editions of the AISC specification raised the allowable stresses as new types of steel were introduced and as further research allowed more accurate calculation of safety margins.

The national steel codes allowed a greater uniformity in the individual specifications written for each project. This was critical to standardization of steel-frame construction, and was recognized in a 1924 detailing text:

For building work, the architects generally write their own [steel] specifications, adapted to the building codes of the cities in which the buildings are erected. The American Institute of Steel Construction has issued specifications which are intended to standardize structural building work throughout the country.[68]

## Present Considerations

The designer looking to modify a steel-framed building built since the 1920s will not find many surprises in the form of the framing connections or members. On the other hand, older buildings must be checked for many possible irregularities, some of which have been discussed above. Obviously, the older the building in question, the more variation from fully evolved steel framing design may exist. In buildings built before 1870, the wrought-iron floor beams may possibly be irregular shapes such as bulb tees rather than true "I's." In buildings this old, the chemical composition and homogeneity of the iron are suspect, and changes in loading are grounds for a chemical analysis of the metal and inspection of its condition.

The other completely obsolete form of framing abandoned by the 1890s is the use of cast-iron beams. These are not found in ordinary locations in buildings dated later than the 1850s, but because of cast iron's well-publicized resistance to corrosion compared with wrought iron, cast beams were sometimes used in sidewalk vaults and other high-exposure locations.

In buildings from the second half of the nineteenth century, especially those dating from the boom period in the middle and late 1890s, the metal composition of beams is still questionable. True wrought-iron beams were still used on rare occasions as late as 1900, and the steel from that period sometimes has a carbon content closer to wrought iron than modern steel. High concentrations of sulfur or phosphorus are also sometimes found, which can result in the metal being brittle and unweldable.

In examining steel beams in general, but especially those used before 1896, the geometric elements of the sections must be carefully measured to aid in specifying the section. Simply knowing the same nominal section depth and weight is not enough, since many different sections were rolled with the same names, and may vary in area and section moduli up to 10 percent in extreme cases.

The forms of old members can cause unexpected loading conditions when analyzed for new loading, and sometimes even under existing load. Combination girders consisting of several I-beams tied together with spacers were designed as if the load would be distributed evenly among the individual beams. This assumption may be reasonably accurate for loads distributed along the length of the beam where the spacers are close together, but it is far less acceptable for concentrated loads such as beam reactions. If new framing is added that includes attaching framing to the outside beams of a group tied into a girder, the stresses in the individual beams cannot be assumed equal. Cast-iron lintels may not be adequate in flexure for concentrated or live loads under modern analysis.

In the same manner, Phoenix columns, built-up box columns, and hollow-section cast-iron columns are inherently subject to gravity moments when loaded through beam connections. The beams are attached to the surface of the column, eccentric from the column axis by half of the column width, creating a moment in the column equal to the beam reaction times that eccentricity. Older columns were not always designed for the effects of this eccentricity, and should not be loaded even to their theoretical capacity based on tributary floor area and the original floor loads unless checked for combined axial stress and bending. Any new loads must be analyzed not only for the axial force they add to the column, but for the direction and magnitude of the gravity moments they induce as well.

The form of the connections on old members can also cause trouble when analyzed in modern terms. Cast-iron columns may be spliced and connected to beams with loose bolts, inadequate in shear for bracing the column section. Wrought-iron and early steel floor beams may be connected with small, loose bolts whose metal does not meet minimum requirements. Early hung lintels and secondary framing may be inadequately braced against movement, and may cause torsion in spandrel beams or moments in columns not accounted for in the original design.

Finally, buildings erected before the building code started specifying material properties in the late 1890s may have been built using material property standards from a specific foundry or other unreliable source. In general, this should not cause unsafe conditions, because the general level of stress allowed was so low, but if possible, the bending and shearing stresses used for beams and the column formula used should be back-calculated if substantial modifications are to be made.

# 7. FLOOR SYSTEMS
## 1890–1940s

The overwhelming popularity of tile arches during the 1880s and 1890s followed the advances in other building components. The substitution of tile partitions for brick walls, the use of increasingly complete metal frames for building support, and the increasing use of machinery on site all led to faster construction of larger buildings. Traditional brick arches, in addition to being extremely heavy and expensive, were slow to build and required carefully built falsework for their segmental shape. Tile arches were a better floor system in every respect but one, the fragility of the individual pieces during construction. In addition to being far lighter, the popular flat arches required only a flat support surface from falsework during erection, and depended on their geometry, not mortar joints, for stability. As soon as a tile arch floor was assembled it could carry its full load, which is never true of traditional masonry.

Tile construction was used for building elements other than flat floors to gain the same advantages listed above. Pitched roofs from the 1880s to the 1910s were often made of "book tile," thin tile plates with two concave edges and two convex edges that interlocked to form a continuous surface inside and out. This type of roof was relatively quick to construct and fireproof, the two advantages of tile construction that also led to the widespread use of tile partitions. These two tile forms were used even in buildings with concrete floors, due to the lack of viable alternates at that time.

Despite the success of tile arches, builders continued to experiment with other materials and other designs for providing floors in steel-frame buildings. Tile arches were a huge step forward

from brick arches, but, to a lesser degree, they shared many of brick's failings. Tile arches still required skilled labor to assemble, required falsework during erection, were constrained in geometry to simple vault shapes supported by parallel beams, were vulnerable to impact from above before their protective cinder or cinder concrete fill was placed, and were too strong. This last does not seem like a problem until the whole process of building design is considered as an attempt to provide the aesthetics and serviceability desired as economically as possible. If the tile arch floors in a given building were capable of supporting three or four times the load that would be required to cause failure in the steel skeleton, extra effort was being wasted in the construction of the floors. While no one has ever taken this logic to the extreme of saying that all building components should fail under the exact same conditions, it is generally recognized that economical construction depends on roughly equivalent levels of service from all components and systems.

## Tile Arches and Testing Alternatives

"The common mode of filling-in between the beams of a floor was the brick arch; but this is largely out of use [in 1893], and in its place a cheap material is now extensively used, which consists principally of burnt fire-clay."[1] At that time, the only alternative to "practically indestructible" brick arches and the tile described was the less-than-appealing "[c]oncrete arch . . . composed of broken stone, fragments of brick, pottery, and gravel, held together by being mixed with lime, cement, asphaltum, or other binding surfaces."[2] From two-floor fireproof systems commonly available in 1890, literally hundreds of minor

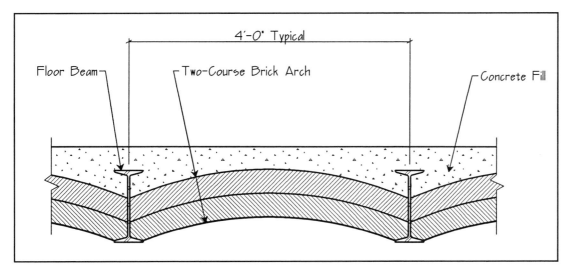

variations arose in the next twenty-five years. The proliferation was not due to perversity on the part of the professionals involved, who can hardly have wanted such a confused situation, but was caused by ambiguity in the New York City building code.

The pertinent code clause read:

All brick or stone arches, placed between iron floor beams, shall be at least four inches thick and have a rise of at least one and a quarter inches to each foot of span between the beams. Arches over five feet span shall be properly increased in thickness, as required by the Superintendent of Buildings, or the space may be filled in with sectional hollow brick of burnt clay or some equally good fire-proof material, having a depth of not less than one and one-quarter inches to each foot of span, a variable distance being allowed of not over six inches in the span between the beams.

This first appeared in the building code in 1891, and remained, in modified form, until 1916.[3] This clause represents an attempt to legislate proper fireproof construction by politicians with little practical building knowledge. By 1892, brick arches had become extremely rare except in sidewalk vault construction, where exceptionally heavy live load and continual exposure to freeze/thaw cycles kept from use both tile arches and the primitive concrete then available. Stone arch floors were practically unknown after the 1870s. Brick arches (figure 7.1) were the primary focus of the law then in force, with the actual floor

systems regularly used relegated to "alternate" status. Since the law as written was devoted to the details of brick arch and tile arch floors, the actual requirements for

"some equally good fire-proof material" as a substitute for brick arches, but the extremely vague manner in which this phrase was worded, the eager competition existing among manufacturers, and the fact that any new floor system, to be approved, must have been passed upon by the Board of Examiners, of which the membership was largely non-professional, brought about a rather chaotic condition in building affairs.[4]

The first alternate allows without exception segmental tile arches meeting certain specific requirements (figure 7.2). Segmental arches were always less frequently used in commercial buildings, and rarely used in residential construction, because of the additional thickness of the floor plate when compared with flat floor arches and the additional cost of providing a hung ceiling. Segmental arches were used relatively more frequently in industrial and warehouse buildings, where a scalloped ceiling was not an aesthetic problem and where the higher live loads were more efficiently resisted by the somewhat stronger segmental form.

The second alternate is vaguely worded, and eventually led to two intertwined industries: the production of patented floor systems, and the load and fire testing of the systems. Since any noncombustible floor system that could pass the

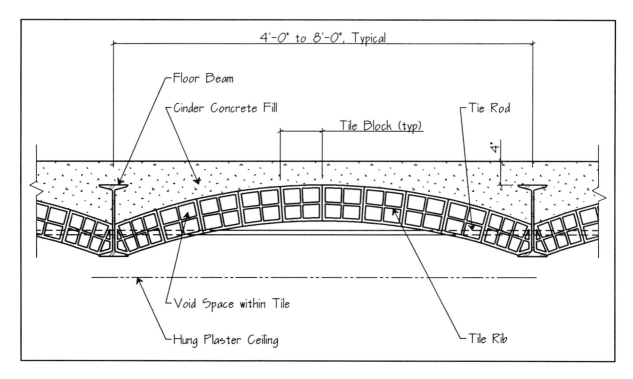

4'-0" to 8'-0", Typical

Floor Beam

Cinder Concrete Fill

Tie Rod

Tile Block (typ)

4"

Void Space within Tile

Hung Plaster Ceiling

Tile Rib

**7.2. Typical Floor Section—Segmental Tile Arch**
**Scale: ¼" = 1' 0"**

Board of Examiners was permitted, any feasible floor system was eventually tested. The extra step in the development of new systems introduced by the tests was inevitably politicized, although it appears from the records that the tests themselves were honestly run. When the tests began in earnest in 1896, many exceptions had been allowed, some for political reasons. These floors had to be tested years after their use to establish a rationale for their acceptance: "The lack of exact, impartial knowledge respecting many of the various systems of fireproofing in use led the Superintendent of Buildings in New York City, Mr. Stevenson Constable," to test all floor systems in current use in 1896.[5] The most pressing need was to provide certification for flat tile arches, which were not yet specifically allowed but had been popular for twenty years.

The basic form of the test was to load a sample panel of the floor system to 150 psf, subject it to a heating of 2000°, cool it with 60 psf streams of water meant to represent firemen's hoses, and then reload it to 600 psf for 48 hours.[6] These conditions were specifically meant to be more rigorous than ordinary service conditions, even those encountered in most fires. This was partly due to the desire to ensure building survival during fires long enough to permit occupants to escape, which has evolved into the modern concept of a timed

fire rating. The overly conservative tests were also based on the realization that the test panels built for the tests received far more care than ordinary floors built on construction sites. Finally, the nineteenth century had seen a number of large-scale fires—the 1871 Chicago fire being the largest—which had created conditions approaching those in the center of a firestorm. In the center of such a concentrated fire, building materials would be exposed to heat greater than that which a normal building fire could produce.

The number of tests conducted far exceeded the actual number of floor systems commercially available. The law was interpreted literally: each "filling-in between the beams" had to be separately tested, even if the circumstances were only slightly different. A floor system that had been tested and approved for 400 psf live loads on 5-foot spans could not legally be used for a residence with 50 psf live loads if the beam-to-beam span were 5½ feet. This irrational method of operation in the Building Department led to the testing of over 190 "different" floor systems between 1895 and 1915. The vast majority of these tests were on a limited number of floor types varying in span and load only. The floor manufacturers and patent holders became more sophisticated later during this period, and submitted more than one example of each type for testing at one time.[7]

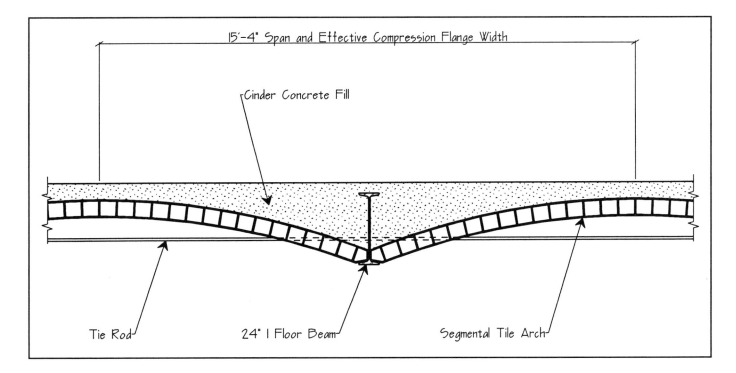

15'-4" Span and Effective Compression Flange Width

Cinder Concrete Fill

Tie Rod

24" I Floor Beam

Segmental Tile Arch

7.3. Typical Floor
Section—Tile Arch
"Composite Beam"
Scale: ⅜" = 1' 0"

The inadequacies of the building code provisions were obvious to engineers. In an article in the *Engineering News* in 1897, the code's bias toward tile arches was said to add 10 to 20 percent to the cost of floor construction in the city. This article, in a respected national publication, also called politicians incompetent to design but willing to legislate design. The Building Department was attacked by the magazine's anonymous contributor for requiring additional tests for concrete floors reinforced with expanded metal and Roebling arches because they did not meet the exact wording of the law.[8]

An item in the engineering press by Gunvald Aus might help explain the long-term stability of tile arch floors. Aus was a prominent engineer, whose company later designed the Woolworth Building. In 1895, Building Department tests were performed on long-span terra-cotta segmental floor arches for use in a warehouse (figure 7.3). As was typical in these tests, one span was built between two beams. The long-span slab, with a 15-foot 4-inch span, required 24-inch steel I-beams as support. The segmental arch began at the beam bottom flanges, and rose 16 inches to midspan. The concrete fill was over 5 inches thick at midspan, and covered the beam webs except for the 6 inches at the bottom where the tile arch was adjacent to the web and bottom flange. The

concrete fill was 3½ inches over the beam top. The test continued to load the arch until the total was 1200 psf over the full arch area. At that point, the bending stress in the steel beams, which were spanning 20 feet, was theoretically 24 ksi. At the time, the accepted maximum bending stress was 16 ksi. While modern codes would allow 24 ksi bending stress, it is not likely that the chemical composition of the beams matched modern steel. More interesting was the lack of measurable beam deflection, despite the high bending stress. Theoretically, the beams should have deflected approximately 0.4 inches at midspan.

Aus's explanation of the unexpected stiffness of the floor beams was that

> this is due entirely to the concrete backing, which, in connection with the beams, forms combination girders, the concrete forming the compression, the beam the tension chord. . . . If this is really the case, it explains, I think, the great rigidity of many an old floor where the strain on extreme fibers of the supporting beams, figured as "beams" would be far beyond the elastic limit. My idea is that the beams act principally in tension, the brick arches and concrete backing forming the [compression] flange.[9]

This view of the facts is the most reasonable explanation, since there was no other structure

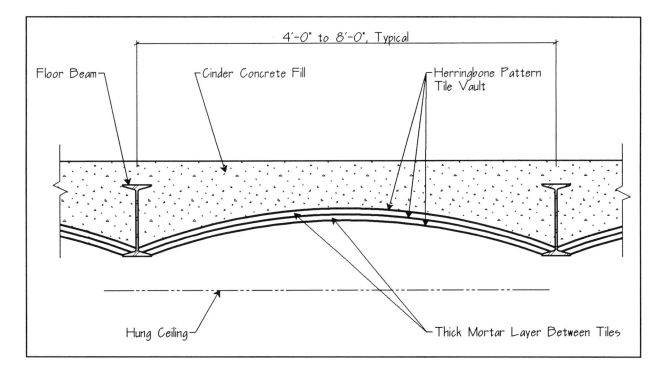

4'-0" to 8'-0", Typical

Floor Beam

Cinder Concrete Fill

Herringbone Pattern Tile Vault

Hung Ceiling

Thick Mortar Layer Between Tiles

available to share the load. A chemical bond does form between concrete and steel that can allow composite action: the bond between small bars and concrete is used in concrete design and the *AISC Steel Manual* has allowed the bond between rolled sections and concrete to be used for composite beam design based on encasement since the third edition in 1937. This action is normally defined as occurring when the beams are entirely encased. Although the possibility of composite action cannot realistically be used in current design or evaluation of this type of floor, the probability is that a bond exists between the concrete and steel, which helps explain the durability of tile arch floors, which are built of a fragile and brittle material.

### Floors Approved by Testing

The following are descriptions of systems definitely approved by the Building Department after 1895 and used in actual construction. The record is not perfectly clear, and other systems may have been used. In addition, there is no guarantee that all floors built during this period met Building Department standards: with so many patented systems and nonpatented, single-source systems available, all of the less-known systems cannot be accounted for accurately. All of the following descriptions, unless otherwise noted, assume the following conditions: steel floor beams parallel to one another and in the range of 5 to 7 feet apart; and New York Building Department approval, meaning that, in good condition, the floors can withstand at least 150 psf live loading. The use of patented systems declined rapidly after the 1916 building code reduced the restrictions on concrete.

Various terra-cotta arches were tested under the provisions for the first alternate. The approved systems included both porous and hard-burned clay, and side, end, and end-and-side combination configurations.[10] Typical examples included the Hard-burned Hollow Tile, side construction, as made by Metropolitan Fireproofing Company, and the Central Fireproofing Company's Floor, which was an end-construction flat tile arch.[11] Tile arches were tested and approved for spans longer than 15 feet and loads up to 1200 psf.[12]

An interesting variation on the tile arch was the Guastavino Timbrel Vault (figure 7.4). The vault was based on Catalan thin shell vaulting and consisted of two or three layers of thin tile, each tile about 1 inch thick and 6 inches wide. The lengths varied, and the joints in the different layers were staggered "securing through this method, by laying in broken joints, a large bonding area making a light and thoroughly cohesive arch." Unlike most tile arch systems, the Guastavino arch could span extremely long distances. The reputation of

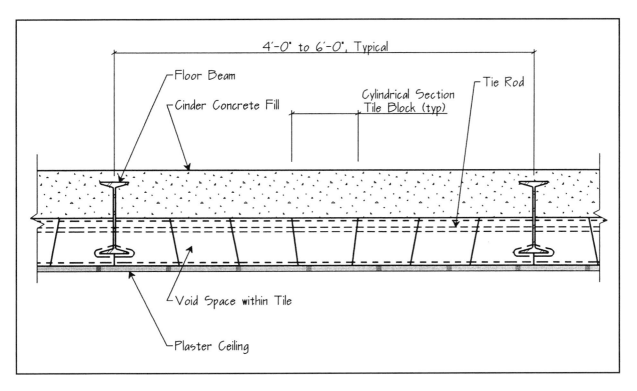

4'-0" to 6'-0", Typical

Floor Beam

Cinder Concrete Fill

Cylindrical Section
Tile Block (typ)

Tie Rod

Void Space within Tile

Plaster Ceiling

7.5. Typical Floor
Section—Fawcett
System
Scale: ¾" = 1' 0"

the Guastavino Fireproof Construction Company rested more on their spectacular domes and vaults left exposed in public buildings than on their occasional use in apartment house floors. This system was the first "special" system used, first appearing in the Arion Club building in 1886, ten years before organized floor system testing.[13]

The Fawcett System was an English proprietary terra-cotta flat arch system (figure 7.5). The tiles were cylindrical in lateral cross section and used in end construction, and were covered with cinder concrete fill. The advantage of an arch consisting of a series of cylinders side by side over a flat top was never clearly stated.[14] The Acme Floor-Arch was a variation on the Fawcett System, where the tiles were trapezoidal in section instead of cylindrical.[15]

The Rapp Floor met the letter of the second set of alternate criteria, although it was totally foreign in spirit (figure 7.6). The load-bearing element in the floor was a series of gauge-steel tees at approximately 8-inch spacing spanning from beam to beam. The beams were usually set at a 4-foot spacing. The tees, 2 inches across the flange with 1½-inch stems, were in the inverted position, with their flanges resting on the top surface of the bottom flange of the floor beams. The tee flanges supported a layer of common brick, which in turn

was covered with cinder concrete fill. The tees were the only stressed element in the system, the brick, cinder concrete, and plaster finish below constituting only fireproofing and a flat top surface. An alternate version had a three-segment arch profile to the tees and brick, with varying depth cinder concrete fill on top.[16] The McCabe Floor was a variation of the Rapp Floor, with terra-cotta blocks resting on small rolled steel tees that rested on the beam bottom flanges 4 feet apart. The tiles were shaped as inverted tees, and the hollow spaces filled with ash. A thin layer of concrete fill provided the top surface. The blocks apparently were as large as 2 feet across, with a corresponding tee spacing.[17]

The Roebling Floor Arch (not to be confused with the Roebling Flat Slab Floor, below) was one of many intermediate forms between arches and concrete slabs (figure 7.7). Arches of dense wire mesh reinforced with steel rods at approximately 2-foot spacing rested on the top surfaces of the beam bottom flanges. The space above the arches was filled with Portland cement stone concrete to produce a flat floor. Wire lath was hung from the beams to permit plastering of the beam bottoms for fireproofing.[18] The Clinton Wire Cloth system was basically identical to the Roebling arch.[19] Floors of this type were originally seen as pure

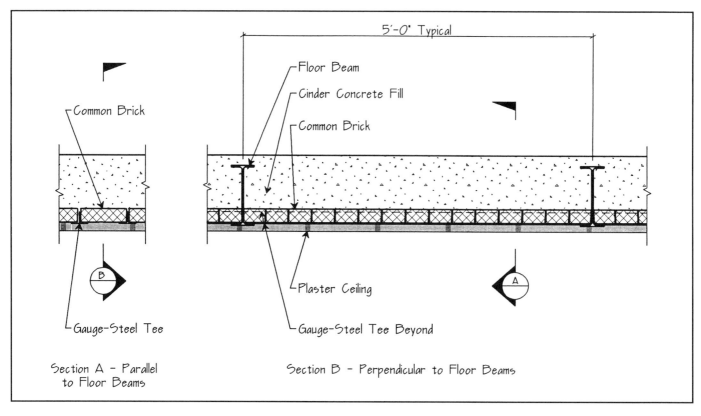

**7.6. Typical Floor Section—Rapp Floor**
**Scale: ¾" = 1' 0"**

**7.7. Typical Floor Section—Roebling Arch**
**Scale: ¾" = 1' 0"**

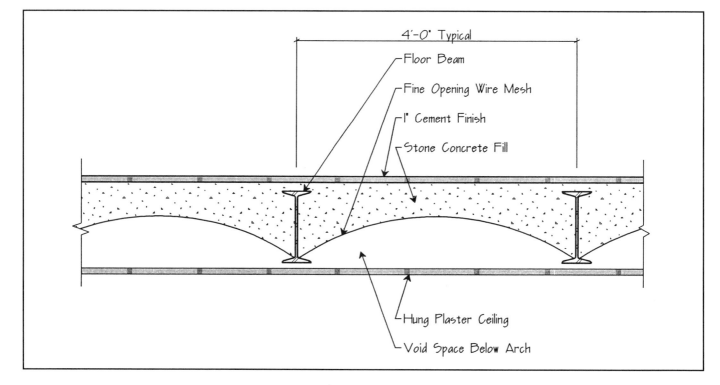

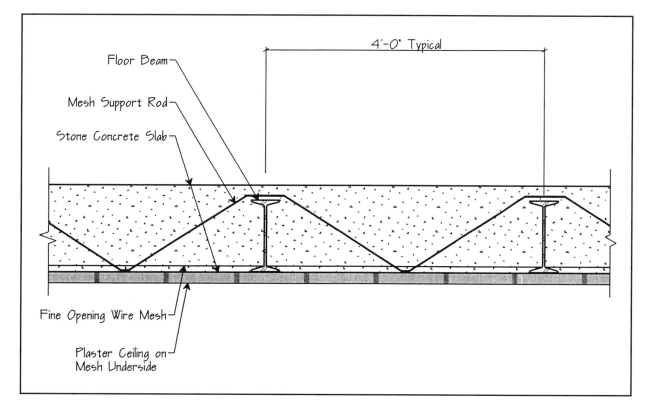

Floor Beam

Mesh Support Rod

Stone Concrete Slab

4'-0" Typical

Fine Opening Wire Mesh

Plaster Ceiling on
Mesh Underside

7.8. Typical Floor
Section—Manhattan
System
Scale: ¼" = 1' 0"

concrete arches, with mesh or expanded metal serving as a permanent form only. The mesh, later referred to as "permanent centering fabric," was used in the 1910s in flat sheets as well as in segmental arches, indicating that designers considered that composite action similar to that in ordinary reinforced concrete was present. Two flaws in this analysis are the lack of a predictable bond between the steel and concrete and imperfect fireproofing provided by the plaster ceiling below. Later systems included Hy-rib by the Trussed Concrete Steel Company, Corr-mesh by the Corrugated Bar Company, Self-Centering by the General Fireproofing Company, Chanelath by the North Western Expanded Metal Company, Ribplex by the Berger Manufacturing Company, and Dovetailed Corrugated Sheets, by the Brown Hoisting Machinery Company and the Berger Manufacturing Company.[20]

The Manhattan System was one step closer to a true reinforced concrete slab than the Roebling Arch (figure 7.8). It consisted of expanded metal mesh supported by the tops of the beam lower flanges and in the midspan by small rods hooked over the beam tops. The space above was filled with stone concrete, which penetrated the mesh openings, and so "bonded" with the metal to pro-

vide flexural strength. The slab was the same depth as the floor beams, which were spaced 4 feet on center.[21] The Expanded Metal Company Floor was extremely similar to the Manhattan System, except that the expanded metal sheets sat on the beam top flanges and the beam projecting below was encased in concrete integral with the slab (figure 7.9). Although the beams were spaced 4 feet on center, the system is notable for its 5-inch slab thickness.[22]

The Multiplex Steel-Plate was an isolated forerunner of modern steel deck (figure 7.10). Steel plates with a squared-off corrugated section, with ribs between 3 and 6 inches on center, rested either on top of the beam top flanges or on top of the bottom flange. The space above was filled with concrete, and a hung ceiling was used below to fireproof the beams and plate underside.[23] The Bailey Floor was a less-modern variation of the Multiplex, using iron "dovetail-formed" sheets spanning 4 feet between beams, with a profile where the ribs were wider at the bottom than at the top, to span between the beam bottom flanges.[24]

The Thompson Floor is an example of construction accepted in the past which makes engineers today very nervous (figure 7.11). It is an

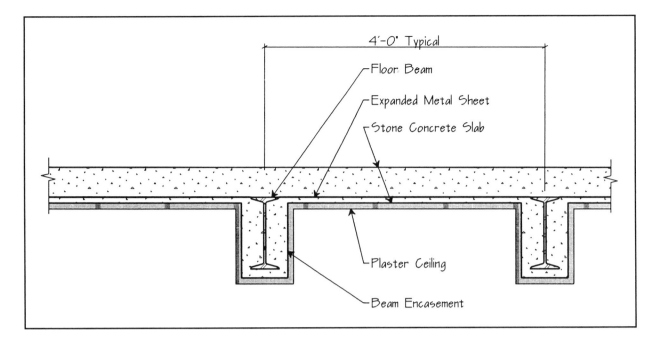

8-inch-thick unreinforced stone concrete slab spanning between beams set approximately 3½ feet apart. Tie rods were provided, similar to tile arch design, at the lower third point of the beam depth, allowing the concrete to act as a series of arches. Wire mesh attached to the beam bottom flanges supported plaster below.[25]

The Roebling Flat Slab Floor was an early reinforced concrete slab (figure 7.12). Flat bars set with their long sides vertical served as reinforcing in stone concrete floors. The bars were twisted

one-quarter turn to rest flat on the beam top flanges and, at the end spans, hook over them.[26]

The Columbian Floor System was another form of reinforced slab (figure 7.13). Cruciform or double-cross-section steel bars were suspended from the beams by gauge-steel straps that hooked over the top flanges and had cutouts the shape of the bar section. The bars acted as reinforcing in a cinder concrete slab. The beams were encased in cinder concrete fireproofing.[27] A far more modern-appearing system was tested as the Monier Floor

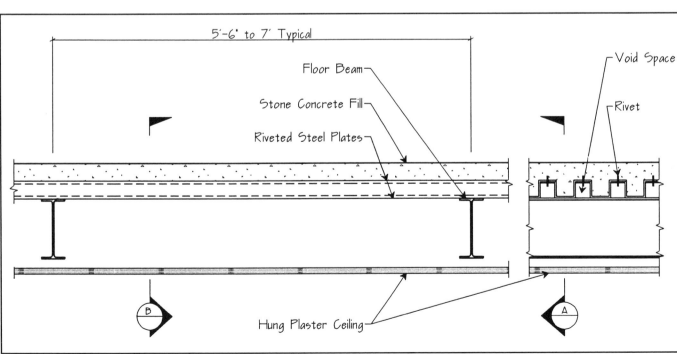

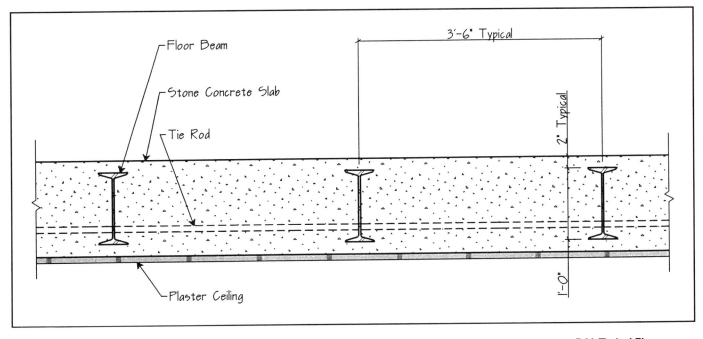

**7.11. Typical Floor
Section—Thompson
Floor
Scale: ¾" = 1' 0"**

System, and was simply a modern-style reinforced concrete slab, using the relatively undeformed reinforcing rods of the nineteenth century.[28]

The Metropolitan System was one of the earliest draped wire floors (figure 7.14). (For a description of this type of system, see the next section.) The floor beams were set up into the 5½-inch thick slab, one inch below the top surface. The reinforcing consisted of pairs of individual wires at 1½-inch spacing, draped over the top of the beams and held down near the bottom of the slab at midspan by a rod parallel to the beams. The slab consists of a peculiar alternative to cinder concrete: three parts plaster of paris to one part wood chips. The beams are encased in extensions of the "concrete" slab.[29]

Most of these systems were declining in use in New York by 1920, because of the growth in popularity of concrete and cinder concrete slabs, especially those with draped mesh. Those systems that retained some popularity typically had one feature that recommended them for a specific use, most

**7.12. Typical Floor
Section—Roebling
Flat Slab
Scale: ¾" = 1' 0"**

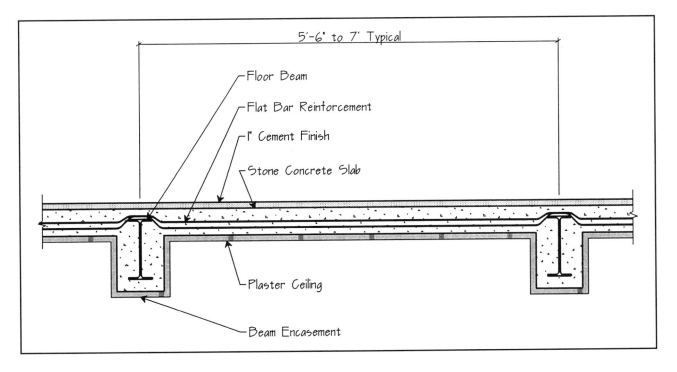

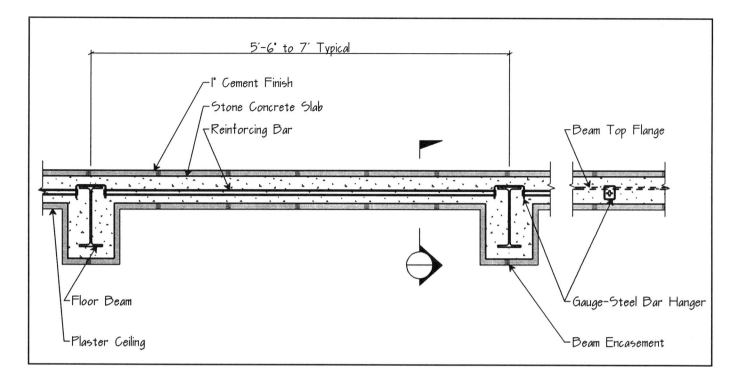

**7-13. Typical Floor Sections—Columbian Floor**
**Scale: ¼" = 1' 0"**

**7-14. Typical Floor Section—Metropolitan System**
**Scale: ¼" = 1' 0"**

commonly durability that promoted their use in extreme loading conditions. A list of factory flooring for heavy loading from 1918 includes brick arches with concrete fill; concrete fill over corrugated sheet iron arches spanning from beam bottom flange to bottom flange, similar to a Roebling Arch; reinforced concrete slabs; Multiplex Floors; and two heavy-duty variations on the Multiplex system, Pencoyd Corrugated flooring consisting of steel sheets riveted together to form a corrugated profile, and Z-bar Floors, consisting of small structural zees serving as vertical webs in a corrugated profile riveted to plates that served as the top and bottom flanges.[30] The Pencoyd system had a brief moment of popularity during the 1890s, before more efficient systems were popularized.[31]

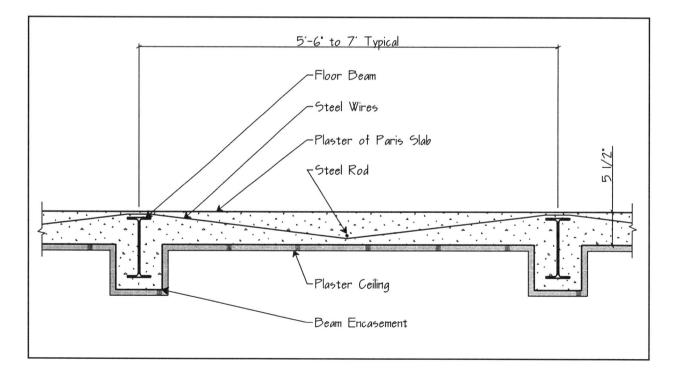

## Tile Arches and Concrete Draped Mesh

The two most common floor systems for iron-and steel-frame buildings between 1880 and World War II were tile arches of various forms and draped-mesh concrete slabs. Each represented the level of building technology at the time of its popularity. Tile arches were used most extensively before World War I, when concrete technology was still relatively new and unreliable. Erecting the arches in a frame building was similar to erecting masonry walls or partitions. The load-bearing tiles were produced under controlled factory conditions, and were set in place in a manner similar to traditional masonry. Draped mesh, which became popular during the 1920s, represented the best development of the turn-of-the-century testing period.

The most important physical differences between tile arch floors and reinforced concrete slab floors is the geometric plasticity and toughness of the concrete. Concrete slabs can be placed in irregularly shaped areas or around obstructions such as pipe sleeves without any special preparation or cutting required. Small holes can be cut through a concrete floor without fear of affecting the overall slab stability. Tile arches are extremely strong under ordinary compressive loads, but are constrained in geometry by the shape of the blocks and the mechanics of arch action. The tile is also extremely brittle, and can be broken during construction by hand-held hammer blows. This makes alteration of a small area without complete demolition of the arch from support to support nearly impossible. Seven of the fourteen floor types examined in the first organized series of tests in 1896 were cinder concrete systems. This type was immediately popular because it was a geometrically adaptable and relatively lightweight form of fireproofing for beams.[32]

Another advantage of concrete over tile arches or other "unit" floors was the chemical bond between cement and steel. This bond, which makes the encasement of structural steel members in concrete far easier than it would be if the materials did not interact, suggested itself to engineers and builders as a method of preventing air, water, and heat from reaching the metal. During the years when skeleton frames replaced bearing walls, engineers were usually aware that they were designing buildings with a much higher percentage of the steel adjacent to the exterior, and therefore vulnerable to corrosion. An article in *Engineering News* in 1904 described extensive experimentation on encased steel, and concluded that steel will not rust if embedded in well-mixed Portland cement concrete and given enough cover to prevent water entry.[33] One and one-half inches was the recommended cover thickness, although the results of these experiments must be viewed as partially suspect, since the embedded steel was unstressed. Stressed steel is more vulnerable to corrosion than unstressed steel, and concrete encasement in service is almost always cracked from flexural deflection, making the test conditions a not-very-accurate representation of the field environment. By 1918, encasement of beams was the preferred method of waterproofing and fireproofing.[34] Other water-protection methods in common use were tar paints, asphalt paints, varnishes, Portland cement washes, and oil-base paints, which included red lead, white lead, and zinc paints.[35]

Draped mesh is a type of reinforcing first developed during the frantic floor-system testing of the 1900s and 1910s. At first, the wires may simply have been draped due to geometric constraints in assembly, but in 1915 tests were performed by the Building Department in cooperation with the School of Engineering of Columbia University on various forms of reinforcing for slabs.[36] The systems tested included proprietary triangular mesh from the American Steel and Wire Company, proprietary expanded metal mesh from the Expanded Metal Engineering Company, ordinary welded wire mesh, ordinary triangular wire mesh, ordinary expanded metal, and plain round and square rods and twisted square rods. The mesh typically was draped, as opposed to the simple straight spans used with the rods. The test results were compared with the then-current working-stress concrete-beam theory. The rod-reinforced slabs exhibited behavior close to that expected, but the concrete compression measured in the mesh slabs was lower than expected. This result could best be explained by catenary action, the automatic tensioning of suspended, anchored, and flexible cables loaded with their own weight and uniformly distributed loads. All vertical loads applied

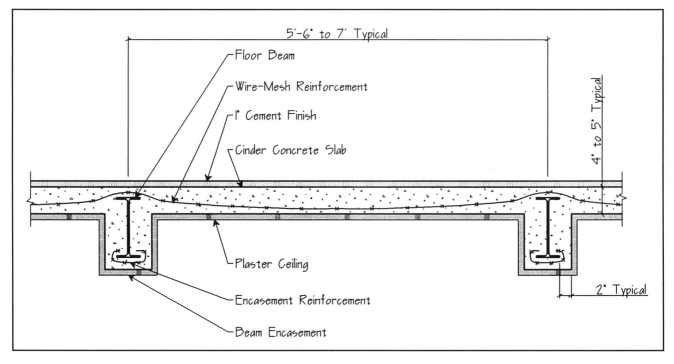

5'-6" to 7' Typical

Floor Beam

Wire-Mesh Reinforcement

1" Cement Finish

Cinder Concrete Slab

4' to 5' Typical

Plaster Ceiling

Encasement Reinforcement

Beam Encasement

2" Typical

**7.15. Typical Floor Section—Draped-Mesh Concrete Slab Scale: ¾" = 1' 0"**

to the wire result in minor changes in the suspended curve and are translated into tension along the wire axis. This structural action is most readily visible in the main cables of suspension bridges.

In a draped-mesh slab, the slab serves only to provide a flat, stable surface for use and to transmit the applied loads to the mesh. The mesh alone spans the beam-to-beam gap (figure 7.15). Because the concrete is not structurally stressed, its composition is far less critical than in ordinarily reinforced slabs where the concrete is stressed. This largely negated the quality-control concerns of engineers and builders in the 1910s and 1920s, and helped explain the immense popularity of this type of floor construction, estimated at being used for over half of the floors built in Philadelphia and close to 90 percent in New York in 1914.[37] For a noncritical role, cinder concrete can be safely used, despite its ordinary maximum compressive strength of 700 psi (as opposed to stone concrete, which rarely has a compressive strength less than 1500 psi). This reduced the weight of the concrete slab to the same range as a terra-cotta arch with cinder topping. In 1913, when cinder concrete was near its widest use, the second Joint Committee Report (see below) stated that "[c]inder concrete should not be used for reinforced concrete structures. It may be allowable in mass for very

light loads or for fire protection purposes." This recommendation was largely ignored, with cinder concrete use remaining popular for both office buildings and high-rise apartment houses into the 1930s.[38]

Catenary action in floor construction was not immediately recognized by the engineering community. As late as 1914, immediately before the New York Building Department established rules for catenary design, an engineer published by the ASCE complained that one of the worst flaws in cinder concrete floors was the lack of truly continuous reinforcing when compared to stone concrete slabs. The comparison is between draped-mesh slabs with the mesh hooked over the beam tops and flexurally reinforced slabs. In some early forms of draped mesh, the actual concrete surface was level with the beam tops, leaving the portion of the wires over the beams embedded in cinder fill. This does not reduce the strength for catenary action, although it is arguably more vulnerable to fire. In the discussion of the article attacking this detail for lack of continuity, one engineer said that "[t]he author ridicules the term 'continuity' thus used without apparently realizing that it is not the continuity of the concrete slab which is referred to, but the anchoring of the reinforcement over the steel beams." This is a clear statement of the catenary principle.[39]

Concrete construction at that time was prone to quality-control flaws, segregation of the mixed components and hidden voids being the most common inherent flaws, and uncontrolled field work resulting in incorrect mix proportions plaguing the practical work. National standards for cement were first written in 1909 for both Portland and natural cements, and by 1917 had been revised limiting structural concrete to Portland cement. The quality of cement available was rarely a problem, and when samples of well-made concrete from this era are tested, their compressive strengths are usually well above the original design values. Segregation is most often the result of lack of adequate vibration during concrete placing; voids are the result of water freezing within the fresh concrete, inadequate vibration, or insufficient plasticity of the unhardened concrete. These problems were finally solved by the use of plasticizing admixtures and modern vibrators, but were common enough that they reduced concrete's acceptance among engineers.[40]

The difference in strength between properly designed tile arch floors and draped-mesh concrete slabs is unimportant because of the relative weakness of the steel floor beams. In ordinary residential or commercial construction, where the highest normally expected live load is 100 psf, the steel beams are designed, as a matter of course, for the actual applied loads. Tile arches were often sized by the beam depth, the arch block depth matching the beam. For a typical layout such as 12-inch floor beams at 6-foot spacing, 12-inch flat tile arches were rated capable of withstanding over 340 psf loading with a safety factor of 7.[41] Anytime a floor system is capable of withstanding load beyond that which will cause flexural failure in the beams supporting it, the exact capacity of the floor is not important to the designer.

In a similar manner, draped-mesh concrete slabs can easily be made stronger than the floor beams. The catenary action of these slabs is inherently stronger than ordinary flexure, in the same way that the arch action of tile floors is. The draped-mesh slabs were typically designed for the actual loads, but since a relatively minor increase in wire size or slab depth could result in a large increase in strength, obtaining adequate slab strength was not a difficulty for the designer. For the example noted above, a 4-inch-thick slab with 0.087 square inches of steel reinforcing per foot of width (#6 gauge wire at 4-inch spacing) was capable of supporting 145 psf.[42]

Since floor strength greater than the floor beam strength was so easily attainable in both tile arches and draped-mesh concrete slabs, strength was never an issue in choosing between them. The price paid for the excess strength was the required labor in construction. Tile arches had to be carefully fit together while supported on temporary centering. The arches, especially the segmental arches, produce a substantial sideways thrust at their supports, the steel floor beams. When the entire floor is in place, the thrust from adjacent arches cancels out, but since the arches are not built simultaneously, the thrust must be compensated for during construction. For this reason, and to compensate for the unbalanced thrust in the last arch on each end of a floor area, steel (or occasionally wrought-iron) tie rods were provided between each pair of beams. Depending on the particular system of arch used, the tie rods fell in the tile longitudinal joints, within specially provided voids in the tiles, or were simply cut into the tile bottoms.

Fire resistance could not be used to choose between tile or cinder concrete. The New York fire tests were used nationally as evidence concerning the ability of various materials and assemblies to resist heat and sudden cooling. The conditions of uniform heating and free air circulation during the tests did not necessarily resemble those in a real fire, but they were useful in broad terms in comparing material reactions. In over forty tests on draped-mesh cinder concrete slabs and flexural reinforced cinder concrete slabs, over twenty on reinforced stone concrete slabs, approximately ten on tile arches, approximately twenty on patented systems such as the Guastavino arch, cinder concrete proved to withstand the direct effects of heat better than any of the other materials. Hard-burned terra cotta can be heated to higher temperatures without any chemical changes, but it is far more vulnerable to secondary stresses created by its thermal expansion in a restrained position such as floor arches.[43]

There was a great deal of debate in the engineering press about the relative merits of using paint, Portland cement washes, Portland cement stone concrete, and Portland cement cinder concrete as rust protection. Cinder concrete had developed a reputation during the 1890s for promoting the corrosion of iron and steel embedded within, specifically because of the acidic nature of the clinker (coal cinder) used as cinder aggregate. Portland cement's protection and bond are based partially on its high pH (basic) nature, and the acidic aggregate is not necessarily the best material to be used in concrete. The practical experience with the protection afforded by cinder concrete is more problematic: how could an engineer examining corroded steel determine if the corrosion was the result of the chemical nature of the encasement or simply excessive water in contact with the concrete seeping inside?

A well-publicized investigation of the ability of cinder concrete to serve as waterproofing was performed during the demolition in 1903 of the Pabst Hotel, built in 1898, at the intersection of Broadway, Seventh Avenue, and 42nd Street, which later became the site of the *New York Times* headquarters. The building was demolished so soon after construction because its structure at the basement level interfered with the new IRT subway under construction as it curved north from 42nd Street to Broadway. The hotel was of typical construction, with steel beams, girders and Z-bar columns, brick curtain walls, and Roebling Arch floors topped with cinder concrete. The report stated:

> The entire structural part of the building is in excellent condition, in fact it seems fully as good as in a newly-built building. Of greatest interest is the condition of the steelwork. This is excellent throughout, in the columns as well as the beams, the surfaces still show the original adherent coat of . . . paint, and there is no indication of deterioration by corrosion. . . . The beams and girders were practically everywhere encased in . . . the cinder-concrete of the floor arches, and this fact should be especially noted in view of a somewhat prevalent view that cinder-concrete, especially when placed so as to have a porous and open texture, is favorable to corrosion of the steel with which it is in contact.[44]

The wire mesh of the arches had light surface rust which was believed to have formed during construction, while the concrete on top and plaster below were wet. Another examination of demolition three years later was under far worse circumstances. The San Francisco earthquake of 1906 left many modern fireproof buildings in such poor shape that they had to be demolished. The report of the Structural Association of San Francisco, examining buildings being demolished, claimed that all buildings with floors using cinder concrete showed heavy corrosion of the floor beams. The report attributed this to the sulfur content of clinker. This view was not universally supported: In the same issue of *Engineering News* containing the report, the magazine published an editorial saying that cinders were not necessarily always harmful, and that the real problem was the unpredictability of the material. More anecdotal evidence turned up in letters to the magazine in response, as engineers wrote to say that they had worked on many alterations that required the removal of cinder concrete, and had not seen rust on the steel.[45]

A more methodical approach to this question was used in 1907, when an experiment was performed to check on the corrosion of reinforcing steel embedded in cinder concrete. Different samples were used, with the concrete mixture mixed wet in the normal manner, mixed dry and set into place before water was added, tamped into place to provide thorough removal of air bubbles and complete mixing of the concrete, and untamped. All of the tests were performed using low-sulfur cinders, since high-sulfur cinders were known from previous experiments to promote corrosion. The conclusion was that cinder concrete by itself was not a cause of rusting—although, to prevent its becoming one, thorough mixing and tamping were necessary.[46] The current assessment of this material agrees closely with the 1907 tests: Completely combusted anthracite clinker is inert and makes a good low-strength, low-weight aggregate; bituminous clinker, or cinders not fully burned, contain free sulfur that will corrode the steel reinforcing.[47]

In 1921, a text on fireproof construction listed: "Types of fire-resisting floors employed in present practice . . . comprise: Mill construction, brick, terra-cotta, concrete, and combination terra-cotta and concrete. In addition to these a great number of patented 'systems' have been used during the past ten or fifteen years, practically none of which have survived the test of time save the Guastavino construction."[48] Mill construction, consisting of heavy timber, was illegal within the fire district, and so was practically unused in New York City. As previously described, brick arches were only used in areas of exceptional load and in vaults. By 1921, terra-cotta use was on the decline, leaving concrete floors as the dominant system.

## Flexural Concrete Slabs

Americans resisted the use of reinforced concrete far more than Europeans at every stage in its early development. Almost all of the early theory came from Europe, and American practice up until the 1920s was more primitive and capable of carrying less load. The construction of tall buildings in the United States had advanced steel construction in everyday building beyond standard practice in Europe and smaller buildings in much of the country were still constructed of wood. Cast-in-place concrete required far more on-site labor than steel frames and tile arches; in this area as in every other, labor costs often were the overriding concern. Despite claims that concrete required less skilled labor than steel erection, the fact that skill was necessary in correctly placing bars, creating proper cold joints separating individual placings of concrete, and stripping forms was recognized nationally after a series of construction accidents after 1900.[49] Overall, Americans were less interested in the advantages of concrete and more conscious of its deficiencies. One peculiar effect of the relative lack of interest in concrete building was that by the time reinforced concrete began to be regularly used, after 1900, modern theories were already available from Europe.[50]

With the isolated exception of a private house with beams reinforced with wrought iron "I's" riveted together, the first reinforced concrete buildings of any importance in the East were built around 1900.[51] As late as 1893, an article describing flexure tests on concrete admitted an "almost total lack of information about these points."[52] The simultaneous beginning of the use of complete concrete frames and concrete floor systems, combined with the repetition of company names

in both areas, suggests linked development.[53] The Building Department did not test concrete columns in the same manner as concrete floor systems; columns were treated as masonry piers by the building examiners and as columns following new design techniques by the design engineers, leaving both parties satisfied.

The Concrete-Steel Company, with its founder and patent holder Ernest Ransome, designed and constructed a reinforced-concrete factory in Bayonne, New Jersey, in 1897. The floor slabs were reinforced with Ransome bars: cold-twisted square-cross-section steel bars. The twists served the purpose of deformations in modern rebar, to provide a mechanical link between the steel and concrete to supplement the chemical bond of the cement.[54] The system was used again in Long Island, New York, for the Nassau County Courthouse of 1901, where the form was called "a class of construction that is certainly ahead of anything known to the building world as regards its fireproof character."[55] Shortly afterwards, there was a small boom in this field, with five all-concrete frame buildings constructed in Manhattan, including a twelve-story office building and a nine-story apartment house.[56] The material was sped on to success by the efforts of the American Society of Civil Engineers (ASCE) and the newly founded American Concrete Institute (ACI) before 1910 and the simultaneous publication of several textbooks.[57]

The design of ordinary beams and slabs of reinforced stone concrete in New York followed the national pattern (figure 7.16). After 1900, elastic design assuming a cracked section, with all tension carried by the reinforcing and all compression distributed linearly from the neutral axis, was the standard. Except for the substitution of a larger compressive stress block and the adoption of higher allowable stresses, this theory and its associated equations are very similar to those in the modern ACI codes.[58] In contrast to steel construction, which was forty years old when the AISC first published a national, nonproprietary specification, reinforced concrete had just entered the mainstream of construction when the ACI was founded in 1905, and when the Joint Committee on Standard Specifications for Concrete and Reinforced Concrete published its first proposed speci-

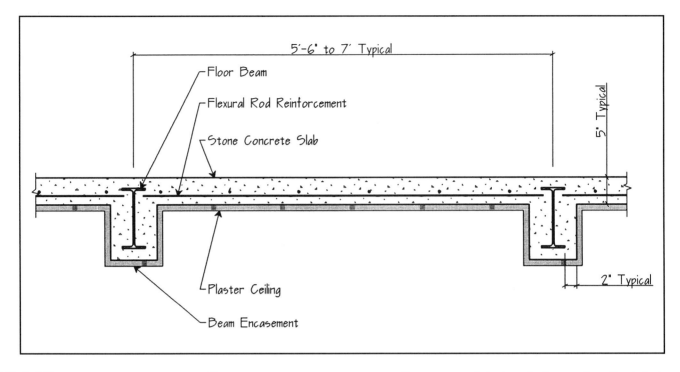

5'-6" to 7' Typical

Floor Beam

Flexural Rod Reinforcement

Stone Concrete Slab

5' Typical

Plaster Ceiling

2" Typical

Beam Encasement

**7.16. Typical Floor Section—Formed Concrete Slab**
**Scale: ¼" = 1' 0"**

fication in 1909.[59] The Joint Committee was a collection of national groups including the ACI, the ASCE, the American Society for Testing Materials (ASTM), and the American Railway Engineering Association. The Portland Cement Association, a national manufacturers' group similar to the AASM, was founded in 1916.

The national, and partially international, nature of concrete theory isolated concrete floor systems from the majority of those tested in New York at the turn of the century. The Monier Floor System, which was nothing more than a modern reinforced concrete slab, was tested for use before 1900.[60] The requirement for testing still existed, but reinforced stone concrete was also subject to nationally recognized stress limits. The dual requirements conflicted and at first produced some nonsensical results. The material allowable stresses were based on rational theory and an attempt to produce a consistent safety factor. The New York tests were simply an extreme load-carrying requirement for a specific design. Hundreds of floor systems carried the specified loads, but there were no tests on stress within the floor component materials. Judging by the descriptions of the designs, the safety factors of the approved floors must have varied widely. Because the stress requirements on reinforced concrete were far stricter than the load test requirements, concrete

floors were relatively safer but underutilized their potential.

The contrast between the two theories of floor design is sharpest when comparing reinforced stone concrete slabs to cinder concrete slabs under the 1911 building code. Stone concrete is ordinarily at least three times as strong as cinder concrete in pure compression tests, but the allowable load on a stone concrete floor was determined by the limit of 650 psi compression in the concrete, 16,000 psi tension in the reinforcing steel, and a ratio between the elastic moduli of the two materials of 15, while the allowable load on a similar-size cinder concrete slab was limited only by its ultimate strength in testing. For heavy loads and long spans these provisions would produce stone concrete slabs thicker and more heavily reinforced than some of the previously approved proprietary cinder concrete systems.[61]

As concrete came to be seen as an ordinary building material, and not as part of a proprietary system, the building code gradually allowed design more in line with the ACI and Joint Committee standards. In 1923, the allowable flexural compressive stress in concrete was 0.325 times the ultimate compressive strength of the mix in New York, opposed to 0.375 times the compressive strength in the ACI code. While there was no rational excuse for the lower allowable stress in

## CASE STUDY: **PATENTED FLOOR SYSTEM**

A nonprofit charity bought a derelict apartment house with the intention of repairing it and converting it to housing for the homeless. The project involved extensive architectural and mechanical renovations, and structural design for such items as replacing the roof-top water tank dunnage. In the course of construction, issues concerning the original construction of the building led to additional investigation, which is described here.

The apartment house is roughly 60 feet by 80 feet, with an irregular outline caused by light wells on the side facades. The floors are wood joists supported on brick bearing walls with steel beams used at gaps in the walls, with the exception of the first and second floors which were known during design to be some form of fireproof construction.

When the first-floor ceiling was removed, the underside of the floor, instead of the expected concrete, was seen to be common brick, with steel flanges exposed in between the bricks. This identified the floor as a Rapp patented system, which made some of the original renovation details impossible to use. Several sections of floor had been scheduled for demolition and replacement, which was still possible, but the upgrade of the mechanical systems of the building required passing new pipes and conduits through the floor. When the floor was assumed to be concrete, core drilling was the preferred method of creating the mechanical openings. Core drilling cannot possibly work in a Rapp Floor, since the bricks are not mechanically fastened in position and would slide out from under the drill bit. In addition, creating holes of any size in a Rapp Floor is problematic, since the loose bricks tend to slide along the surface of the "T" flanges when disturbed.

The solution was to group the pipes and conduits as much as possible, to create larger chases that could be built by removing entire sections of floor. This allowed most of the floor to remain undisturbed. The large openings for the new chases were carefully cut, with the "T"s spotwelded to prevent sliding during construction. All new sections of floor at the two fireproof levels were built of concrete slab on metal deck.

---

one place, the absolute cap of 650 psi no longer existed. Engineers regularly specified 3000 psi concrete, which the building code would allow to be stressed to 975 psi, a gain of 50 percent over the old maximum allowable design stress attained without any other change.[62]

Scientific design of concrete slabs coincided with the development of the slab constituents. Standards were developed for cement, aggregates, reinforcing, and details to ensure that the material calculated to bear stress was capable of it. The first ASTM standard for cements, in 1906, gave allowable ranges for the elements and compounds making up natural and Portland cement. Portland, made by combining specific proportions of the elements, was by its nature more readily subject to quality control, and by 1917 the Joint Committee was specifying the use of Portland cement in stressed concrete.

Aggregates could develop only in limited ways, given that inert stone is inert stone. The previously described use of coal clinker as lightweight aggregate was one of several attempts to produce concrete that was between half and two-thirds the weight of stone concrete. The most successful effort during the 1910s produced modern lightweight aggregates. The need for a vastly increased number of ships during World War I led the United States Shipping Board to build reinforced concrete freight ships in an attempt to ease the pressure on the domestic steel industry. In buildings, the weight of concrete was not a serious handicap, since the load was ultimately all transferred to the earth. The material weight was far more important in shipbuilding, where every extra pound of ship weight reduced the ship's freight capacity and thus negated some of the advantage in building "steel-less" ships. In 1918, the Concrete Ship Section of the Shipping Board tested numerous lightweight aggregates for use in shipbuilding. The best found was burned shale, which was far less dense than ordinary stone but nearly as strong. In the 1918 tests, concrete produced with the shale was said to be equal in strength to stone concrete, producing concrete compressive strengths in the range of 3500 to 5500 psi, but at a weight of 110 pounds per cubic foot instead of the standard 150.[63]

Before 1930 the variation in steel reinforcing bar types was as wide as the variation in floor systems, and for some of the same reasons. Many of the rebars were experimental in nature because there was no direction in design that was clearly best; many were minor variations on familiar forms created to allow their producers to obtain patents, and many were part of proprietary floor systems. For use in New York, most were tested in slabs along with the other systems described above. The first specifications specifically for reinforcing were developed by the Association of American Steel Manufacturers in 1910, hoping to repeat the success in standardization the group had achieved with rolled beams.[64]

The peculiar shapes of bars were all the product of one idea: to create the maximum possible bond between the steel and the surrounding concrete. There is a chemical adhesion of Portland cement to steel, which in some old reinforcing designs was assumed to provide all of the bond necessary. In buildings built before 1920, plain bars of simple square or round cross section can sometimes be found. On rare occasions, plain bars with "structural" cross sections, such as channels or tees, were used. These were specially rolled sections under 2 inches in depth.[65] Plain flat bars were even more rarely used, because early experiments indicated that the chemical bond stresses were lower than for other shapes.[66] There is no reason why this should be, and the experiments were most likely revealing the tendency of flat bars under stress to split the concrete, and thus destroy the existing bond. Plain bars were phased out during the 1910s and early 1920s in favor of deformed bars, "the principal object of which is to furnish a bond with the concrete independent of adhesion, a mechanical bond as it is usually called."[67] The primary advantage of mechanical bonding is that it is less dependent on the surface condition of the steel and the exact proportions of the concrete mix than are chemical bonds.

Two types of deformation were used, longitudinal and radial. Radial deformations produce bars with complex cross sections that remain the same down the length of the bar. This type of design was created to increase the surface area of the bar relative to its area, and so increase the potential for chemical bonding. The strength of chemical bonds can be expressed in pounds per square inch, and so the strength is directly increased by increasing the area over which the bonds act. Types of radially deformed bars include wing bars, which have a complicated "+" cross section; Monolith bars, with a "figure eight" cross section; Girder bars, with "I" cross sections, and Hanger bars, with "+" and "+-+" cross sections (used in the Columbian Floor System).[68]

Longitudinally deformed bars are the type familiar to us today. The cross section of the bar varies repetitively along its length, providing indentations that fill with concrete and so "lock" the two materials together to transmit force. Outside of laboratories, this is more dependable than chemical bonding and is the primary bond used in modern codes. The oldest consistently used longitudinal reinforcing was the Ransome system twisted bar. This was a square bar of the desired cross-sectional area, cold-twisted along its long axis until the bar edges perform a complete revolution every foot or so. These bars were originally patented by E. L. Ransome for use in Ransome Company–built buildings, but were later used by others in ordinary floor slabs.[69]

Other forms of longitudinally deformed rebar included Thatcher bar, which was square with cross-shaped deformations on each face; lug bar, which was square bar with small round projections at the corners; Inland Bar, which was square with "stars" similar to diamond-plate-type deformation on each face; Herringbone, Monotype, and Elcannes bars, which had a complex cross section similar to the radial deformed bars but with longitudinal deformations; Havemeyer Bar with round, square, and flat cross sections with deformations similar to diamond plate; Rib Bar with a hexagonal cross section and "cup" style deformations consisting of shallow oval depressions in the bar faces; American bar with square and round cross sections and low circumferential depressions; Rib bar

## CASE STUDY: DRAPED MESH

The managing agent of a luxury apartment house became concerned with the safety of a section of floor after a plumber had repaired a nearby terrace drain. The plumber had cut into the structure to replace the horizontal length of pipe below the drain, and an engineer was called in to investigate the seriousness of the damage. The following is a summary of the investigation and course of repair.

The building dates from the late 1920s, and has a seventeen-story steel frame with a brick curtain wall. The living rooms, main bedrooms, and other major spaces in the apartments have hung ceilings, making identification of the slabs impossible. From the appearance of the ceilings in the kitchen and the former servants' bedroom where the damage was, the floors appeared to be concrete draped mesh. The encased beams were visible, as were the board form marks on one area of concrete where the plaster had been removed.

Upon close examination, it was apparent that the plumber had inadvertently destroyed the ability of approximately 60 square feet of floor to carry load. The drain pipe, after passing through the terrace slab, turned and was "hid-den" next to the web of one of the floor beams. To remove this pipe, the plumber had removed the beam encasement for roughly 10 feet and, more importantly, had cut almost all of the wires in the welded mesh where it passed over the top flange of the floor beam. The mesh in a draped-mesh slab provides both moment and shear support, and the encasement provides additional shear support. With the demolition performed by the plumber, the panel of slab adjacent to this beam was effectively unsupported on one end.

In order to actually repair the damage, the terrace above would have to be demolished, as well as part of the apartment next to the terrace. This was not practical in the occupied apartment, so the repair design concentrated on re-supporting the affected areas of slab. The cut wires were welded to the beam flange to ensure continuity for the next panel of slab, and small beams were placed below the slab, framing into the sides of the existing floor beams. To fireproof the new construction and hide the alterations, the entire room received a dropped ceiling at the elevation of the bottom of the existing beam encasement.

with basically square and round complex cross sections and low circumferential ribs; Scofield bar with an oval cross section and discontinuous circumferential ribs; Corrugated bar with round, flat, and square cross sections and cup deformations; Slant rib bar with a flat cross section and low projecting diagonal ribs on the flat faces; Cup bar with round cross sections and cup deformations; and Diamond bar with round cross sections and low circumferential ribs.[70] The last two greatly resemble modern deformed rebar except that the deformations on the older bars do not project as far from the bar center. The general trend in rebar since the 1940s has been to reduce the assumed strength of bonding, and so increase the surface provided for mechanical bond and increase the length of bar required to consider the bar fully developed.

The 1924 Joint Committee report legitimized the modern designations of #3 to #8 for round cup or diamond deformed bars of ⅜ inch to 1 inch (⅞) diameter. This report still mentioned square bars, although most manufacturers of nonproprietary bars produced only twisted squares, not otherwise deformed ones. Until World War II forced simplification of production, most manufacturers also produced (¼ inch) #2 bar, often undeformed.[71] The small undeformed bar could be dropped from production without hardship because its heavier applications could be performed by #3 bar and its lighter ones by wire fabric.

Wire fabric first appeared in New York in proprietary tested floor systems. Presumably some form of systematic design took place in the offices of the patent-holding companies, but it was not considered by the officials in New York who tested these systems. The oldest ancestor of wire fabric was expanded metal. This reinforcing was created by slitting a sheet of iron or steel in a staggered pattern, and then pulling the sheet at right angles to the slits to expand the openings into diamond shapes. By 1920, at least five companies were manufacturing variations of this material. It was used in several floor systems, most notably one type of Roebling arch. Depending on the degree to which the sheet was pulled, the openings could be kept small enough to allow the metal to serve as a permanent form for wet concrete. A similar sheet reinforcing was the Rib Metal produced by the Trussed Concrete Steel Company, which consisted of sheet steel semicircular ribs produced by fastening together a corrugated sheet on top of a flat one, and slitting the sheets between the ribs and pulling out the sheets to create ribs separated by thin straps. This was meant for use as slab flexural reinforcing.

Individual wires had been used before 1900 in floor systems such as the Metropolitan, and it was a short step to using wire fabric. The wires set at right angles to the slab span do not carry load, but serve to tie the main wires into a single sheet, such as was already familiar from expanded metal. This simplified construction and helped ensure correct location of the reinforcing within the slabs. Since the load-carrying, longitudinal wires in all of the proprietary wire fabrics were the same standard gauges, the only differing details were the form of connection to the transverse wires. The connections varied among small wire staples at each joint in the Unit Wire Fabric by the American System of Reinforcing Company, gauge-steel washers locking together the wires in the system produced by the Cargill Manufacturing Company, simply wrapping the transverse wires several turns around the longitudinal at each intersection in the Page Special Process fabric and lock-woven fabric by W. W. Wight & Company, the wrapping of two sets of transverse wires each at approximately 20° from right angles in the Triangle-mesh Wire Fabric manufactured by the American Steel & Wire Company, and welding the wire fabrics produced by the Clinton Wire Cloth Company and the Witherow Steel Company.[72] Welding the wires together became the design of choice by the 1930s.

Reinforcing details vary with the theory used to explain the slab strength, the gross shape of the concrete, and the type of reinforcing used. A flexurally reinforced slab interior span does not need any mechanical anchorage for the rebar, but it was occasionally provided in the form of hooking the bars around the floor-beam flanges. This was a carry-over from early concrete theory of the 1890s that treated slabs as tied arches—for example, the Thompson Floor. In arch theory, unanchored bars are incapable of carrying load. At the same time that engineers were learning that mechanical anchorage was not necessary in flexural slabs, it remained crucial in draped-mesh slabs, whose catenary action is of course merely arching upside down.

Concrete slabs resembled tile arches in one aspect of fireproofing: for economic use, the floor material had to provide protection for the floor beams so that hung ceilings would not be needed. Slabs were placed at various elevations relative to the beams, with the most common design being a slab that extended several inches above and below the beam top flanges. Since the slabs all had to be created with wood plank forms, and there was no structural reason that the slab thickness should be anywhere near as large as the beam depths, this configuration did not require any more formwork than any other. The beams were encased in extensions of the slab, poured monolithically with the rest, and often reinforced with expanded metal or wire fabric stretched under the bottom flange and hooked over its edges.[73]

Depending on the top finish requirements, cinder fill, cinder concrete fill, and solid cement finishes were often placed on top of the structural slabs for both reinforced and draped-mesh construction. The fills were used as fire protection and as an easy way to embed sleepers for wood finish; the cement as waterproofing and a wearing surface.

## Concrete Joists

Reinforced concrete theory assumes cracks in the concrete on the tension side of the neutral axis, throwing all of the tensile force into the rebar. While the tension-side concrete is necessary to provide a connection between the steel and the compression concrete to transfer stress and to cover the bars to protect them from fire damage

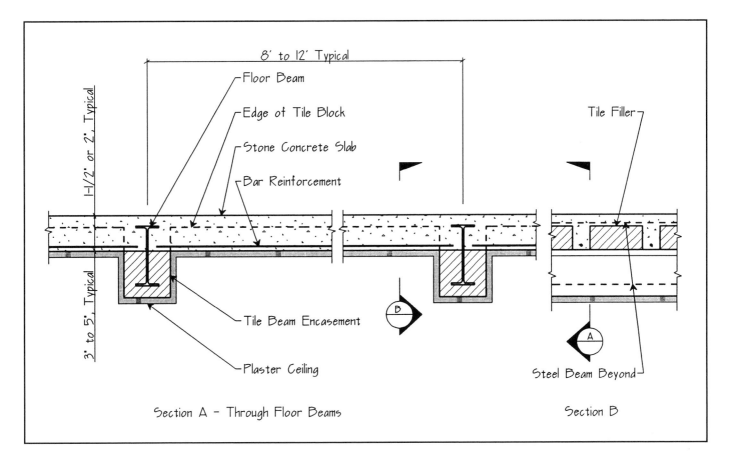

8' to 12' Typical

1-1/2' or 2' Typical

3' to 5', Typical

Floor Beam

Edge of Tile Block

Stone Concrete Slab

Bar Reinforcement

Tile Beam Encasement

Plaster Ceiling

Tile Filler

Steel Beam Beyond

B

A

Section A - Through Floor Beams

Section B

7.17. Typical Floor
Section—Tile and
Concrete Joists
Scale: ¾" = 1' 0"

and corrosion, the vast majority of this material is wasted structurally. In the same way that a search for more efficient forms led engineers to the wide-flange beam, it created the concrete-joist floor slab. The analogy breaks down when examining the efficiency of construction: once four roller mills were installed there was no loss in changing from American Standard to wide-flange shapes; the creation of concrete-joist or waffle slabs required either the creation and placing of permanent tile forms or the use and removal of steel forms, both of which required substantially more labor than the simple flat forms of an ordinary slab. It is for this reason that the majority of concrete floors constructed were not joists or waffles, despite the advantages of those forms.

The two most common forms before the 1940s were one-way slabs and two-way slabs with structural tile embedded in the bottom. Clay tile, gypsum block, concrete tile, or lightweight concrete-like tile such as "slag block" was laid on top of the flat wood base forms, either in continuous rows to create a one-way system of concrete joists or separated individually to create a two-way waf-

fle. After the reinforcing required in each joist was placed between the blocks, the concrete was poured over and around the tiles (figure 7.17). In this type of construction the thin slab over the tiles was most often reinforced with wire mesh to span the 1 to 2 feet between joists, although other designs existed. Especially during the 1930s and 1940s, systems such as the Natcoflor one-way floor system did not use a top slab. The clay tile had corrugated sides designed to transfer stress to the 2-inch-wide concrete ribs.[74] The general theory of concrete- and tile-joist construction justified the assumption of shear transfer: "The one-way combination floor or roof construction . . . is essentially a flat slab in which a large percentage of solid concrete has been replaced by structural clay tile. . . . The tile units between reinforced concrete joists become an integral part of the slab and act in combination with the concrete, contributing to the flexural, lateral, and shearing strength of the floor."[75]

Standard one-way systems existed before 1910, and most commonly were created with 1½ to 3 inches of top slab poured monolithically with 4-

to 6-inch-wide ribs separated by 12-inch-square tiles. Small pieces of tile the thickness of a single web were often placed in the forms to finish the bottom of the joists and to provide a continuous fire protection and furring surface of tile across the entire slab underside.[76] Minor improvements were made into the 1940s, most notably the use of 16- or 20-inch-wide tile. Where heavy live loads existed or long spans were required, shear strength could control the design. This rarely happens with solid slabs because of the additional material, but it is often a factor in joist design. To provide additional shear strength without using steel shear reinforcing, ribs were often widened near the slab supports by switching from the wider tiles back to 12-inch-wide tiles, and so widening the concrete ribs.[77]

An interesting variation on the standard that was used in the 1910s but died out later was the Merrick Floor, which substituted dense wire-mesh cages for the blocks, and had 2-inch-thick solid slabs at its top and bottom. Presumably this system failed to achieve any popularity because of the additional effort required to create the bottom "slab," whose purpose was not clearly defined.[78]

Two-way joist systems, or waffle slabs as they gradually became known, were most often used in concrete-frame buildings.[79] The design of the reinforcing was always performed in a similar manner to that used for solid concrete two-way slabs, while the details resembled those used for one-way joists. An early waffle system, the Schuster Floor patented in 1915, used tiles 12 to 20 inches square without a top slab, or groups of four tiles to create 30-inch rib spacings with a 1½ -inch top slab.[80] The Corr-tile Two-way Floor, at approximately the same time, used "U"-shaped blocks as forms at the concrete ribs to present a continuous tile ceiling below, with minimum joist widths of 3 inches and a total depth of 5 to 7 inches.[81] The Smooth Ceiling System is a two-way concrete-and-tile waffle system without floor beams and with short steel channels to serve as shearheads within an all-concrete panel at each steel or concrete column.[82] This system was most often used with concrete columns, as it most resembles all-concrete-frame buildings. The Sandberg System was similar, except for the substitution of rebar for the rolled sections as shearheads.[83]

Many patented systems of precast beams, joists, and slabs using tile as permanent forms for concrete existed between 1925 and 1950, but few were used in New York. These systems were substitutes for precast concrete or cast-in-place concrete, two less-used systems in the New York area.[84] One bizarre combination system used in the New York area—although mostly in low buildings—was the United Floor Construction, consisting of open-web steel joists at 30 inches on center, carrying tile arches that spanned between the lower chords of the joists and a continuous concrete slab over the tile and joist tops. The arches spanned the short direction; steel trusses acting compositely with the concrete slab spanned the other.[85] Many light slabs were used with light-gauge or open-web steel joists where fire rating was not an issue; this type of construction was prohibited in the New York fire zone. A similar, but older, detail was the use of deafening with wood joists, where small nailers were attached to the sides of the joists to carry planking which in turn carried a concrete slab.

## Present Considerations

So many floor system variations exist that no short list can cover all of the possible conditions a present-day designer may encounter while making renovations. Just cataloging the known styles of tile arch would involve listing dozens of minor variations that are structurally indistinguishable. As with many other elements of building construction, past performance is often the clearest indicator of future safety: a floor slab that has seen continuous use for eighty years and shows no signs of distress is probably adequate for the same use for an indefinite period of time. The use of plaster ceilings in older buildings provides an easy way to check for sag or water penetration: simply by examining the condition of the floor's finish from below.

Many of the older systems are more fragile than modern concrete slabs or concrete on metal deck. This becomes important whenever previous renovations have been made. It is not uncommon to see utility lines hung from tile arches by expansion bolts drilled or shot into the bottom tile flange. This is a worthless connection, even if the hangers appear secure, because there is no way to judge if the tile is cracked internally and any tile failure will be without warning. The only safe hanger details with tile arches are those that attach directly to the floor beams and those that reuse the original hangers, which were attached to the floor beams or, rarely, the tie rods. When hangers are attached to the floor beams, the amount of tile removed from the beam bottom flange edge and top surface must be minimized, since this area is the seat for the skewback tile, and thus for the entire arch.

All the proprietary floor systems described have quirks that must be dealt with in renovations. For example, the Rapp floor's gauge-steel tees are not directly fastened to the floor beams. If a small piece of floor is demolished—for example, for new plumbing risers—the adjacent tees can move sideways towards the hole. The amount of movement needed to dislodge the brick layer is small, and can lead to progressive movement and failure of the floor. The safest way to deal with this problem is to demolish the entire bay, but this is rarely feasible. Other solutions, such as welding the tees in place, must be considered.

Draped-mesh concrete slabs are vulnerable to the destruction of the wire anchorage. The catenary action designed in creates tension in all of the reinforcing; if the wire is cut loose from its end anchorage, usually consisting of the last span being wrapped around the last beam or the last beam's top flange, the entire slab's stability is threatened. This can occur naturally where the anchorage is at a spandrel beam. Water entering through the facade can penetrate the slab, especially if it is cinder concrete, and corrode the wires right at the anchorage.

The potential problems with concrete slabs—spalling, rebar corrosion, and water penetration—are well documented. Any portion of an old concrete slab that is expected to carry high stress should be examined for honeycomb voids caused by lack of proper vibration during the original concrete placement. If any concentrated loads are to be placed on slabs from before World War II, or if the spans are short and highly loaded, the anchorage of the bars should be checked, since the old-style deformations (or in some cases, lack of deformation) can increase the length of embedded bar needed for development of the full rebar strength.

# 8. CURTAIN WALL SYSTEMS
## 1890s–1950s

For sixty years, the development of curtain walls meant, more than anything else, a search for thinner exterior walls. The original curtain walls were supported versions of the traditional bearing wall, and, except for the presence of spandrel beams, the two types were indistinguishable. True skeleton frames include curtain walls that are loaded only by their own weight and wind loads and are supported at every floor height. Curtain-wall thicknesses do not need to increase towards the base of a building the way bearing walls did, and recognition of this fact led to increasingly relaxed code provisions for wall thicknesses.

Bearing-wall thicknesses had been partially determined by empirical examination of existing buildings, with some gravity-load stress analysis. The use of thinner walls meant that the wind pressures used to design the overall lateral-load resistance of the building frame also had to be analyzed for their effect on the exterior walls. Once the design of curtain walls for lateral loads shifted from empirical rules of thumb to calculated stress, there was no longer any rational minimum thickness for walls other than that determined by bending and compressive stresses. For masonry this eventually was standardized at approximately 8 inches for a normal floor-to-floor height of between 9 and 12 feet.

The use of metal as decorative sheathing in the 1920s and 1930s had its origins more in traditional copper and tin roofs than in the technology of cast-iron facades. Those roofs had been built as a skin of metal over a framework of wood or metal. This "skin-and-bones" construction was similar enough to skeleton framing to provide an almost seamless transition to the mansard and tower roofs on early skyscrapers. The early skyscraper peaks led to the metal-skinned spires on the Chrysler and Empire State buildings in 1930 and 1931. The implications were broad: if a thin metal skin backed with a few inches of tile or brick was acceptable for an area as large as Chrysler's spire—which is taller than an eight-story building—then the technology could be used for entire facades.

Glass-and-metal curtain walls developed from two sources: the stress analysis of metal-fronted facade walls and the more empirical design of glass storefronts. Again, the individual components of the wall, glass panels, metal spandrel panels, and mullions had to be designed for the lateral loads and supported from the building frame. These were basically the same issues that had previously turned up in masonry curtain-wall design, only now applied to walls with a total thickness in the range of 3 to 4 inches instead of 12 or more.

## Thick Curtain Walls and Thinner

The early curtain walls were barely distinguishable from bearing walls. Writers from the 1890s through the 1910s who discussed the history and future of skeleton framing inevitably contrasted "thick" bearing walls with "thin" curtain walls, but this distinction is based more on the theory than the reality.[1] The allowance in the 1892 New York building code of a 4-inch reduction in thickness for a supported curtain wall when compared to a bearing wall, the first such reduction permitted, is minor in terms of structure. Because skeleton frames were considered to be "high technology," they were used first in large commercial buildings. New York designers were faced with the problems of tall building construction in the first group of frame and curtain-wall buildings constructed. The height of these buildings led to walls more than 2 feet thick, which today are often mistaken for bearing walls.

The irrationality of requiring curtain walls to increase in thickness at a building's lower floors was recognized immediately by the advocates of skeleton frames. Presumably the code requirement was based on an attempt to introduce gradually the new idea, but not to stray far from the familiar bearing-wall building. William Birkmire, whose writings in favor of skeleton construction in the early 1890s were influential because of his position with the Architectural Iron Works, stated in 1894 that:

> [t]he New York building law imposes certain conditions upon the use of curtain walls, in that the curtain walls of the lower floors shall be built thicker. Why this is so the author has been unable to determine.[2]

The building laws of 1887 were a major revision taking many developments into account, but they did not allow for skeleton frames and curtain walls.

> After a year or two's experience with that law it was seen where it could be improved in many respects; indeed it became absolutely necessary to make certain additions thereto. A new method of constructing tall buildings came into use subsequent to the date of passage of the new law. The skeleton of iron and steel simply surrounded by thin brick walls, used to-day in nearly all the great buildings now in the course of erection, the law had made no provision for. Application has to be made to the Board of Examiners in each case.[3]

In 1892, the newly revised building laws recognized the existence of curtain walls. The language used in the code was a little confusing, and as late as 1898 there was a reference to "curtain walls . . . not supported on steel or iron girders," a form which is found only in cage buildings.[4] The original provision for supported curtain walls gave a minimum 12-inch thickness for the topmost 50 feet of a building, increasing 4 inches every 50

Chrysler Building, 405 Lexington Avenue, east facade. The spire above the round-topped center bay and all of the three-dimensional ornaments are stainless steel. This was an early use of the material for so large an area. The extent of stainless steel at the spire is clearly visible.

feet down. An alternative was simply to subtract 4 inches from the specified thickness for bearing walls, with the exception that no wall could be thinner than 12 inches. Several provisions in the code indicate clearly that it was written by people not comfortable with skeleton frames. Where the walls were required to be 16 inches thick or thicker, vertical support was only required at every other floor, so long as the wall was tied back into the frame at every floor to prevent horizontal movement, and the wall was not used for support of interior floor beams. Where the walls were required to be 20 inches thick or thicker, floor beams were allowed to rest on the wall, as long as iron or steel ties were embedded in the wall to tie the columns together at each floor level.[5] These provisions, when combined with the required thicknesses, greatly diluted the concept of a skeleton frame supporting a curtain wall, although the buildings produced were functionally sound.

The code provisions changed very slowly, despite the construction of increasingly tall buildings. The 1901 code required a 12-inch-thick curtain wall for the topmost 75 feet of a building and an increase of 4 inches for each 60 feet down.

**55 Liberty Street, south and east facades. Close-up shows damaged terra-cotta blocks being removed during repairs. After the Woolworth Building, one of the largest all-terra-cotta facades.**

Cage-enclosure walls were specified at 12 inches for the topmost 60 feet with a 4-inch increase for each 60 feet down.[6] The 1916 building code still had the exact same requirement for curtain walls, despite the construction of buildings over 600 feet high.[7]

The New York City Building Department's conservative approach was not shared by everyone in the field. In 1898, Birkmire was proposing curtain walls no thicker than 16 or 20 inches, with the caveat that

> [i]t is, of course, understood in recommending this saving that the skeleton-frame is rigidly built, and sufficiently strong to support all loads and sustain, without any possibility of derangement of its parts, all manners of strains to which it may be subjected.[8]

Birkmire pointed to Chicago's more liberal laws regarding curtain-wall thickness as being closer to his ideal.[9] At this time, the practice of constructing curtain walls starting at the third or fourth floor was already established.[10] This practice, instituted to prevent logistical conflicts at the ground on a construction site, constitutes tacit acceptance of the nature of the walls, highlighting the pointlessness of heavy curtain walls.

Ten years after Birkmire's proposal, another prominent engineer voiced the same complaint. O. F. Samsch, the structural designer of the Singer Building, investigated the practical maximum height of a skyscraper. He found that under the highest rock-bearing stress allowed by the New York code it would be possible to build 2000 feet high on a 200-foot-square lot, but that the height would in fact be limited by the required wall thicknesses. A description of his research concluded:

> Structurally considered, this provision of the building code is unnecessary, since the weight of the walls at each story is carried directly upon the steel framework at the base of that story; and there is, therefore, no structural necessity for making the walls at the base of a 20-story building more than 12 to 18 inches thick.[11]

The categorization of some walls as bearing walls and some as curtain walls based on the presence or absence of spandrel columns and beams is unclear to modern engineers. Both sets of walls were ordinarily built of solid masonry, most often common brick tied together wythe to wythe with numerous headers. The curtain walls were supported either by hung lintels positioned at an elevation where they would fit in the brick coursing or by the brick resting on the floor slab above the spandrel beams with the floor-to-floor height an even multiple of the brick course height. This type of curtain wall had no provision for horizontal expansion joints. A modern supported wall would have a gap below the spandrel beams and hung lintels to create a soft line for an expansion joint. The walls also were laid up continuously around the perimeter of the buildings, without provision for vertical expansion joints.

The lack of expansion joints in the curtain walls enabled them inadvertently to carry structural loads. In full-frame buildings with skeletons designed to carry all applied loads, this accidental load capacity of the facade has had several well-documented effects. The facade masonry "stacks," with the curtain wall at each level supporting the wall above instead of having the independent action for which skeleton frames are designed. Stacked curtain walls often carry a substantial portion of the perimeter floor load because the walls loaded in axial compression are far stiffer than the spandrel beams loaded in bending. The spandrel beams deflect more than the walls under their respective loads, and so the beams come to rest on the wall and be supported by it, reversing the intended design. The presence and design of the hung lintels intended to support the wall are of little importance as stacking occurs because, without expansion joints to allow deflection of the lintels, the load will again be carried through the masonry. The compressive stresses in the facades at the base of this type of building have been measured to be as much as two-hundred times as high as the value expected for a true curtain wall.[12]

The masonry facades, if considered as shear walls, are stiffer in resistance to lateral load than all steel moment frames and many braced frames. As described above, the stiffer element will absorb load. The facades of many older buildings provide significant stiffening against lateral movement, a fact that was not fully realized until glass-curtain-wall buildings were constructed and their lateral load responses measured. Without the unintentional stiffening provided by masonry facades, the drift and lateral acceleration of steel frames later became primary concerns for designers.

The use of decorative terra cotta tended to amplify stacking and make its consequences more dangerous. Any material made from baked clay is subject to expansion from water absorption, but the degree of expansion is far greater for terra cotta than for common brick. Terra cotta "is as small after firing as it is ever going to be in its entire life. When in service it will have a tendency to grow."[13] The popularity of terra-cotta facades between 1880 and 1940 was based on aesthetics and economics, the mass-produced clay being the only method available at that time that could economically supply complicated ornament.[14] Large facades of terra cotta without control joints, such as were used at 55 Liberty Street or the Woolworth Building in 1909 and 1913, are subject not only to thermal and building-movement-related cracking but also to compressive stress caused by expansion. Terra cotta's fragility means that such a facade, more likely to be highly stressed than a brick facade, is

also more likely to crack and spall under stress.[15] By the 1920s, when expansion joints and support at every floor were recognized standards, the era of greatest terra-cotta use was already past. These standards were also sometimes ignored even when known.[16]

As described in the context of column fireproofing, the curtain walls usually wrapped around the spandrel columns.[17] Before the use of concrete floor systems, the curtain-wall masonry was usually built into the space between the spandrel-beam web outer face and its flanges. Once concrete floors became popular, this detail was used less often, because the spandrel beams could easily be encased in concrete. But it remained in use as long as solid two-wythe brick walls were used with steel frames, approximately through the 1940s.

The main effect of encasing the spandrel framing in brick or tile anchored into the curtain wall was to create "hard points" in the wall susceptible to cracking. In modern curtain-wall design, hard supports are locations specifically designated for expansion joints. The rationale is simple: since there is no way to prevent cracking at these locations, the crack should be constrained in form to an intentional control joint. In the forty years following 1890, hundreds of tall buildings were built without any control joints. The cracks that have

**Woolworth Building, 233 Broadway, east facade. Close-up showing the facade construction, which was originally all terra-cotta block, and has been repaired with modern replacements. The high ratio of window to wall in a 700-foot-high building is an indication of the steel columns hidden behind each major terra-cotta pier.**

developed over the years from building sideway and facade thermal movement have been concentrated directly outboard of the steel members. The most common exceptions to this pattern are those buildings where fashion dictated the presence of masonry quoins. The different material properties of brick and stone concentrated movement in these buildings at the quoin joints, thus protecting the remainder of the wall from cracking.

The Potter Building of 1886 at 38 Park Row is an interesting transition between bearing walls and curtain walls. The building's exterior walls are brick and terra-cotta true bearing walls carrying the wrought-iron floor beams. The building, as mentioned earlier, was erected on the site of a spectacular fire, and was the most elaborately fireproofed building in the city at the time of its construction. Only noncombustible materials were used, although not necessarily in ways that meet modern fireproofing standards. Portions of the facade consist of cast-iron bays between masonry piers. Although cast iron's vulnerability to fire was well established by 1886, these bays are closer in design to a cast-iron curtain wall backed up with a thin layer of brick than to the iron bearing walls used in the low cast-iron facades of the previous two decades.[18] Cast-iron or copper-skin bay windows in vertical bands were used extensively until the 1910s, when changing architectural fashion and the use of true masonry curtain walls reduced their relative attractiveness to designers.[19] These tall bay windows were usually supported at each floor by a pair of cantilevered beams, which were, along with the metal and brick skin, the notably advanced features. These bays represented the first cantilevered curtain walls and the first use of a consistently thin curtain wall.

Cantilevered curtain walls, where the entire wall system is outboard of the last column line, evolved at the same time as constant-thickness curtain walls. Cantilevering the wall is not necessary, but it does have various architectural advantages. The issue of waterproofing exterior columns can be totally eliminated if there are no columns closer to the exterior than the interior face of the curtain wall. That idea was taken to its logical end after 1950, when glass curtain walls were often cantilevered beyond the spandrel beams as well as the columns.

Besides the isolated vertical bays, another ideological source of cantilevered curtain walls was cage construction. The St. Paul Building on Park Row was designed by George Post and completed in 1899. Post was an advocate of cage frames instead of skeleton frames, but by the late 1890s the mainstream of design had moved so far from cages that even Post had to accept supported curtain walls. The St. Paul Building had a true curtain wall that in many ways resembled a cage frame's exterior wall. The wall was of increasing thickness and supported on double spandrel beams which were connected to columns by cantilevered brackets. The columns were entirely inboard of the wall and fireproofed with tile block independent of the wall.[20] This construction was consistent with Post's stated distrust of column waterproofing in curtain-wall buildings. Later in his career, Post allowed his feelings about skeleton framing to dominate his structural design. In 1907, his design for Shepard Hall at City College, a low building with an extremely large footprint that included a slender tower over 100 feet high, used primarily bearing-wall construction. This was unusual enough for a nonresidential building after 1900, but in addition, the walls were in places entirely terra-cotta-face tiles, which were not ordinarily considered for bearing.[21]

As described earlier, part of the revolution in building design caused by skeleton frames was the conscious design for lateral loads. This design was applied as it became clear that the nineteenth century's engineering structures could not be built without more analysis than their predecessors. Design wind pressures were not subjected to much rational analysis until the collapse of the Tay Bridge in Scotland in 1880. This disaster, the collapse of a high bridge under combined train and wind loads, was the impetus for new research into wind loading in the United States as well as in Britain. The formulas used in the 1890s were simplified versions of Newton's formula relating the pressure exerted to the square of the wind speed. The United States Signal Service gave the wind pressure as one two-hundredth of the wind velocity squared, which was adopted by many building code writers across the country.[22] The local weather records provided the expected maximum wind velocities.

While the Signal Service formula provided a basis for overall wind pressure, the design of facades for wind was still not necessarily practiced. Early curtain walls were so thick relative to the floor-to-floor height that modern analysis reveals them to be stressed far below their capacity, a condition that allowed them to be built fairly safely without design. An examination of damage to steel-frame buildings after the 1897 St. Louis tornado found that certain common detailing practices were not adequate.[23] Among the recommendations were the exclusive use of rivets instead of unfinished bolts in order to provide connections stronger than the connected members; stronger direct connections between curtain walls and the building frames; and the use of a minimum 30 psf wind load on facade elements, increased to 50 psf at building tops.[24] The use of riveted, welded, and slip-critical bolted connections has of course become standard practice; direct connection of facades spread in the twenty years following the tornado as curtain walls became thinner; and most modern building codes now contain minimum wind loads for facade elements higher than those used for the building frame.

Some of the provisions in the New York building code and descriptive literature at the turn of century were forerunners of more sophisticated practices that came into use later. As early as 1891, the building code contained provisions specifically addressing the use of hollow walls.[25] The allowance was written in the same language as that for the use of piers instead of a continuous wall: hollow walls were allowed on the condition that the amount of total wall area was the same as required for solid walls. Brick or iron ties connecting the different parts of the wall were also required.

There are two main types of hollow wall: multi-wythe walls continuously separated into two parallel segments in the manner of modern cavity walls; and walls containing voids specifically dedicated to some purpose such as carrying chimney flues or serving as chases for plumbing. The latter was an attempt to put the large amount of floor area lost to walls to active use, was more common in bearing-wall buildings, and was most common in small bearing-wall buildings, when compared

with curtain-wall buildings. The use of the wall floor area to carry mechanical services became less frequent as the walls grew thinner and more adjustable in layout, and as architects developed dedicated vertical mechanical cores and shafts within buildings.

The central idea of a cavity wall is to use an air gap to provide insulation and water protection for the building interior. A gap of 2 inches could be created by separating two wythes and bridging the gap with metal ties; gaps of 4 inches could be created by laying out brick into a gapped pattern one wythe wider than the actual amount of brick would warrant. The second method, referred to as a "row-lock" or "Rolok" wall, required only a variation in the brick-laying pattern. It was similar to the manner in which individual chases were created, but it surrendered more floor area than was necessary and, in a thin wall, would require substantially more brick than a mechanically tied wall.[26] Old drawings tend to be very unclear about this type of detail, but it appears that cavity walls did not become standard practice until the late 1930s, and mechanically tied walls did not become the dominant form until the 1940s.[27]

Cavity walls are used today primarily for their ability to trap water outside of the occupied building space. The insulation provided by the air is discounted for temperature change, with rigid board or fiberglass batt insulation often being placed within the wall; but the insulation provided by the distance for water control is extremely effective. It is accepted that water will penetrate the outer wythe, either through gaps in joint material or through generalized seepage. Flashings, and occasionally vapor barriers, are provided at the face of the inner wythe to guide water down the back of the outer wythe and to weep holes that take it back to the outside air.

Curtain walls built to minimum code standards reached a thickness of 12 inches in 1916.[28] In the late 1910s and 1920s, the most common form of curtain wall consisted of three wythes, the innermost structural clay tile used to provide a furring surface for plaster. The outer two brick wythes were tied together in standard bond with a course of headers every six or seven courses, usually spaced so that a header course was laid directly above the interior floor slabs. The two interior

30 Rockefeller Plaza, south and east facades (background); 610 Fifth Avenue, south and east facades (foreground): Two steel-framed buildings of Rockefeller Center with limestone-veneer curtain walls. The spandrel panels in 30 Rockefeller Plaza are cast aluminum. This was the first masonry curtain wall in New York to be a consistent eight inches thick top to bottom.

wythes rested on the slab and the spandrel beams directly below; the outer wythe was "hung" from the middle wythe by the header courses. This wall construction was used with minor variations until the 1930s, when a two-wythe version became standard, either with a tile interior wythe or all brick.

The first building with an 8-inch curtain wall was the RCA Building, the centerpiece of Rockefeller Center, completed in 1932. At that time, with the full effects of the Depression first being felt, the Building Department was perhaps more flexible about its rules than it had been, especially where the only large private construction project in the city was concerned. The facades at Rockefeller Center are limestone panels in vertical strips between the windows, with aluminum spandrel panels in the window-line vertical strips.[29] Rather than requiring enough backup to make the total wall thickness equal 12 inches, the building examiners interpreted the language of the code, which said that walls supported at each floor "might" be 12 inches thick for the full height of building, to mean that a thinner wall could be allowed.[30]

While overturning twenty years of practice in a backwards manner is not necessarily good, the use of thinner walls was unquestionably a step forward.

## Increasing Window Area

The progression of the masonry curtain wall from multiple wythes of brick to the more modern two-wythe brick or brick-and-block cavity walls was largely complete by 1930. The drive for thinner curtain walls and architectural requirements for increasing large windows resulted in the detailing changes that separate earlier masonry walls from modern ones. A structure such as the Little Singer Building on lower Broadway, built in 1904, showed how much leeway the building code actually could allow. The rules about wall thickness did not specify directly how much of the wall surface area had to be masonry as long as the total wall cross-sectional area was maintained. Little Singer's window area is approximately three-quarters of the wall surface, set behind a decorative wrought-iron and steel screen, with spandrel panels of masonry covered in terra cotta.[31] The code-specified wall cross section is concentrated into piers between the windows. While this building is an obvious predecessor to glass curtain walls, it also shows the type of architectural effects that would be pursued in the next thirty years in glass and masonry curtain walls.

As window areas increased, the use of interior wythes of masonry to support the exterior wythes and the use of individual loose lintels over each window became less practical. Hung lintels, which had previously been used primarily for the support of large projecting elements such as cornices and occasionally for the floor-level support of facades, became necessary to support the band of masonry at each floor level that separated the bands of windows and masonry piers above and below.[32] Another architectural trend that fit this pattern was the use, starting in the late 1920s, of corner windows. Unlike planar windows, windows at the corner of a building require a braced hung lintel regardless of the window width.

The masonry curtains used in these buildings were far thinner than those used previously. This reduced the total wall weight, shrinking the required size of the spandrel beams and columns.

It also forced engineers to begin considering the structural design of the wall for lateral, out-of-plane loading. In the years in which wall thicknesses had been prescribed, the only reference to the effect of wind load on wall surfaces had been indirect: the required increase in thickness when the unbraced area of wall exceeded certain limits.[33] For a solid rectangular section, such as a brick wall, the bending stresses are inversely proportional to the square of the thickness. Reducing a wall from 16 inches to 12 inches thick increases the bending stress induced by wind load by 77 percent. Reducing that wall to 8 inches thick increases the stress to four times the original value. For the ordinary range of floor-to-floor heights, 12-inch walls can be used freely, while 8-inch walls may require additional bracing or reinforcement depending on the exact conditions.

The use of large areas of glass increased the difficulty of designing a thin masonry curtain. A common architectural design of the 1940s was the use of horizontal strip windows. Since these cut off the top support of the wall, analyzing the wall as spanning vertically from floor to floor was no longer possible unless a secondary system of framing was introduced. Secondary framing consists of posts and beams set within the wall plane and oriented for horizontal load to brace the masonry panels and window frames. In structural form, it is little different from the mullion systems of glass curtain walls, and perhaps it helped suggest the technology of that next step.

## Glass Walls

Despite the frequent discussion of glass curtain walls as a twentieth-century phenomenon that grew out of storefront design, there is a history of all-glass walls in limited use at least as far back as the 1850s. The London and New York Crystal Palaces and Paddington Station in London contained large wall and roof areas of iron framing and glass panels. The design methods of that time, particularly in the United States, were not adequate for this type of structure without confirmation through load testing. Until the 1870s, when design calculations caught up to construction, the early iron-and-glass structures had little influence. The construction of large, open interiors in that decade, most prominently in the train sheds of

**Little Singer Building, 561 Broadway, east and south facades. Close-up view of a fairly typical loft building, fireproof on the inside, with a steel, glass, and terra-cotta curtain wall.**

railroad stations and the generation of federal armories built after the Civil War, produced lighting problems that were not entirely solvable. The best interior lighting of the time was gaslight; the brightest lighting of any kind was either limelight or the newly developed arc lights. None of these was very successful at lighting a space such as the train shed at Grand Central Depot, built in 1871, which was 600 feet long and 200 feet across.

At Grand Central, New York got its first permanent, large, iron-framed open space. The lack of adequate light in the train shed, which was a full block in area and crowded with locomotives releasing coal smoke into the air, was addressed by skylights that made up over 50 percent of the roof area. The roof was a balloon shed carried on thirty-two wrought-iron arches. The plain roofing material was galvanized corrugated iron sheeting attached directly to the arches and I-beam purlins. The skylights were carried above the line of the arches by secondary iron framing: a center strip with a gable section and two side strips with shed sections. In addition, the rear wall of the shed was an iron-and-glass curtain wall which covered the full 200-foot width and 100-foot height of the shed.[34]

While the construction of the glass portions of Grand Central may have lacked certain features of a modern glass curtain wall, most notably weep

**Grand Central Depot, before 1898, interior, facing north from the station building proper. The balloon shed with its wrought-iron arched trusses can be seen plainly at the skylights.**

holes and other waterproofing details, there is no question that this construction can be defined as a true glass wall. Unlike earlier mostly glass walls such as rose windows in churches, the construction at Grand Central was designed by engineers for specific loads instead of being designed by rules of thumb and builders' knowledge. Isaac Buckhout was the engineer for the building as a whole, and the designer of the balloon shed was Wilhelm Hildenbrand. Hildenbrand, who was 22 years old in 1871, later became one of the chief assistant engineers for the Brooklyn Bridge.[35]

The Grand Central shed had a fairly short life due to the functional inadequacies of the railroad operations in the building, and was demolished during the construction of Grand Central Terminal in 1910. The general form of a trussed roof with skylight monitors and glass curtain walls at one or both ends was used on many other train sheds now demolished, and on many armories that are still standing. The most notable of these may be the 1879 Seventh Regiment Armory at 643 Park Avenue, which was probably the second balloon shed built in the United States, and is almost certainly the oldest still standing. The general form of the drill shed portion of the building

is extremely similar to the train shed of Grand Central, which the armory's designers consciously copied. The shed is 180 feet wide and 300 feet long, and the main roof covering is built of wood, but otherwise the description of the Grand Central roof applies. The armory is more advanced than Grand Central in its use of two hinged trussed arches of varying depth to equalize stress in the different portions, but a step backward in its lack of a glass curtain wall at the end and its use of a wood plank roof.

In 1880, the company that eventually became Pittsburgh Plate Glass began producing large sheets of smooth and transparent plate glass. While plate glass was not newly invented in 1880, PPG was the first company in the United States to produce it cheaply enough to make a profit. Its immediate use in building construction was in cast-iron storefronts and facades. The next stage in storefront development came after 1900, when steel replaced cast iron and wood for storefront frames. The Kawneer Company, founded in 1906, was one of the pioneers in steel storefront manufacturing. Individual panes of glass as large as 80 square feet were used, showing that the changes needed to produce glass curtain walls were in frame and panel design, not in the glass-making technology.[36]

The combination of storefront glass and glass-holder technology with the earlier skylight frame design produced glass curtain walls. Glass-wall construction proceeded overseas, but was prevented in New York and the United States in general by code provisions for exterior walls. In 1929, Harvey Wiley Corbett made a proposal for a hundred-story tower for the Metropolitan Life North Building at 24th Street and Madison Avenue. The building actually erected was much shorter but shared the original design's full-block footprint. Corbett was concerned about providing adequate light in such a massive building, and had originally proposed curtain walls entirely of metal and glass. This idea was abandoned because of the building code's rules requiring masonry exterior walls.[37]

In addition to purely architectural concerns such as facade aesthetics and interior lighting, glass walls were also favored for construction issues. With the almost-exclusive use of steel framing and concrete floors in large commercial and residential buildings in New York during the 1930s and 1940s, masonry's role had been reduced to curtain-wall and partition construction. By switching to glass or metal curtain walls, the trade could be practically eliminated from base-building construction, leading to faster erection times. In addition, the new walls were thinner than masonry curtain walls, squeezing a few extra net square feet of floor space from the same building footprint.[38] Glass curtain walls brought the separation of enclosure from the building frame to its final extreme. Not only did the exterior wall play no role in the design of supporting the interior floors, but it could not, under any circumstances, support load.[39]

The first applications of the all-glass curtain wall in New York were the Secretariat Building of the United Nations in 1950, the Lever House at 390 Park Avenue in 1952, and the Manufacturers Hanover Trust bank at 510 Fifth Avenue in 1954.[40] All three used the most basic form of glass curtain wall: individual mullions mounted to the building frame and glass panels that were sealed on the exterior to provide waterproofing. This system was first used in the United States at the 1947

Lever House, 390 Park Avenue, south and east facades. The second glass curtain wall in New York, stainless-steel covers over steel mullions. The Racquet and Tennis Club, at left, is a 1918 steel-frame building with a curtain wall.

Manufacturers Hanover Trust Bank, 510 Fifth Avenue, north and east facades. Early glass curtain wall with exceptionally large individual sheets of glass from mezzanine to second floor.

Equitable Building in Portland, Oregon.[41] Structurally, the systems were almost identical, despite the difference in metal used. The Equitable Building used aluminum mullions, the Lever House carbon steel with stainless-steel covers, and the Secretariat and Manufacturers Hanover carbon steel.[42]

The Chase Manhattan Building, completed in 1961 at Cedar and Nassau streets, was typical of early glass curtain-wall design in New York. The walls were structurally identical to those at the Lever House, exceptional only because they were the first over 800 feet high. The walls of this building have functioned well despite their fairly primitive expansion joints largely because the building frame itself was designed to be extremely stiff, limiting the total drift to approximately half that of similar, newer buildings. The relatively small building movements reduced the amount of movement the curtain wall's expansion joints were asked to accept.[43] A more important advance for New York in curtain-wall design was the use of aluminum mullions and spandrel panels in 1953 at 99 Park Avenue and at 460 Park Avenue, and in 1954 at 112 West 34th Street without masonry backup.[44]

## Modern Curtain Walls

The majority of refinements in glass curtain-wall design since 1960 have been architectural. Double glazing, pressure equalization, new sealants, and new mullion configurations are all methods of improving the thermal insulation, waterproofing, and material life span of the walls. These are important issues, but they rarely have any influence on structural design and rarely require structural engineering to develop. They are not discussed here for the simple reason that they are complex architectural subjects well covered elsewhere.

Curtain-wall design is an area where architectural decisions influence structural performance, most notably in mullion section selection. The modernism of the 1960s and 1970s included a movement towards totally flat curtain walls, where the mullions would be pulled back into the wall plane to present as flat and seamless an exterior as possible. In some cases the mullions are entirely within the building and the glass is secured with mechanical fasteners and adhesives; in others a single thickness of steel or aluminum extends from the mullion within to cover the edge of the glass panels. In several cases, windows have "popped out" of the mullion frames, endangering people in the street below. The glass supports had not been adequately designed to resist the wind suction that exists on the leeward side of a building.

Other structural developments have been the refinement of glass-panel design, recognition of the more complicated requirements for expansion joints in glass curtain walls when compared with masonry walls, the use of new materials for glass curtain-wall mullions, panel design for masonry curtain walls, and the use of metal studs as masonry backup.

The last item on this list, metal-stud backup, was first introduced in the mid-1960s, as a variation on the brick-over-wood-stud walls previously used in private houses and tenements.[45] The basis of the system is a series of light-gauge metal studs spanning vertically from floor to floor, anchored to the top of the floor slab below and the spandrel beam or floor slab above, designed to resist lateral loads applied to the wall. The enclosure is most often built of a single wythe of brick—although precast panels, concrete block, and other materials have been used. Although only used in recent years, there are no elements of a brick-and-stud curtain wall that did not exist in the 1890s. The use of this system is predicated on the studs being cheaper to install than masonry backup, which was not true before World War II.

Stud-backup use requires certain assumptions of questionable accuracy.[46] With the ordinary assumption that the brick shares in load when the wind blows in towards the facade (where the brick would be in compression), it is a simple matter to design studs that can withstand the stress created by code wind loads. The brick cannot share the suction loading that exists on the leeward side of the building because there is no provision for loading it in tension. While the recommendations for allowable deflection vary, it is not particularly difficult to meet the most stringent suggestions (one six-hundredth of the span length) with 4- or 6-inch-deep studs. However, regardless of the structural adequacy of the studs, water penetration through a wall of this type is inevitable. Even using the strictest deflection standards, some cracking of the mortar under suction loading will occur.[47] Masonry cavity wall design, which is regularly used and accepted by the majority of designers, assumes that water will penetrate the joints of a single wythe of brick and must be drained away within the cavity. There is no form of waterproofing that can be put between the brick and studs that will not be punctured by the wire ties holding the brick in place; so there is no way of preventing water transmission.

The examinations of facades and their supporting structure required by New York's Local Law 10 (see below) have clearly shown the vulnerability of steel in close proximity to the outside face of the enclosure. Rolled steel beams with ¾-inch-thick flanges are often found to be seriously corroded, with more than half of the metal lost. Given the evidence of water transmission through brick masonry, the relative speed with which a 16-gauge steel stud will rust, and the impossibility of flawless vertical membrane waterproofing, stud backup must be considered as a lower quality, shorter-lived method of constructing curtain walls.

The widespread use of aluminum was a result of industrial advances during the 1920s and 1930s. The metal's application to construction has always been limited by its relative weakness when compared to steel (allowable bending stresses in the range of 10 to 12 ksi instead of 20 to 24 ksi) and its more expensive smelting process. When Pietro Belluschi designed an aluminum-and-glass curtain wall for the Portland Equitable Building in 1947, he proposed a structural aluminum frame, but this was rejected as unbuildable with the technology of the time.[48] Aluminum has, however, been regularly used for facade elements, where the forces to be resisted are orders of magnitude lower than those in building frames and where close tolerances are required. The Empire State Building, in 1931, was built with aluminum for the decorative mooring mast at the building's top, exterior ornamental trim, and decorative spandrel panels below each window. Similar aluminum spandrel panels were used at around the same time in the original Rockefeller Center buildings.[49] In 1957, the Tishman Building at 666 Fifth Avenue became the first building to have an all-aluminum curtain wall.[50] This precedent has only rarely been followed, mostly because of architectural styles

Empire State Building, south facade. Close-up shows limestone panels for typical wall, aluminum spandrel panels below windows, and stainless-steel vertical strips covering limestone and aluminum unfinished edges. Steel strips are the direct ancestor to curtain-wall mullions.

dictating either glass or masonry curtain walls. One prominent example of the aluminum wall is the Citicorp Center at 53rd Street and Lexington Avenue, built in 1977, which has an aluminum facade over 900 feet high.[51]

The use of aluminum paneling followed the same pattern as stainless steel, which had gone from decorative elements to large surface areas with the Chrysler Building and, in 1955, was used for the entire facade of the Mobil Building on East 42nd Street.[52] The aluminum panels were treated as simply finish, and were supported by the code-mandated 4 inches of masonry block backup.[53]

Ever since exposed metal has been used in building facades, designers have been dealing with the problem of corrosion. Cast-iron facades have survived reasonably well because of the metal's inherent resistance to rusting and the thorough painting of the facades. The Empire State and Chrysler buildings used stainless-steel facade elements, which became popular for glass curtain walls.[54] Stainless steel and aluminum resist atmospheric attack better than any other materials, but they are also the most expensive metals used in construction. The brief vogue for "rusting steel" was an attempt to find a facade material that was no more expensive than ordinary steel but did not require constant maintenance. This material is a steel alloy that rusts under similar conditions to ordinary steel but whose corrosion product is dense and remains bonded to the base metal. This behavior, which is exhibited naturally by aluminum, prevents the corrosion from spreading deeper into the metal as in classic rusting and delamination.

The Ford Foundation Building on 43rd Street, built in 1967, and the Annenberg Building of Mount Sinai Hospital, completed in 1976, both use Cor-ten steel, one of the more popular rusting steels. Martin Luther King, Jr. High School on Amsterdam Avenue, completed in 1975, uses Mayari-R, a slightly different alloy.[55] Rusting steel never achieved the popularity of other facade materials for both architectural and engineering reasons. The protective layer of rust is dark reddish brown tending towards black, a color that many people dislike, especially in a building as large as Annenberg, which is over 400 feet high. This color also "spreads" over portions of the facade not made of rusting steel: small pieces of rust break loose from the base metal and stain adjacent masonry or concrete. Finally, unless the facade is detailed to prevent any standing water, the rust protection breaks down and harmful corrosion can result.

The opposite end of the material spectrum from glass curtain walls was the increasing use after 1960 of new forms of masonry. The most widely used has been precast concrete panels. Large concrete panels spanning vertically halfway or more between floors allowed architects to use the visual weight of masonry without the expense of laying up finish-quality brick or block. The rough look of concrete was also a popular style feature at that time, as a number of exposed, cast-in-place concrete-frame buildings were built. Two of the more prominent examples of exposed precast cladding were the Pan Am Building at 200 Park Avenue, finished in 1963, and the American Bible Society building on upper Broadway, completed in 1966.[56] Precast panels were designed as ordinary reinforced-concrete simple beams vertically oriented, with the natural panel-to-panel joints serving as expansion joints. This application was structurally successful enough to encourage the use of precast panels as backup for veneer stone facades in place of the more common concrete block. This application shows the genealogy of precast, which is related more closely to glass-and-metal curtain walls than to unit masonry.[57]

Brick, which had never entirely fallen out of architectural fashion, has been modified in various ways since 1950. Most of the variations, such as the use of jumbo sizes, have no structural implications. One common variant that has proven structurally unsound under certain conditions is silica-glazed brick. This brick, while available in many colors, was used so often from 1950 into the 1970s in light gray or white on high-rise apartment houses that it is usually referred to as "white glazed brick."[58] The danger in the material comes from its nearly impervious exterior surface. Most water that finds its way into a facade of this material through joints or flashings is trapped, because it cannot make its way back through the outside face of the brick. In winter, trapped water can freeze and spall the outer face of the brick.[59] The extensive use of glazed brick in residential construction, where stud backup was also most common, had the unfortunate effect of combining the worst aspects of both materials.

Another variation on precast construction was glass-fiber reinforced concrete (GFRC) panels. These panels are used in the same locations that ordinary precast concrete might be used, but the special nature of the concrete, with polymer additives, glass fibers scattered throughout to bind the mass together, and only small aggregate, permitted the use of extremely thin webs. GFRC panels typically are $\frac{5}{8}$- to $\frac{3}{4}$-inch thick, compared with a practical minimum thickness of 2 inches for ordinary precast. In the 1980s, GFRC began to be used commercially for decorative elements such as cornices and ornamental balconies and for repairs. The panels are typically hung from the building structure by light-gauge steel, and joint-sealed with GFRC slurry.[60]

The extreme light weight and flexibility in configuration made GFRC a natural choice for replacement of deteriorated terra cotta. In the largest application of this type, over 200,000 terra-cotta panels on the facade of the Woolworth Building were reanchored or removed and replaced with GFRC.[61] The facade restoration of the Woolworth Building, conducted in the mid-1980s, was one of the largest initiated by Local Law 10, the code-mandated examination of facades.

# CASE STUDY: CURTAIN WALL FAILURE

The owner/manager of a large office building in midtown Manhattan called in an engineering firm to investigate a serious ongoing problem with the curtain wall. The problem, in short, was that the facade shed glass into the street during high winds. The building staff developed a close watch on weather conditions, and would close portions of the pedestrian plaza at the base of three facades when sustained winds over 30 miles per hour were expected. The following is a summary of the investigation.

The building is a fifty-five-story commercial building built in the mid-1960s. The facade is an aluminum-mullion and glass-panel curtain wall, with individual "sticks" of aluminum attached directly to the building's steel frame. The majority of the vision windows are 4 feet wide and 8 feet high, ⅜-inch thick. Very few of the smaller spandrel windows had failed.

The building fills a city block, and is therefore relatively stocky, even for its height. While many tall buildings from the 1960s are overly flexible, this building does not have a history of perceptible motion. One common cause of facade failure—excessive shortening of the leeward facade causing outward buckling—was therefore unlikely. There was no observable pattern to where the lost panels came from beyond the loss being more common in the upper floors. This behavior is to be expected if the wind pressure is the direct cause of failure.

The panels are somewhat thinner than modern practice would use for a new building, but the stress level in a panel under the expected wind load was not enough to cause flexural failure. The lack of a pattern to the loss locations eliminated wind "hot spots" caused by the adjacent buildings as a probable cause.

The 4-foot dimension of the panels is horizontal, with the panels assumed for design to deliver all of their load to the vertical mullions. The main portion of the verticals is an "I" shape adequate to carry the tributary wind load. Expansion joints were correctly located above the every-other-floor spandrel-beam attach-ment. The panels were placed into the frame opening from the outside, against a strip of adhesive tape on the outside face of the mullions, and then secured with snap-in mullion extensions on the sides. The entire panel edge was then caulked. The adhesive was meant to prevent minor movements of the glass only; the glass-panel weight was borne by the horizontal mullion at the panel base and the snap-in mullions were designed to resist suction.

The snap-in mullions required very little rotation (although a reasonable amount of torque) to reach the position where they snapped loose; since the adhesive was not strong enough to restrain the panels under the full leeward-side suction load, the snap-ins would be loaded both in shear and torsionally by the deflected glass panels. The thinness of the panels, while not directly threatening, allowed larger deflections of the panels than the recommended glass would, increasing the rotation at the glass edge.

The basic design flaw found was that ordinary loading of the panels could cause enough deflection to snap loose the portions of the mullions that secured the panels. In most curtain walls, these rotations are in opposite directions, so the suction rotation of the panels' edges is not an issue.

Given the occasional nature of the glass loss, adding structural steel to the rear of the mullions to stiffen them or changing the snap-in design to prevent this type of failure would be too expensive. The building's manager began a program of replacing the glass with thicker panels that would deflect less, and therefore would be less susceptible to failure. In addition, a small channel section was designed that could be attached to the spandrel beams and the mullion rear, which would stiffen the main portion of the mullion, and thus reduce a secondary movement that contributed to the snap-in's movement. The decision was made to not install the channels until the effect of using the heavier glass could be assessed.

**2643 Broadway, east facade. Typical fireproof apartment house. The original sheet-metal cornice was removed after a facade inspection some years ago. The new, white cornice is GFRC. The black cornice on the building to the right is the original sheet metal.**

## Local Law 10

An accident in 1979 triggered the most comprehensive review of curtain walls, and facades in general, in the history of construction in New York. While the problem of deteriorating facades was not new, the sensational nature of the accident served to focus the attention of both the public and the Building Department.

On May 16, 1979, a Barnard College freshman was killed by a four-pound chunk of masonry that fell from an upper-floor decorative window lintel at 601 West 115th Street. This building is a twelve-story apartment house owned by Columbia University. While the facade was found to be noticeably deteriorated, it was not any more so than those of hundreds of other "prewar" buildings in the city. In retrospect, the absence of any guidelines for the inspection, maintenance, and

repair of facades makes negligence difficult to prove. The outward displacement of the lintel masonry from water infiltration behind was probably not visible from either the street or the apartment inside.[62]

This accident was unfortunately not an isolated incident. During the 1970s, the deteriorating condition of facades was becoming well known. One of the most spectacular incidents was the 1976 failure of connections in a new stone-and-glass curtain wall, resulting in 6-foot-long stone panels falling twenty stories into the street. Even more frightening in its implications was the discovery of deteriorated structural steel during a facade repair. The building in question was sixty-six years old and receiving brick and terra-cotta repairs, but the project engineer, Peter Corsell, found that the structure of the top five floors of the building was in danger of collapse from rusting caused by facade water infiltration.[63]

The eventual reaction was to ensure that inspection, maintenance, and repair of such facades were clearly defined in law. Local Law 10's accident-driven popularity is shown by the fact that it was introduced into the city council at the mayor's request, and was sponsored by nine of the thirty-four council members.

The new law, incorporated into the building code but still commonly referred to as Local Law 10, requires all buildings of more than six stories to receive a "critical examination" by a licensed architect or engineer at regular intervals. "[A]meliorative work" specified by the professional performing the inspection for any deficiencies noted has to be performed immediately. The only exception for buildings in this class was allowed for those buildings undergoing continuing maintenance as defined by the Department of Buildings.[64] The requirements for this type of maintenance are stringent enough to ensure that very few buildings qualify for this exception.

The practical effects of the law have been extensive, although not necessarily in the ways intended. There have been many examples of building owners stripping damaged ornament off building facades rather than repairing it. One of the most prominent examples of stripping is the Mayflower Hotel at 15 Central Park West. In 1983, all of the terra-cotta cornices and window heads were removed and replaced by brick-red stucco. This form of "simplification" has been repeated on numerous apartment houses and row-houses around the city.

A far more beneficial effect of the law has been the education received by the engineers and architects performing the inspections. A professional in New York can learn about the aging of masonry curtain walls in the course of performing inspections in what is arguably the world's largest laboratory for the destruction of brick. The inspections have shown up the extent of water damage to steel members behind the facade, the widespread presence of cracks at the corners of buildings without facade expansion joints, and the fragile and damaged condition of much of the existing terra-cotta ornament. These conditions are not new or surprising, but their existence is far better known now than before the Local Law 10 inspections began. Since 1980, similar inspection laws have been proposed or passed in other cities.

Because the law specifies the class of building to be examined by the number of stories, and not by the height, most cast-iron buildings are exempt. The majority of cast-iron-front buildings are five stories high—although those stories can vary from 12 to 15 feet, making the buildings as a whole as tall as most modern seven-story buildings. When cast-iron facades are examined, extensive damage can often be found, most often cracks in the connecting flanges, rusted bolts, and missing broken segments. Unfortunately, the exemption of cast-iron buildings from the regular scrutiny of Local Law 10 inspections has meant that fewer repairs have been made and less development in repair methods has taken place. These lacks may be partially offset by the relative scarcity of iron facades stripped and mutilated.

Besides the height limitation, the other aspect of the law most often criticized is the use of visual inspection. Inspections conducted from the sidewalk, setback roofs, and adjacent buildings, even when performed with binoculars or telephoto camera lenses, cannot be as thorough as those conducted from a hanging scaffold climbing the facade. Emergency facade repairs on the New York Municipal Building triggered an *Engineering News-Record* editorial that stated, in part:

A facade law that does not require physical inspection is seriously flawed. Instead of protecting the public, it can actually create danger by giving even responsible building owners a false sense of security about the condition and safety of their facades. A physical inspection often is required to pinpoint deterioration, such as rusted connections and supports that are hidden from view.

How did that visual-only provision find its way into the law? Powerful real estate interests, says one New York City-based consulting engineer. The cost of a physical inspection, which typically requires hanging scaffolds, can exceed the cost of visual examination by $5,000 to $20,000, depending on the age, configuration and design of a building, says the consultant.[65]

## Present Considerations

The body of an early masonry curtain wall, if composed of brick or stone, rarely requires structural attention. These walls are so heavy compared with the loads that they must carry, and are so water-resistant through their sheer mass, that they have held up extremely well over the last hundred years. Problems with old curtain walls are most common with those portions built of terra cotta and with the areas near structural support and corners.

Terra cotta's problems are relatively easy to understand. The material is extremely brittle and was usually laid up without provision for movement or expansion. Depending on the building's status with the Landmarks Commission, the owners' and architects' plans for the facade, and the owners' ability to pay for expensive repairs, damaged terra cotta is removed and replaced with common brick or stucco, removed and replaced with cast stone or GFRC, or repaired and pinned to the backup masonry.

The structural concerns near steel supports are more complicated. The lack of expansion joints in old curtain walls almost inevitably has led to cracking at external corners. Over the years, the thickness of these walls has fooled many people making alterations, who have used the supposedly non-load-bearing walls as supports for new stair beams and other renovation framing. Another common alteration that is inherently dangerous is the use of flexible sealants in the joint directly over steel window lintels. This was done to many high-

**Two Park Avenue, north and east facades. All the ornamentation visible is glazed terra-cotta block, laid up with the brick curtain wall.**

rises in the belief that it would prevent water entry into the curtain walls, while in reality it has served to trap water within the walls. Misapplied sealants are probably the most common cause of destructive rusting of lintels, to the point where few Local Law 10–inspired facade repairs proceed without some lintel replacement. Finally, brick stacking has sometimes created enormous pressure within facades that can cause bowing out of entire areas, spalling, and local cracking.

The steel structure behind the facade was not waterproofed in any effective way until after World War II. When the masonry has stood undamaged, the steel is usually in good condition, often with its original paint intact. At areas where the facade is damaged, at cracks, loose parapets, and open joints, the spandrel beams and columns can be rusted so heavily that there is literally no solid metal remaining. Often the rust itself contributes to the facade deterioration: as the steel rusts, it expands, pushing out the masonry and creating new cracks for water to enter.

Obviously, when any of these conditions are encountered during renovations they must be carefully examined as to their extent and seriousness. In New York Local Law 10 has provided a mechanism by which larger buildings are theoretically checked for all of these problems on a regular basis. Once organized inspection starts, a minor surface condition may be found to be the outward sign of extensive damage.

Glass curtain walls present the extremes of structural functionality. The vast majority of them are well designed for wind loads and have expansion joints adequate to prevent glass-panel stacking. The exceptions are always clear: they are the buildings that shed glass into the streets during high winds. On the rare occasions when this condition exists, the curtain wall must be retrofitted with additional structure to stiffen it, and sometimes with new expansion joints as well. These unfortunate buildings are of more use as examples during the design of new buildings. Given the forty-year history of glass walls in the United States, a designer today should always be capable of correctly detailing mullions and glass panels to prevent failure.

Building materials and systems can be tested in wind tunnels, and mock-ups built for exposure tests, but no testing is infallible. It is the rule that long-term testing is performed through the weathering of actual buildings. The use of glazed brick was popular for approximately thirty years, until enough evidence built up about its flaws to discourage architects. The debate about stud backup is still in progress, and GFRC is so new that no long-term evidence exists to debate. In such cases, the best that the designer can do is question the details and uses of new materials and apply the same standards, either architectural or structural, that would be applied to the material being replaced.

# 9. THE FIREPROOF BUILDING (II)
## 1910s–1960s

Fireproof, before the Chicago fire of 1871, often meant incombustible. The first fireproofing in the modern sense was the patented tile column covers and tile arch floors of the 1870s. These tiles were made of porous terra cotta, an incombustible material intentionally used to protect building structure from the heat of a fire regardless of the flammability or heat-sensitivity of that structure. The development of a new type of fireproofing dependent on the use of a new floor system occurred again in the 1950s and 1960s. The new floor was concrete slab on metal deck and the new fireproofing was sprayed-on plaster. This combination is now the most common in steel-framed buildings—although other options, such as the use of fire-rated ceilings, still exist.

In the nineteenth and early twentieth centuries, buildings were sometimes advertised as "perfectly fireproof."[1] Experience has taught us that nothing is totally fireproof, and that even attempting to make a building safe against all fires is impractical. Efforts in the twentieth century have focused more on the safety of the building's occupants. Fire-protection ratings are spoken of in terms of time: a given floor system has a rating of two hours, or three. This represents not the time that the component will take to burn or be rendered useless by heat, but an approximate measure of the amount of time the area protected or supported by that component may be safely occupied. Given this change in emphasis, it is no surprise that the majority of innovations in fireproof construction have been architectural or mechanical apparatus to aid the occupants. Items

such as fire-rated stair enclosures count as "fire protection" rather than "fireproofing," since the building can survive without them but the tenants cannot.

There are two approaches to fireproofing based on the properties of the materials used. One approach is the successor to buildings built of monolithic masonry: the use of structural materials that can be shaped to provide integral fireproofing. Concrete is the most prominent example, although terra-cotta tile arches also qualify. The other, more modern, approach is the use of freely applied materials that meet all of the criteria for fireproofing, even if they cannot be left exposed because of their appearance or they cannot be subject to any stress because of their fragility.

Terra-cotta tile fireproofing for columns, the first modern fireproofing patented, is definitely in the second group. While a concern for usable floor space in buildings dictates that the tile should be as thin as possible and conform to the column shape as closely as possible, these preferences are not requirements. Certainly, the occasional round column enclosure containing an "I" or square column is proof of tile's free application. Tile fireproofing for beams, in the form of tile arch floors and soffit blocks, and for girders, in the form of square blocks fitted in and around the lower flange, belong roughly to the first category. Because the beam fireproofing is integral with the structure of the floor arches, its shape must meet the requirements of transferring the arch thrust to the beam. The distinguishing feature of integral fireproofing is a geometric or constructability constraint based on structural stability.

Concrete encasement of beams and girders is the best integral fireproofing system developed. Encasement is a feature of draped-mesh slabs, the floor construction standard in the 1920s, 1930s, and 1940s. The floor beams in draped-mesh slabs were typically set into the slab, with their tops 2 inches below the top of the slab (see figure 7.15). This position automatically created the proper high mesh position at the slab supports, and allowed the mesh to "droop" down at the slab midspan during placement. Unreinforced concrete is weak in shear, and cinder concrete especially so. As a result, the slabs needed more extensive vertical shear support at each beam than the 2 inches of slab that continued over the beam top flange. (This support is not needed for slab safety, because catenary action entirely supports the slab. Without vertical support other than the mesh, however, cracking would develop at the edges of the beam flanges. This would affect the usefulness of the slab top surface, lead to continued spalling of the broken slab edge, and lessen the *appearance* of safety.)

The most logical choice for slab shear support in this case is a concrete haunch extending down from the bottom surface of the slab to the beam's lower flange. For support purposes, the haunch could rest on the top surface of the beam bottom flange, but by extending the haunch below the bottom flange, and fully encasing the beam, fireproofing could be provided at minimal additional cost. In order to keep concrete suspended from the underside of a steel beam, extra pieces of wire mesh were usually added, wrapped around the tips of the bottom flange and continuously across its underside. (The bottom-flange encasement reinforcment later became a requirement if the encasement was to be used as a method of ensuring composite action between the beam and the slab. When composite design became popular during the 1940s, it was another reason to use integral concrete fireproofing; see chapter 10.)

Draped-mesh slabs with beam encasement were very popular for many reasons, of which structural efficiency was not the most important. Fireproofing in an integral system disappears as an issue, simplifying construction sequencing and reducing cost. Slabs made of cinder concrete, which could not be safely used in ordinarily reinforced slabs, were thinner and lighter than any other noncombustible floor system that had been built previously, thus saving on steel weight and cost. Finally, in most circumstances the underside of the slab and beam enclosures could be plastered and left exposed, removing the cost of building a ceiling. The penalty for all of these positive features was the requirement for building complex wood forms around each beam and below each span of slab. In 1931, it was noted that concrete "[m]ixing and placing costs are decreasing so fast that figures a few years old are almost worthless. It costs less to pour concrete today than it did before the war, when wages were about half."[2] The cost of providing forms is excluded from that statement, because unlike mixing and placing costs, which were reduced through mechanization and technological development, the cost for carpentry work rose steadily with wages. In an early study of metal deck, Bengt Friberg clearly presented the cost of wood forms. In 1954, a 4-inch-thick concrete slab cost approximately fifty cents per square foot for materials, mixing, and placing. The cost of the formwork more than doubled the cost of construction to over one dollar per square foot.[3]

In the 1950s, steel building construction costs were being drastically reduced through the replacement of riveting with bolting and welding and through the replacement of masonry curtain walls with glass and metal. To complete this "revolution," engineers began searching for cheaper and easier ways to build floor systems. The sequence of information presented below is to some degree arbitrary: the use of metal deck and the development of new fireproofing methods were connected, and neither would have succeeded entirely without the other.

## Metal Deck and Composite Metal Deck

There is no clear lineage for the development of metal deck. Corrugated iron sheeting had been used as siding for temporary and industrial buildings as early as the 1850s, and several of the patented floor systems of the 1890s and 1900s contained some form of steel sheeting. Daniel Badger's patent on cast-iron construction included iron sidewalks and floors, but few cases of this type of floor were built.[4] Iron daylighted sidewalks over vaults were used in front of many cast-iron-front buildings, but the vast majority of fireproof floors built before the development of tile arches were brick arches. A few examples of iron arches covered with concrete fill from before 1890 may exist, despite the fact that this construction was never common. Iron and steel sheeting has always been thin enough to have its thickness measured by gauge, rather than attempting to designate it as "³⁄₃₂ of an inch." Thicker sheets are not practical for use because of their weight (a ½-inch steel plate weighs 20 pounds per square foot), so corrugations were used as a way of increasing the strength and stiffness of the sheets.

Around 1900, corrugated iron deck was used as a permanent form for concrete slabs. The rise of encasement and draped mesh precluded any further development along those lines at that time. Development of beam fireproofing also slowed because of concrete encasement.[5] Column fireproofing was still required; most often it took the form of brick or tile enclosures before 1920, and gypsum block or plaster enclosures after 1920.[6] The gypsum block was used the same way as other masonry, its primary advantage being its light weight and ease of working. Unlike brittle tile, gypsum block can be worked with carpentry tools without the danger of shattering. The block was sometimes reinforced with very-light-gauge wire mesh and attached to the columns with sheet steel clips. Plaster enclosures were very similar in effect and appearance, the difference being in installation. Wire lath would be placed around the columns and covered with 1 or 1½ inches of gypsum plaster.

The use of deck as an integral part of floor slabs was contingent on advances in the technology of fireproofing. Using only the fireproofing methods available in the first half of the twentieth century, a modern composite slab on metal deck can only be protected from fire below through the use of a plaster ceiling, either hung from a light steel grid suspended below the structure (black iron) or on metal lath attached directly to the deck soffit. The use of plaster on metal lath as fireproofing continued through the 1930s, but almost entirely for column protection. Because hot air rises, ceiling fire protection must withstand substantially higher temperatures than vertical column protection.

Plaster had been gradually eliminated from this role because of its fragility and the possibility of the black iron creating vulnerable hot spots. During the 1930s and 1940s, a period when nearly all fireproof floor construction in New York was some form of reinforced concrete, the simple and obvious beam fireproofing solution was complete encasement. Concrete is not an ideal fireproofing material, but it can easily achieve the two-hour fire rating that most floor structures require.

As long as concrete encasement was being used to protect the floor beams, there was no practical method of installing metal deck that did not increase the amount of effort involved in building a floor. Pouring the concrete encasement first, and then placing deck for slabs has never been done; the amount of labor in placing the concrete encasement without the work platform provided by the adjacent slabs or forms, placing the deck, and then placing concrete for the slabs is almost exactly twice that for the traditional method of building forms and placing concrete. Once gypsum-board enclosures and spray-on fireproofing allowed easy installation of fireproofing separate from the structure, the way was open for metal deck.[7]

The form deck used early in the century served merely to support the wet concrete. Modern composite deck eliminates the need for slab reinforcing by using the deck as reinforcing. Form deck did not develop directly into composite deck. An intermediate type of steel floor deck, used briefly around 1950, did not support a structural concrete slab, nor did it act compositely with concrete. Rather it was the floor structure, which was then covered with concrete on top and plaster on the bottom. The concrete acted as a wearing surface and vibration dampener and the plaster was finish and of course fireproofing. H. H. Robertson popularized "Q" deck after World War II. This was the closest approach to an all-steel floor ever designed. The deck cross section was a series of enclosed hexagonal cells of light-gauge steel sheet linked by a short, single thickness of sheet. The cells were a series of tubular "beams" spanning from beam to beam. The steel was covered by a nonstructural concrete "finish" or fill between 2 and 3 inches thick. This floor had the advantage of eliminating formwork, enabling fast erection of

The Millennium, Broadway and 67th Street, Lincoln Square, south facade. Close-up shows construction of brick curtain wall with block backup. Wall construction beginning in midair is the hallmark of skeleton frames. Storefront top hung lintel and interior structure are visible. Steel beams are covered with spray-on cementitious fireproofing. Deck is not fireproofed because the combination of slab and deck thickness meets required fire rating without additional fireproofing. Deck corrugations for composite action can be seen on each deck rib.

the lightweight panels, and serving as its own work platform during erection. These advantages were at the cost of expensive deck fabrication, the inefficiency of wasting the concrete's strength, and a relatively short maximum slab span imposed by the deck capacity.[8] Because the deck provided all of the slab strength, Q deck was even more sensitive to heat from a fire below than other forms of deck construction. Many of the earliest buildings built with spray-on fireproofing had Q-deck floors, not composite deck.[9]

Friberg's 1954 study of metal deck was the first to examine the significance of using both the metal deck and the concrete slab, acting together, to span from beam to beam. He referred to the deck as "re-form," meaning reinforcing and form, a term that quickly and mercifully was replaced with "composite metal deck." The idea is a straightforward extrapolation from previous deck use: the bare deck section would still have to be strong enough to support the construction loading of wet concrete, but it would act as the reinforcing for the concrete slab to allow the combination to support all load once the concrete hardened. Some form of temperature and shrinkage reinforcing was required in the slab to prevent the top surface from cracking excessively. Friberg's study was based on Granco Steel Products Co-far deck, which had small reinforcing bars running at right angles to the deck ribs and welded to the rib tops.[10] These bars served a dual role, acting as shear-transferring "deformations" to bond the concrete to the slab in the direction of the deck span

and as temperature and shrinkage reinforcing in the perpendicular direction.

Co-far deck was a proprietary product, but the general idea of composite deck relied on concepts too old and well known to be patented. All composite deck produced since 1954 shares certain characteristics, including squared-off or half-hexagonal ribs wider than they are deep, and some form of repeated deformation to create bond between the deck and the concrete above. Since the 1960s, the characteristics of deck have been mostly standardized. The primary characteristics of deck from different manufacturers are similar enough to allow engineers to act as if the differences did not exist. The ordinary method of specifying deck is to designate a specific deck size, type, and manufacturer, but to allow the substitution of any "equivalent" deck. Equivalency is usually defined as deck that has the same depth and gauge, a bare steel moment of inertia and a composite moment of inertia within a set tolerance from the moments of inertia of the deck specified, and the same finish.

Co-far deck was hot-dipped galvanized.[11] Galvanizing and painting are both popular finishes for the metal left bare when the deck is not sprayed with fireproofing. Sprayed deck is often left unfinished, to save money and aid the adherence of the spray-on to the steel. Tests performed by the United Laboratories as products are introduced to the market determine the fire rating of each particular combination of deck, concrete, and fireproofing. A number of systems have been given ratings without fireproofing. These systems have enough concrete to act as a heat sink to keep the deck safe for the fire-rating time during heating. In cases where the heavier slabs would be provided anyway, such as when longer spans are used or when soundproofing is required, the fireproofing can be eliminated at no extra cost. In many projects where a thinner slab could be used, a cost analysis is performed during schematic design, balancing the cost of spraying the deck underside against the cost of additional concrete and slightly heavier steel framing.

Composite metal deck is laid directly on the bare steel frame; it serves as a work platform, as a permanent form for the concrete slab, and as flexural reinforcing for the slab. These three functions

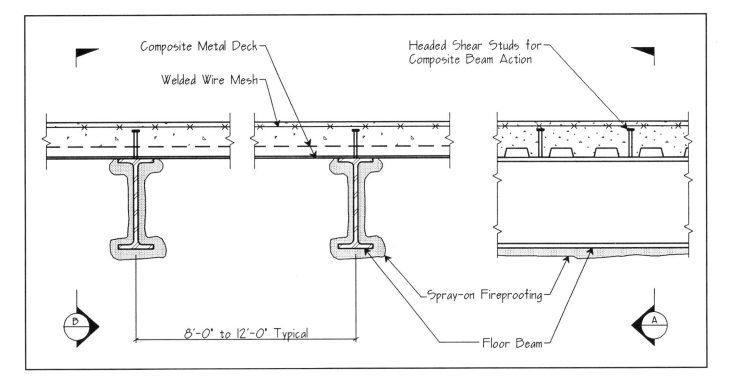

**Composite Metal Deck**

**Welded Wire Mesh**

**Headed Shear Studs for Composite Beam Action**

**8'-0" to 12'-0" Typical**

**Spray-on Fireproofing**

**Floor Beam**

had never been combined before in one material or one physical object. The almost-universal use of metal deck in modern steel buildings and the overwhelming percentage of that deck that is composite deck are the result of those advantages.

Composite deck is bonded to the concrete slab through small deformations similar to those on the surface of rebar. Once the deck has been laid over the floor beams it is fastened down, either through puddle welding of the deck to the beams below, shear studs welded to the beams through the deck, or less-commonly mechanical fasteners shot or screwed into the beams. Welded wire fabric is usually provided as reinforcing near the top of the concrete fill. This serves to reduce the size of cracks resulting from the drying shrinkage of the concrete and from the areas of "reverse" bending stress over the top of the floor beams. In an ordinary concrete slab, the latter would ordinarily be analyzed to treat the series of spans as one continuous slab, but since the composite deck and slab sections are designed as simple spans stretching only from one beam to the next, the top reinforcing need not be designed for any actual stress (figure 9.1).

The bare steel deck is designed elastically to support the weight of the wet concrete.[12] While it is possible to shore the deck while the concrete is

curing, to allow longer spans or thicker slabs than would be possible otherwise, this is not ordinary practice. One of the great advantages of metal deck is its elimination of the shores required with wood-formed concrete slabs, and few designers are willing to add that expense back into a project. If stronger bare deck is required, the normal solution is to increase the depth of the deck section (e.g., from 2-inch deck to 3-inch deck) or to increase the deck thickness (e.g., from 20-gauge to 18-gauge).

When concrete fill is placed in the deck and hardens, it is mechanically bonded to the deck through the metal's deformations. Under load, the deck serves as reinforcing for the concrete slab. The maximum strength of the composite is controlled by the shear bond strength between the deck and slab.[13] Since the bond strength can be increased by changing the deformation pattern, the rest of the slab and deck can be designed for maximum efficiency. This final result is the same thin, one-way concrete slab spanning from floor beam to floor beam that has been, in broad terms, the most commonly used floor system since 1900. It shares some of the same limitations as other reinforced concrete floors: the requirement for support at any opening, the lack of an easy method of upgrading the capacity, and the messy and

**9.1. Typical Floor Section—Composite Metal Deck**
**Scale: ¾" = 1' 0"**

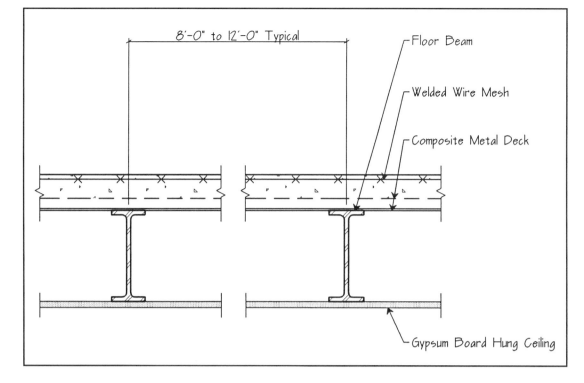

slow placing operation. In addition, where the slab and deck combination is heavy enough to meet the required fire rating without other fireproofing, the deck is usually left exposed within the floor plate, making it more vulnerable to corrosion than ordinary reinforcing, which is fully encased in concrete. These limitations are far outweighed by its advantages, which include the ordinary ones for concrete of a stable, noncombustible, and high-strength floor, as well as the absence of formwork and shoring, and the prominence of lightweight concrete. Lightweight can be used in almost any application, including the construction of entire buildings, but its lower shear strength and modulus of elasticity have limited its applications. It is perfectly acceptable as composite deck fill, however, where the stresses are low and the steel provides shear capacity. The use of lightweight also reduces the dead load of the floor system, and therefore allows the use of smaller floor beams.

The widespread use of metal deck floors required a return to the same fireproofing issues that had been widely examined between 1890 and 1910. At that time, plaster ceilings had been the most common solution to fireproofing exposed beam soffits with segmental tile arches or Roebling arches. The reason that this solution was used

with less-than-perfect enthusiasm in the 1950s was that the labor eliminated from the slab formwork was returning in the form of plasterwork.

## Fireproof Enclosures and Ceilings

When Q deck was first introduced, designers faced a problem that had almost disappeared after 1920. The exposed bottom surface of the deck and the exposed beams had to be fireproofed. This was actually more important than fireproofing the bottom flanges of beams had been during the tile-arch era, since the thin deck was more vulnerable to heat than the large structural sections used as beams. The initial solution was the same system used previously: coating the underside of the deck with plaster. The normal method in the early 1950s was to create boxes around all beams and columns of expanded metal or gypsum board-lath, attach lath to the deck underside, and then coat all of the surfaces with gypsum perlite plaster.[14] This is reasonably effective fireproofing and provides for architectural finish. The great drawback to the system is the amount of labor involved in plastering the ceiling. This may not have appeared as an extra item at the time, since almost all ceilings were plastered anyway, but it was prime for improvement.

Calculated fire resistances had been introduced

in the 1940s, and had changed some fireproofing assumptions.[15] Gypsum board, which had been used primarily as lathing and backup for plaster, turned out to be a more efficient and more consistent fireproofing material than wet-applied plaster. Gypsum-board panels are extremely simple to install and can be manufactured so as to be acceptable visual replacements for plaster. Within a few years of its introduction, gypsum board had displaced plaster as fireproofing and greatly reduced its role in creating finished surfaces. Gypsum-board ceilings are used as fireproofing in many circumstances, especially in buildings with few mechanical services in the ceiling space and with joists. The most efficient fireproofing for open-web or light-gauge joists is a gypsum-board ceiling attached to the joist bottoms; nearly the only possible fireproofing for wood joists is gypsum board (figure 9.2).

A less-common variation on gypsum-board enclosures is rigid panel fireproofing for columns. The product is a set of mineral-fiber-based panels approximately one inch thick with widths determined by steel column section depths and flange widths. The panels have integral gauge-steel clips that can be fastened from outside. The panels are attached to form a continuous box around the column. The biggest advantage to the system is its low profile, which results in finished column perimeters smaller than for any other fireproofing material. This is obtained, however, by using a proprietary system that can cost more installed than generic systems.

Gypsum-board enclosures are at their most efficient when they are used for flat ceilings and square column boxes, but they can be freely adapted to almost any shape desired for aesthetics or space planning. The only absolute requirement in the use of enclosures as fireproofing is that they be continuous, so as not to provide entry for flame or hot gases. This freedom has meant that gypsum-board ceilings have a tendency to have peculiar box-outs in corners, usually concealing plumbing or mechanical ducts.

## Spray-on Fireproofing

The use of metal deck slabs has coincided with sealed buildings dependent on mechanical ventilation. Fireproof ceilings become vastly more difficult and expensive to build when they are penetrated every few feet by air registers. Even in buildings with operable windows, sprinkler heads and light fixtures can turn a fireproof ceiling into a more complicated exercise than is readily apparent. The solution to this problem is to return fireproofing to a position close to the beams and slab, where it is separate from interference caused by nonstructural construction. This solution returns to the beam box-out problems, unless a new method of installing the fireproofing is used.

Two well-publicized buildings built in California in 1958 show the beginning of this new method. An apartment building near San Francisco had the more advanced floor system, consisting of lightweight concrete on composite deck, but it was fireproofed with a seventy-year-old system: plaster over wire lath. The most advanced feature of the fireproofing was the use of vermiculite plaster, which contains expanded mica particles that are poor conductors of heat.[16] The other building, an office building in Los Angeles, had floors consisting of concrete fill on Q deck, but its vermiculite plaster was sprayed from nozzles directly onto the deck underside. Beams and columns were enclosed in wire mesh and iron frames, which were then sprayed.[17] Once the step of spraying fireproofing had been taken, all other changes to the system were relatively minor. Vermiculite plaster was replaced by the late 1960s by other cementitious products, with gypsum, perlite, or vermiculite aggregate in a chemical binder. Mineral fibers were developed as a more efficient alternate to cementitious products. The sprayed fiber is an extremely poor heat conductor and entirely inert, but is more fragile and vulnerable to being washed away by water than cementitious products. The most effective mineral fibers previously used contained asbestos, and are no longer used.[18]

Spray-on fireproofing has no inherent shape, simply adhering to the material below in a reflection of the base shape. In cases where an architectural ceiling will be hung below complex structure, spray-on is a simple solution to fireproofing cheaply. Almost all sealed buildings, and most steel-framed buildings of all types, use spray-on because of its flexibility. The ability to simply spray an entire area, regardless of the details of the

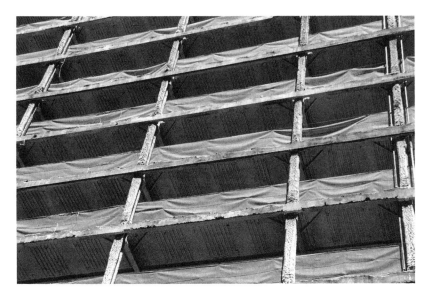

**320 Park Avenue, south side. Steel framing exposed during building reconstruction. Interior beams, columns, and deck underside are covered with original cementitious spray-on fireproofing. Spandrel beams are fireproofed with concrete poured with slab. Areas without deck ribs are reinforced with bottom sheet for utility trenches within slabs. Short diagonal beams are supports for deck corners unsupported at column box-outs.**

framing or finishes, contrasts sharply with the precision required in assembling a rated ceiling.

## Other Solutions

In contrast to the bulky and obvious fireproofing solutions developed since the 1870s, two new techniques require little space. These techniques take the idea of a timed fire rating one step further than traditional materials. Intumescent paint is used to deflect heat from the steel surface for a specified period of time; reduction of spray-on thickness for heavier members uses the structure's material properties to withstand heat.

Intumescent paint shares a characteristic of all paint—charring under exposure to heat—but uses it to provide a fire rating. The paint acts sacrificially, absorbing a large amount of heat in the chemical change as it chars and blisters away from the steel surface. The energy absorbed in this manner and the physical distance created by the paint blistering protect the steel below long enough for a timed fire rating.[19] This method works better on heavier members, which need less protection because they act as heat sinks.

Traditional fireproofing materials depend on poor heat conductance to protect the structure behind. Terra cotta or gypsum are noncombustible under ordinary circumstances, but just as important is their ability to insulate steel from rising temperatures. Concrete is a much less efficient insulator, but it functions reasonably well as fireproofing because the mass of concrete beam encasement and connected slab serves as a heat

sink.[20] Rather than keeping the heat at the surface, as other materials do, concrete spreads it out to keep the average temperature change low. Both effects will eventually fail under prolonged exposure to a fire, but before failure both work acceptably well. The thirty-year-period when concrete was the most commonly used fireproofing for steel may have suggested that heat sinks are worth investigating as an alternative to insulators.

The most readily available heat sink is the steel frame itself. Steel is an extremely good heat conductor, a property that had always been considered a drawback in fireproofing design. Local heating applied to steel tends to spread out through the material. The temperature gradient from the location where heat is being applied to the location farthest away will be far more even than for any other structural material. To use this property practically, the piece of steel involved must be fairly thick, both to prevent local temperatures from rising too quickly to avoid destructive effects and to provide enough mass to absorb enough heat to have any meaning. Fireproofing assemblies are rated by the United Laboratories. The UL ratings allow thinner coatings of spray-on for heavier members, but do not pursue this issue very far. Most of the producers of spray-on fireproofing have spent enough time on research to know that the fireproofing can be reduced greatly when the steel members are heavy enough. In new construction, the heaviest steel is typically in the lower columns, where flange and web thicknesses of 3 inches are not uncommon. There is not much of a need to reduce the fireproofing thickness on the columns, and so not much interest has developed. In renovations, new columns or transfer girders heavy enough to act as heat sinks are sometimes introduced into extremely cramped quarters, where saving one inch of fireproofing may be important.

## Present Considerations

Existing fireproofing from the 1920s or later is a rarity in renovation. As long as it is undamaged, it is almost certainly adequate, regardless of its component material. In many cases, the existing fireproofing is damaged by the renovations in progress; terra-cotta-block protection is inevitably destroyed by any attempt to work near it. Given the ease with

## CASE STUDY: **MODIFYING METAL DECK FLOORS**

A large new hospital building in New York was designed and built fast-tracked, with construction starting before the mechanical, architectural, and structural plans had been coordinated. During the late stages of construction, coordination revealed numerous locations where duct openings had been cut through floors without proper structural reinforcing, or where conduit embedded within the slabs reduced the slab capacity. Despite the fact that this was new construction, correcting these conditions took on most of the properties of a renovation project, as several hundred individual penetrations or weakened slabs were addressed. The following is a summary of the types of problems noted.

The slab penetrations in question were typically for mechanical equipment added for individual areas in the hospital, such as certain advanced examination rooms, that had above-average requirements. In order to install additional ventilation or electrical service to these rooms, holes were often cut through the floors. The typical floor was 3¼ inches of lightweight concrete fill on 2-inch composite metal deck, which is a system that does not require additional fireproofing. The additional holes were typically located without regard to the slab supports or areas of high moment.

In a typical case, the slab penetrations were small enough so that the cut edges did not have to be re-supported because the slab surrounding the opening was strong enough to carry the additional load, if the load could be transferred laterally. Depending on the size of the opening, gauge-steel sheet or more commonly small rolled angles were fastened to the bottom of the deck. The additions serve as transfer beams to carry the vertical reaction at the cut edges to the

slab on either side. In the small percentage of cases where the adjacent slab did not have the capacity to carry the cut portion, small wide-flange or channel beams were added to span to the original floor beams. In the typical case, the continuous floor adjacent to the opening was still a composite of concrete and steel deck spanning from beam to beam, and the cut portions were composite spanning from one beam to the reinforcement.

The other common problem was the density of conduit near the electrical closet on each floor. This area contains the "home-runs": the portion of conduit leading from the switch boxes in the closet to the various portions of the floor where the service is required. Because of this condition, common to any large building with centralized electrical service, the deck in this area was originally designed as "electrified deck." This form of deck has an additional gauge-steel sheet connecting the bottoms of the ribs. The hollows created by the bottom sheet and the underside of the corrugated deck can be used to run conduit. In this case, electrified deck was used not to run conduit parallel in the hollows, since the conduits ran in a fan pattern out from the electrical closets, but rather because it has a higher section modulus both when bare and when filled with concrete. The final density of conduit in some areas was so great that composite action could no longer be assumed between the fill and the deck. Reinforcing was added to the bottom of the deck in the form of additions as described above or as new beams to reduce the span. In both cases, the deck was assumed to be the load-carrying element, and the concrete fill merely a bearing surface around the conduit.

which fire-rated gypsum partitions and soffits can be worked into new construction, often the old fireproofing is removed and replaced as a matter of course. In the case of plaster or gypsum-board ceilings, the difficulty sometimes can be simply in recognizing the fireproofing for what it is.

There are types of fireproofing used in the past that are not rated by the current standards. The most common of these is a covering of plaster attached directly to steel members. Obsolete practices that can create dangerous conditions are the lack of adequately sealed joints in otherwise rated suspended plaster ceilings, the lack of firestopping to seal openings in concrete slabs, and unreinforced concrete encasement which can easily collapse during a fire. Another common problem is the previous removal without replacement of fireproofing materials during renovations.

Structural materials from the past are very similar to the materials used now. This is not true of other materials, often because of health hazards now known to exist. For years, all steel was painted with red lead, which did a good job of corrosion protection but cannot be removed without special precautions to prevent dust from spreading. Since new steel is usually attached to existing members by welding, which requires surfaces free of paint, this issue is a constant problem. A similar problem is presented by older mineral-fiber spray-on fireproofing, which was often asbestos-based before 1975. Newer versions are asbestos-free, but anytime that existing fireproofing is to be touched, it should be checked for asbestos content whether or not the check is legally required.[21] Unlike asbestos pipe insulation, which is limited in area and relatively sturdy, asbestos spray-on often cannot be paint encapsulated. The most vulnerable aspect of spray-on fireproofing is its adherence to the base metal, making encapsulation less attractive than removal.

# 10. "MODERN" STEEL CONSTRUCTION
## 1920–PRESENT

The basics of steel framing have not changed since the adoption of true curtain walls before 1920. Structural systems have evolved, as low unbraced frames and higher braced frames have been followed by tube frames and bundled tubes for taller buildings, but this change is a reflection of an analysis theory that does not necessarily affect the design or detailing of individual members. Other changes, such as the customary use of composite metal deck with concrete floor slabs, have reduced the effort involved in constructing the remainder of the building after the steel skeleton is in place but, again, do not directly influence the design of the steel.

Changes in framing systems and details have been aimed primarily at more efficiently using the high strength of structural steel. For example, the average flexural stress in a steel beam composite with the adjacent floor slab is higher than that in a non-composite beam. The higher the average stress in a steel member, the more efficiently the metal's capacity has been exploited. The same logic, coupled with the detail simplification inherent in welding, has led to increased use of continuous beams and other moment-connected forms.

Steel construction competes with other materials, as it has all along. By the time that steel frames had effectively killed off large bearing-wall buildings, concrete frames had become competitive. Steel has remained the dominant material in the Northeast, largely because of the high cost of labor-intensive concrete construction, but there are numerous exceptions, especially in smaller buildings. This has been true for years, as even the AISC had to admit in 1929

> that the use of structural steel as frames for buildings of twelve stories or less, is comparatively rare, particularly in that portion of the country west of the Atlantic seaboard, and competitive interests have taken many of the fifteen, sixteen, and eighteen story structures. . . . Factory buildings, warehouses, garages, automobile plants and countless other structures are being erected in other materials than steel.[1]

Entirely concrete buildings are most popular for residential use, where the exposed bottom of flat two-way slabs can be plastered as ceilings, close column spacing is possible, and the relatively more soundproof floors are a selling point. Small apartment houses and rowhouses are often modern variations of old masonry construction, with block and brick bearing walls and precast-concrete plank or wood-joist flooring. Small commercial buildings are also built in this manner, occasionally using all light-gauge metal stud and joist construction. Except for concrete frames, none of these other types lends itself readily to structural renovation, and few major renovations are performed on these buildings.

### Floor Systems

Few steel-framed buildings are now built with any floor system other than concrete-filled composite metal deck. Areas of exceptionally heavy load, such as X-ray rooms in hospitals, may sometimes be built with formed concrete slabs, and small buildings are occasionally built with precast concrete plank laid over steel beams, but these are small percentages of the total.

In almost all cases where metal deck is used, the role of the designer has been reduced to calculating the construction dead load to select a suitable bare steel section and calculating the total floor load to pick the final slab depth. These decisions are influenced by architectural requirements for slab depth and, occasionally, vibration and by the desire to avoid additional cost for fireproofing. The most popular metal deck floors are those thick enough to not require additional fireproofing, but made of lightweight concrete so as to reduce the floor's dead load. The lighter the floor load, the more money can be saved on steel for floor beams, columns, and foundations.

Two types of non-composite deck are still common: form deck and roof deck. These types are defined more by usage than by any distinguishing characteristics. While there are deck styles sold specifically as form deck and roof deck, composite deck is capable of acting as either. The reverse is not true: the less-expensive deck sold as form deck or roof deck lacks the deformations that make composite action possible.

Form deck is used in areas where the loads are too high for composite deck to act as reinforcing, or where condensation is expected that might corrode external reinforcing, even if galvanized. An example of use that often combines these two problems is the construction of new roofs for sidewalk vaults. As the buildings from New York's turn-of-the-century building booms age, the vaults tend to require more reconstructive work than any other portion of the buildings. Given the increase in truck weights since World War II and the fact that any accessible sidewalk area must be treated as a place where trucks, especially fire trucks, might someday go, new vault roofs are generally far heavier construction than the roofs being removed. With the vault removed, building forms from the street is simple. Form deck serves to eliminate the problem of removing forms from the vault once it is re-enclosed. Garage floors have service conditions and loads similar to vaults, but

are more accessible during construction and so are more often built with all-concrete construction.

Roof deck is used at the other extreme in conditions; it is a logical result of the lighter loads applied to roofs. The minimum roof load has been gradually reduced for much of the country during the twentieth century as accurate analysis of probable snow loads has become possible. In New York, the load was reduced from 50 pounds per square foot in the 1890s to 30 psf in the 1950s. The interior of a building will have floor live loads ranging from 40 psf to 100 psf or more for warehouses and some industrial uses. The interior floors also usually carry a distributed partition load for stud partitions. The result is that the roof total load, excluding the slab itself, can be as little as half the ordinary floor load exclusive of the slab. If the slab loads do not reduce this discrepancy substantially, lighter framing can be used for the roof, saving expense for steel there and in the columns for their full height.

Roof deck remains a minor feature of construction in New York for two broad reasons, economics and functionality. In a city dominated by multi-story construction, the percentage of the total building cost saved by using roof deck instead of concrete-filled composite deck is usually small. This savings is bought at the expense of increased noise transmission, increased vulnerability to damage from imperfect waterproofing membranes, and a decreased capacity for roof renovation or secondary use. Roof deck is used in much of the country for buildings such as schools, shopping malls, and factories; these uses are restricted in New York. Schools tend to be three or more stories in height and heavily built, and the number of strip-mall-type buildings is relatively low. The most common uses of roof deck in New York have been in industrial buildings or similar shed-type buildings such as freestanding school gymnasiums and auditoriums, but these are a small fraction of the total buildings built.

## Composite Construction

The term "composite beam" is ordinarily used to indicate a rolled steel section that acts in conjunction with a concrete slab, although technically it also refers to a reinforced-concrete beam or any other beam made of two different materials acting together, such as a flitch plate. The principle is the same in both cases: a design using steel only in tension, to eliminate buckling concerns, in combination with concrete used only in compression. This use of the materials in ways specifically suitable to their properties became commonplace with the spread of reinforced concrete after 1890, but did not appear as part of steel-beam design until the 1930s.

In 1937, the third edition of the AISC code contained the first steel-based design provisions for composite beams. The specification allowed for composite action between a concrete floor slab and a steel beam encased by a haunch integral with that slab to carry live loads applied after the concrete had hardened. The steel beam had to be designed to carry all dead loads by itself, the haunch had to provide a minimum of 2 inches cover on the bottom and sides of the beam and 1½ inches on the top, and the bottom of the haunch had to be mesh reinforced to prevent spalling. The specification did not explicitly restrict the width of the concrete flange that would act with the steel beam, but the general circumstances of construction at that time acted to restrict the possibilities.

Ordinary steel-framed buildings in the late 1930s and early 1940s with concrete floor slabs had short one-way slabs spanning closely spaced filler beams. The slabs were board formed and usually reinforced with draped mesh, although the form of slab reinforcing does not affect the ability of the slab to participate in composite action. The logical guess for the width of the concrete flange of the composite beam would be the distance from beam to beam, with the flange spanning on each side of the beam centerline to the adjacent slab span midpoint. For a typical case, where the beam span is approximately 12 to 15 feet, the slab 4 or 5 inches thick, and the beam spacing 6 to 7 feet, the current code provisions for the flange width are all in the neighborhood of 35 to 40 inches on either side of the beam, approximately the same as the intuitive guess of half of the slab span.

The AISC third edition allowed a bending stress in the steel of 20 ksi—the same allowed for ordinary bending or axial tension—but did not specify allowable compression in the concrete, leaving that to the national concrete specifications. These provisions remained unchanged through the fifth edition of the specification, issued in 1947. This specification was so well adapted to the typical construction of the time, with its short-span slabs and concrete-encased beams, that it probably could not have been improved upon without the use of new construction techniques introduced later, most notably metal deck and shear studs.

The first AISC composite specification depended on chemical bonding created by the cement to join the steel beams and the surrounding concrete for shear transfer between the steel and the concrete flange.[2] The general trend in concrete design has been to move from cementitious bonding to mechanical bonding, resulting in the use of deformed rebar and the use of epoxies and grouts for binding fresh to existing concrete. Practical considerations prevented both of these methods from being used in composite beam construction. Using epoxies in new construction to create bond would complicate the process of creating the concrete floor, and would thus be a step backwards in the general trend of simplifying field work. Providing deformations on the beams of the type used on rebar would complicate all fabrication, as well as require a post-rolling manufacturing process, again complicating the work. The solution that evolved was to provide "deformations" consisting of separate pieces of structural steel welded to the top surface of the beam, which would mechanically interlock with the concrete, providing a known amount of shear transfer regardless of the bond strength of the cement.

Composite beam design was initially used far more extensively in highway and bridge construction than in buildings. The heavier loads and longer spans common in road construction gave early composite design an advantage that offset its additional cost. By the time composite construction use increased in buildings, during the late 1940s and early 1950s, its features had been tested and improved in road design.[3] In road design, where the AISC code does not govern, mechanical shear fasteners were first used. The two oldest types are short lengths of light channels or zees and short helices of steel rod, both welded to the beam top flanges at regular intervals.

The helices, called "spirals," are more effective

in shear transfer because they are stiffer than the web of a zee or channel in bending, and therefore reduce the amount of slip between the concrete and steel. No slip would occur in ideal composite action since the steel and concrete would deflect together at all points along their length. Some slip is inevitable because of the discrete nature of mechanical shear connectors: having pinned the two materials together at specific locations, cracks in the concrete allow differential movement between those locations. Keeping the amount of slip to a minimum is one of the prime criteria in the theoretical design of shear connectors.

Despite their theoretical advantages, spirals were difficult to fabricate and install, and probably delayed the introduction of mechanically fastened composite beams into building design. Channels and zees were easier to fabricate, but still required extensive field welding for installation. The solution to these problems was the introduction of an entirely new connector: the headed shear stud.[4] Studs are cylinders made of steel rods with their bottom end welded in a circle to the beam top flange and their free upper ends enlarged. This form is stiffer than channel connectors, is easy to manufacture in bulk, and, most important, lends itself to mechanized installation. During the 1950s, the Nelson Stud Welding Division of Gregory Industries developed not only the stud form but also an electric welding machine that holds the stud in place and welds the entire perimeter at one time. A patent was awarded for the process in 1961, but in a significant change in attitude from the patent holders of floor systems and some types of concrete construction sixty years earlier, Nelson Stud chose not to charge royalties.[5] This was not based entirely on altruism, since encouraging the use of technology in a field where Nelson Stud had a huge competitive advantage was a profitable business practice.

Composite design use increased greatly from the introduction of headed studs in the 1950s and changes in the AISC specification in 1961.[6] The "stud gun" welding machine made rapid installation of shear connectors possible, since "[t]he studs are light and easy to handle, and a competent crew can install 600 studs per day."[7] The code changes allowed the use of composite construction that relied entirely on shear connectors by

removing the requirement for beam encasement. This code change was further supported in the sixth edition of the AISC manual, first issued in 1963, which included design tables for shear connector composite beams.[8]

The use of metal deck made composite construction simpler. In theory, shear connectors could be shop welded to the beams, but in practice this is rarely done, since the studs or other connectors are relatively delicate and prone to break off during steel shipment. With metal deck, the steel could be shipped and erected in the ordinary manner, the deck placed and then used as a work platform to weld the connectors to the top flanges. Composite construction actually saved labor using this scheme, since the welds that attach the connectors, being "burned through" the deck, serve to connect the deck to the steel. The extra step of welding the deck down is thus eliminated.

Because composite construction became popular at the same time that composite deck use was growing, the use of both techniques together was considered experimental. The use of deck meant that there was a vertical gap between the beam top flange and the average bottom elevation of the concrete slab, increasing the bending force on the shear connectors for a given shear level (see figure 9.1). In 1962, a load test of 27-inch-deep wide-flange beam with a 4-inch lightweight concrete slab on Co-far deck and zee shear connectors welded through deck was performed and showed behavior almost identical to the theoretical values.[9] With the almost-universal use of metal deck in steel-frame buildings after the early 1960s, the combination of composite beams with deck has become common.

The current AISC specification allows two designs: encased beams and beams mechanically fastened with studs or channels. Encased beams can be designed either with the composite section carrying live loads with the allowable steel stress specified as 66 percent of the yield stress and the allowable concrete stress defined by the concrete codes or with the steel section designed to carry all loads with an allowable stress of 76 percent of the yield stress. The latter is not as efficient, but in cases where encasement is used for reasons other than to specifically provide for composite action it allows an easy method of design. Mechanically fastened design is more complicated but provides for the most-efficient use of the materials of all three options. This method is used for nearly all of the composite design in new buildings.

## High-Strength Steel

The specification for ASTM A36 steel, the most common alloy used today, is fairly similar to that for A9 steel in the 1910s. A9 was defined as having an ultimate stress of 55,000 to 65,000 psi and a yield stress of half the ultimate stress, or 27,500 to 32,500 psi.[10] A36 steel, eighty years later, has an ultimate stress of 58,000 psi and a yield stress of 36,000 psi. The current definition of high-strength steel is any alloy with a yield stress higher than 36,000 psi.

Higher strength steels as used in buildings do not directly affect design or detailing. The design of any individual member in A36 or A572 steel is performed in the same manner, although deflection analysis becomes more critical as the use of higher strength steels gives designs of smaller beam sizes and higher actual stresses. The elastic moduli of all commercial steels are very similar, regardless of the yield point or ultimate stress. Higher allowable design stress values therefore lead to beams with greater deflections if all other contributing factors are the same. In many designs using high-strength steel, the limits on beam depth, acceptable deflection, and vibration force the use of beams larger than required for stress alone. In some cases, these constraints produce the same beam sizes as designs using ordinary A36 steel.

The constraints on moment of inertia described above are less critical in column design, especially columns with high ratios of axial load to moment. Deflections in columns from axial load are far smaller and less noticeable than those in beams from bending. One response to this difference has been the construction of buildings using high-strength steel for columns and bracing members and A36 steel for floor beams.

Another way for designers to avoid the limitations deflection design imposes on the use of high-strength steel in beams has been to use high-strength composite beams. The addition of the concrete flange results in a much larger moment of inertia than the bare steel section, reducing the deflections proportionally.[11] This type of development can sometimes seem futile, however, since the increased beam strengths encourage the use of longer spans and smaller beam sizes. The deflection and vibration problems that discourage the use of high-strength steel beams recur at longer spans when the beams are composite. This type of floor system may not be the most serviceable, but it is often the cheapest to build, leading to the noticeably bouncy floors in many speculative retail and office buildings. The liberal modern live-load reduction provisions in building codes (see "Code Changes" below) only exacerbate the problem.

## Connections

The modern steel frame did not develop until riveting had made rigidly joined connections possible. Since 1950, riveting has been supplanted by welding and high-strength bolts. These two technologies create rigid joints with less labor than riveting, making them less expensive. In addition, welding allows simpler details, reducing the amount of connection material and connection fabrication required. Most bolted connections are almost identical to riveted connections of fifty years ago; welded connections at first were versions of those riveted connections, but they have evolved to take advantages of welding's special advantages.

Structural ironwork was bolted before riveting and welding became possible, but bolting was abandoned because of the weak and loose joints it created. Wind frames require joints that are more rigid than the members they connect, a rigidity that unfinished bolts cannot provide. The era of modern bolting left research for practical applications in 1951, when the first specification for high-strength bolts was issued and endorsed by the AISC.[12] High-strength bolts are made of tempered steel capable of withstanding high levels of combined tensile and shear loading. When used in slip-critical connections, the bolts are tightened past the point where they could be considered ordinarily tight in order to prestress them in tension. The bolt tension serves to clamp together the connected pieces between, preventing slip at shear values below the maximum design load. The use of bolts spread quickly because of the vast labor savings over riveting. A bolt crew of two men, using air-powered impact wrenches, can install bolts faster than a four-man rivet crew, who need both air-powered hammers and coal-fired salamanders to heat the rivets.[13] Bolts also offer slightly quieter installation and a known, experimentally determined slip value, as opposed to rivets' empirically determined design values.[14] By the late 1950s, a number of buildings were successively the "world's tallest all-bolted skyscraper," although many used bolting for field connections while still using shop riveting. The increase in efficiency was far greater in the field, where pneumatic presses for installing rivets could not be used.[15]

By 1963, rivets were hardly considered for use in field connections, having been replaced almost entirely by bolts. Wind-frame connections and any highly loaded connections were made by welding, or more frequently with high-strength bolts in "friction" connections, which are the pretensioned, slip-critical form. Other field connections were made with high-strength bolts in ordinary bearing connections, or unfinished bolts in bearing connections. Comparative costs for equal-strength connections were $1.22 for rivets, $0.96 for high-strength friction bolts, $0.83 for high-strength bearing bolts, $0.73 for unfinished bolts, and $0.64 for welding.[16]

Welding's introduction to routine use was slower, starting in the late 1940s, and not becoming common until the late 1960s.[17] The New York Stock Exchange Annex was the largest welded frame in New York in 1956, at twenty-two stories, which was considered an achievement worthy of notice.[18] In 1960, the Western Electric Building at Fulton Street and Broadway, designed by Purdy & Henderson, was the tallest steel frame in the eastern half of the country with welded connections; buildings nearly 60 percent taller with bolted connections had already been built in New York.[19]

Connection forms have been influenced by developments other than the connector used, most notably the evolution of wind systems. Skyscrapers built as late as the 1920s still had column grids approximately 20 feet by 20 feet. Such close spacing made internal bracing, usually located around the central service core of the building, easy to fit into the floor plan; close spacing also gave each floor enough beam-to-column connections to encourage the use of unbraced frames built with partial moment connections. These designs, which were and still are the overwhelming preference of engineers for the wind system in steel-framed apartment houses, are practical only when the moment to be resisted at each joint is relatively low. Since the overall moment comes from the wind load and is independent of the building's internal structure, partial moment connections are eliminated from use in slender buildings and buildings with wide column-grid spacings—two cases where the number of joints at each floor has been reduced by the frame geometry.

In older buildings, where partial moment connections were used, beam-to-column connections are typically large riveted brackets. These connections resemble stiffened seat connections with a second stiffened seat turned upside-down sitting on top of the beam. The seats are capable of developing the required wind moment and taking the beam's gravity load. The obvious drawback to using these brackets is their extreme complexity, with up to a dozen separate connection pieces requiring fabrication and more than a dozen rivets requiring driving. Even the simplest bracket of this type would require driving eight rivets: two each into the top and bottom beam flanges and two each from the top and bottom angles to the column.

Partial moment connections are still used in residential construction, where the relative permanence of architectural layouts has meant that columns can be hidden in partitions and so are still closely spaced. Modern top and bottom angles are usually bolted, which is some gain in efficiency over riveting (figure 10.1). Larger and more complicated connections are usually shop welded, reducing the old need for numerous filler plates and eliminating the need to use angles to create outstanding attachment plates. One of welding's greatest advantages in simplifying connection design is its ability to attach two plates at right angles, instead of requiring that attached elements be parallel.

The spread of less-expensive and more portable welding also encouraged the use of full moment connections in building frames. Connections to develop the full moment capacity of large members were extremely difficult with riveted construction because of the relatively low shear capacity of each individual rivet. The large flange forces in a beam in such a connection are resisted by rows of rivets in shear, transferring the forces to connector angles, tees, or other pieces which are then riveted to the column. Welded full moment connections can be as simple as a beam butted to a column.

Full moment connections were not the only form dependent on welding. The use of shear tab (or "strike plate") connections is obviously impossible without cheap welding. Welding prices had fallen far enough by the 1970s that the difference in price between punching holes in the column and connection angles for bolted framed connections and shop welding the plate for a shear tab connection had become negligible. In connections to heavy column sections, where connections are often welded because of the expense of drilling bolt holes through the thick material, the cost comparison is more direct. The two connection angles have been replaced by one plate of comparable size and thickness, and the total amount of weld between the connector pieces and the column is almost exactly the same, with the two vertical welds on the outstanding angle legs being replaced by the two vertical welds on the plate sides.

Just as strike plates represent an improvement on clip angle shear connections made possible by welding, yield plates and end plates are alternates to top and bottom flange angles to obtain partial moment connections. Unlike strike plates, which can reduce the total amount of cutting, punching, and drilling, yield and end plates can require more fabrication effort than flange angles, and so have not gained widespread use.

Yield plates are conceptually the simplest form of partial moment connection. They consist of plates welded to the column and welded or bolted to the beam framing in, with a narrowed "neck" in the middle. The beam's vertical end reaction is carried by a separate shear connection, usually a strike plate. The end moment specified for the connection is provided by the force couple of the two plates times the beam depth. The neck of the

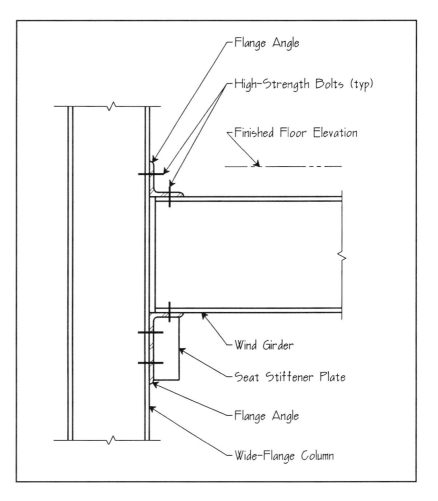

plates is designed to yield, to stretch without the application of additional load once the plate force matches that created by the desired moment. The result is a connection that resists applied moments until a certain level is reached, and then ceases to increase its resistance. This is the behavior needed for unbraced wind frames, but it comes with some expensive fabrication and erection problems. The plates must be cut into the necked shape in addition to standard fabrication. If the plates go to the site welded to the columns, their placement must be extremely precise to ensure connection to the beams, but at the same time, such a tight fit complicates placing the beams between the plates. The other option, to weld only one plate to the column in the shop, substantially increases the amount of welding required compared with flange angle connections.

End plates are far more complicated to design. The same plate, welded vertically to the end of the beam, is bolted to the column, providing both a vertical shear and moment connection (figure

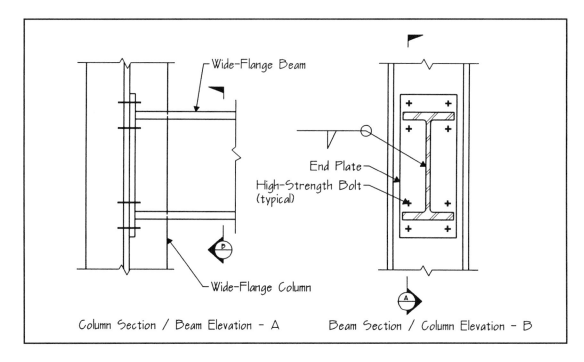

Wide-Flange Beam

End Plate

High-Strength Bolt
(typical)

Wide-Flange Column

Column Section / Beam Elevation – A          Beam Section / Column Elevation – B

10.2). The basic idea, to use the vertical plate to provide a connecting element parallel to the column flange or web, is straightforward, but the design of the plate and bolts is not due to the plate's rotation under load and prying on the bolts. Leaving aside the design, end plates have two complications in erection that have limited their use. Because slip-resistant bolted connections must be used for bolts with axes parallel to the beam's axis, there is far less allowable variation in the beam length than with other connections. A beam cut too long or too short by a fraction of an inch may be unusable unless it is shipped back to the shop for additional work. Secondly, the exact fit of the beam between the columns can make placement of a beam between two column webs extremely time-consuming, because the beam cannot be swung in from the side. Variations on the basic end plate connection that involve a plate welded across the column flange tips can eliminate this problem, but at the cost of another element to be designed, fabricated, and welded.

In other cases, welding has influenced the sequencing and configuration of field construction. One prominent example is the use of "trees" in tube frame structures. The details of tube-type structures are not necessarily different than those for traditional moment or braced frames, but engineers have taken advantage of the actual distribution of forces within the various members to

simplify erection. The tube "walls" perpendicular to the direction of the applied wind load are subjected to simple axial loads, those parallel to the applied wind are shear walls, composed of discrete elements rather than the traditional masonry or concrete continuous shear wall. The walls are most commonly unbraced frames, or Vierendeel trusses, of moment-connected beams and columns. In such a frame, there are points of inflection, or zero moment, at some location along each beam. This location shifts depending on the direction and size of the applied load, but in general the moment in the middle of the beams can be assumed to be lower in magnitude than that at the ends.

Tubes are commonly erected in sections called trees, consisting of a two-story-high column with stub beams on the wall plane sides. The length of the stubs varies from around one foot up to half of the column-to-column spacing. For any length less than the maximum, a center beam section must be placed between each adjacent pair of trees; if the erection of trees with half-bay-length arms is practical, the step of erecting the center beams can be eliminated, cutting the number of individual steel sections to be erected by approximately one-half and the number of connections to be made by approximately two-fifths. The connections required, at the column and beam splices, often do not need to be designed for the full section moments, and so are reduced in size from the theoretical maximum. Trees have been used on occasion for moment frames other than tubes, but ordinarily have extremely short arms, used more to provide clearance away from the column to make field connections than to take advantage of reduced moments. Without cheap shop welding, tree construction would be at a disadvantage when compared with ordinary beam and column construction, because of the additional expense of the beam center connection. Today's economics encourage the use of trees, because the total of two shop-welded and one field-bolted connections is cheaper than two field connections of any type.

## Miscellaneous Detailing

Modern building design often includes sealed windows. The development of practical, large-scale air conditioning and ventilation, combined with building owners' desire for larger floor areas, has led to buildings where occupants might be almost 100 feet from the nearest window. This is a marked contrast to architectural design in the 1930s, when large buildings usually were designed with occupants no further than 30 feet from a window. In return, the amount of ductwork required has expanded tremendously, since the habitability of large-floor-area buildings depends on continuous mechanical ventilation.

The vertical openings required for ductwork do not usually create structural difficulties. Any reasonable amount of vertical duct space can be created simply by framing an opening in the floor with small beams. Structural engineers prefer to cluster these openings in buildings incorporating composite floor beams, since each hole destroys the continuity of the concrete beam flange, and requires that the steel beam be designed non-composite. The horizontal runs of duct can create more serious problems. The trend in large commercial construction has been towards fewer columns in order to create more flexible space, requiring longer floor beams that often must be deeper than those used in older layouts. The trend towards efficient use of the building volume, leading to shorter floor-to-floor heights, combined with the code-minimum ceiling heights, has squeezed the depth of the floor plate. Column transfers are now more often built-up plate girders

than the previously popular transfer trusses. Finally, the use of sealed windows and large floor areas has increased the number and size of horizontal duct runs within each floor plate.

The inevitable result of these trends has been spatial conflict between the building structure and mechanical systems. Often, HVAC engineers are able to organize their ducts so that the majority of runs are parallel to the floor beams, but the conflicts can only be reduced in this manner, not eliminated. In cases where there is not enough vertical room, the structural engineer must provide openings through the beam webs to permit ductwork to continue through unimpeded. This is a fairly expensive detail to build, but to many building owners, the one-time expense of additional steelwork is more than offset by the additional floors that can be squeezed into a given zoning-determined building height by reducing the floor plate depth.

Web openings are ordinarily provided by reinforcing the area around the hole to be cut with both vertical and horizontal plates welded to the web. The area is analyzed as a Vierendeel truss, with the portion of web above the opening one "beam," the area below another, and the areas to either side the truss vertical columns. In this analogy, the horizontal reinforcing plates serve as the top flange of the beams below and the bottom flange of the beam above, and the vertical plates serve to increase the minor-axis (out-of-web plane) buckling resistance of the columns. To lessen the stress concentrations that exist at a Vierendeel's corners, the opening's corners are usually cut round and the reinforcing plates are continuously welded to the web and to one another.

In the United States, web openings are fighting an uphill battle economically. They are used when there is no alternative, simply because of the cost of welding the reinforcing. The welding, even when performed in the fabricator's shop, substantially increases the cost per ton of steel. While this condition is true for individual floor beams anywhere in the world, castellated beams represent an option used in Europe that is extremely rare here. Castellated beams are ordinary wide-flange sections that are cut in two along a square wave line through their webs. The two halves are then welded back together after being shifted to align the wave "peaks." The result is a deeper beam with a series of square, round, or hexagonal openings through the web. Because the openings are relatively small, the beam can be analyzed as an ordinary truss, much like an open-web joist. The primary advantage of castellated beams is the possibility of using robotic cutting torches and welding for the extremely simple and repetitive work. This is outweighed in the United States most of the time by the inability to fit main ducts through the small web openings provided.

The opposite end of the fabrication cost spectrum is the general trend towards larger and larger sizes of steel shapes that have been rolled over the years. Bethlehem Steel's innovation of wide-flange beams did not immediately take hold; the majority of buildings built in the 1910s still had built-up columns and American Standard beams; the 1923 Carnegie Steel Company handbook, published simultaneously with the first edition of the AISC manual, was dominated by heavy American Standard beams and built-up columns.[20]

All of the economic and practical advantages lay with the use of wide-flange beams and columns, which in the 1938 Bethlehem catalog were available in sizes up to 36 inches deep and 300 pounds per foot (36WF300) and 14WF426, respectively.[21] American Standard beams had reached their practical limit at beams in the range of 27I90 and 24I120. The maximum sizes from the 1938 catalog remained the maximum sizes in the AISC manual through the fifth edition in 1947 and the sixth edition in 1963.[22] There was no great impetus for larger sizes because the majority of buildings did not require larger sizes. In places where these maximums were not sufficient, such as transfer girders in the lower floors of high-rises or lower column shafts in extremely tall buildings, built-up plate girders and columns continued to dominate, although they were welded more often than in the past.

The architectural trend towards wider column spacing, rooted in the desire for large, open floor spaces in commercial buildings, led to an increase in the heaviest column section available, up to W14x730. This maximum held for the seventh edition of the AISC manual in 1970 and the eighth edition in 1980.[23] Foreign manufacturers began offering heavier shapes in the 1980s. These

took two forms: heavier sections in existing depth and flange-width series, and entirely new sizes such as 44-inch-deep beams.[24] Even though these new sections are not available from most domestic manufacturers, many were added to the ninth edition of the AISC manual in 1989, whose tables extend up to W44x285, W40x625, W36x848, W33x619, W30x581, W27x539, W24x492, W21x402, and W18x311, in addition to W14x730.[25]

The heaviest wide-flange shapes are referred to as the jumbo sections, currently defined as sections containing flanges or webs 2 inches thick or thicker (figure 10.3). These shapes came into use before they had been fully researched. Experimental analysis of jumbo shapes has established that the metal grain structure in the center of thick elements can be coarser than that in other steel, and so more brittle.[26] The initial use of jumbo shapes was to replace built-up members as columns carrying exceptionally heavy loads. Since such columns are unlikely to carry moments large enough to ever create tensile stress and are not usually connected with full penetration welds, the brittleness caused by crystallization remained a minor issue.

As soon as engineers and steel fabricators became comfortable with jumbo shapes, they began to find other ways to use them. In areas where a beam must carry heavy loads and is severely restricted in depth or allowable deflection, a jumbo shape can provide a solution easier to build than a plate girder or truss. In addition, the rise of tube bracing for the lateral load system of tall buildings and column-free interior floor plans has created a class of columns carrying heavy axial loads and heavier moments and shears than columns in traditionally braced frames. These uses put jumbo shapes in tension, and often require large fillet welds or penetration welds for connections.

A supplement to the eighth edition of the AISC code issued in 1989 and incorporated into the ninth edition later that year refers to the increase in "non-column use applications" in its restrictions on welding and fabrication technique and provisions for minimum material toughness.[27] The toughness restriction requires that the steel used in jumbo sections subject to tension or flexure that

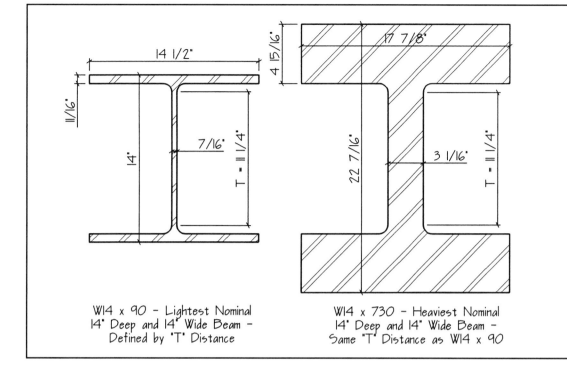

W14 x 90 – Lightest Nominal
14" Deep and 14" Wide Beam –
Defined by 'T' Distance

W14 x 730 – Heaviest Nominal
14" Deep and 14" Wide Beam –
Same 'T' Distance as W14 x 90

will be connected with full-penetration-welded joints pass a specific ASTM-defined test to limit the possibility of segregation under load. The test "shall also be considered" for jumbo sections subject to tension or flexure connected with partial-penetration and fillet welds.

The fabrication of welded connections has also been specified in much greater detail for jumbo sections than otherwise. At any location where two welds meet at an angle, such as the juncture of the web weld and flange weld in an "I" section splice, weld access holes are required. These holes, also called "rat holes," are small openings cut in one of the flat planes at the weld bend to prevent cooling shrinkage in two directions at the bend. The typical detail used in beam splices is a teardrop-shaped hole in the web on one side of the splice. The same detail is occasionally used for beams smaller than jumbos.

To prevent discontinuities in the steel that cause stress concentration, tension splices are required to be ground smooth, and all portions of connections, such as access holes, must be cut or ground smooth. These requirements do not apply to compression (ordinary column) splices if they are made using details that do not induce large amounts of shrinkage, such as fillet-welded or bolted splice plates.[28]

The use of jumbo sections has also led to concern over the effect of certain weld details on the surrounding metal. Full-penetration welds have traditionally been given less attention in design practice because they represent a connection whose strength is exactly equal to that of the member. They have long been listed as "pre-qualified" for the full capacity of the base metal when formed correctly. The possible danger in their use with jumbo sections is in the heat of their creation. No matter how big a weld becomes, it is still created by building up weld material in pass after pass with an electrode that ordinarily deposits ⁵⁄₁₆ inch or less on a single pass. The most common full-penetration weld is the groove weld, shaped like a "V" with a 45° root angle. In 2-inch-thick base material, a groove weld splice is 2 square inches in cross-section area, and will require approximately twenty-six passes to fill with electrode metal. Even allowing for cooling-down periods between each pass, this represents an enormous amount of heat being added to the steel.

All steel beams contain along their main axis minor compressive and tensile stresses that arise during cooling after the rolling process. These "residual stresses" are ignored during design because of their relatively small size compared with the material yield stress, the safety factor used

Roughly one year after a new commercial tower was completed, the partners of a law firm leased two complete floors for their new offices. The structural firm that had designed the tower was hired by the speculative builder who owned the building to review the design of the lawyers' engineer for alterations. The review process is described below.

The building is a large, steel-framed building, fairly typical except for its fifty-floor height. The floors are lightweight concrete on composite metal deck, with the slab and deck thickness sufficient to provide self-fireproofing. The steel beams and girders are composite with the slab, and covered with spray-on cementitious fireproofing. The full permissible live-load reduction was used in design of all girders and columns.

The proposed alterations included cutting an opening through one floor slab to provide an interior tenant stair and the construction of a library in which all materials would be stored in high-density shelving. All other proposed alterations were architectural additions to the bare space that had no implications for the building structure.

The high-density shelving covered an area of roughly 1200 square feet with a live load of roughly 150 pounds per square foot (psf). The floors had been designed for a 50-psf live load, reduced, and a dead load consisting of the slab and hung ceiling weight and a 20-psf partition allowance. Even taking into account the absence of partitions, the floor beams and one girder directly below this area were overloaded, and had been reinforced with bottom flange plates by the design engineer. The details of the existing connections were unknown to the designer, and were hidden from view by the fireproofing. The connections could not be exposed during design because the lease was not yet final, so the additional load on the double-angle beam-to-girder connections and the strike-plate girder-to-column connections was examined by the reviewer checking the capacities from the original steel shop drawings.

The stair was more difficult to deal with, since its location adjacent to a column destroyed a portion of the slab "flange" for one beam and one girder. Since there are no ordinary design provisions for a beam composite with a slab that does not extend the full length of the beam, these two members were re-analyzed with a flange on one side of the center line only, as "L" beams instead of "T" beams. They both required reinforcing with WT shapes to be adequate, despite the reduction in load from removing part of the slab.

in steel design between maximum allowable applied stress and yield stress, and their symmetrical nature. The last comes from the origin of residual stress in cooling: the portions of the shape farthest from its center and with the most air circulation—the flange outer tips and the center of the web—cool first. In cooling they contract. This creates no stress, since the still-hot metal adjacent is incapable of resisting the contraction. As the remaining portions of the beam—the "T"-shaped sections consisting of one end of the web and the center of the adjacent flange—contract, they are placed in tension and the already-cool areas are placed in compression. This pattern of stress remains in the beam permanently. Since each flange and the web "average" at zero stress, the residual stresses can be ignored.

Welding creates residual stresses through the same mechanism, but with the important exception that those stresses may not be balanced. Even small fillet welds can warp a section if improperly placed, although with thin sections and small welds the cooling is usually close enough to simultaneous to reduce the effect. When jumbo sections are welded, the retention of heat and differential shrinkage during cooling are of paramount importance. The AISC limits these problems through a requirement for a relatively hot pre-heating of the base metal for jumbo-section welds.[29]

## Code Changes: Column Formulas

The sixth edition of the AISC code, issued in 1963, changed the column formulas used exclusively in building construction for twenty-five years. The code introduced a cutoff slenderness ratio, $C_c$, which represents long columns from short columns. Long columns, with slenderness ratios between $C_c$ and the upper limit of 200, were designed around elastic buckling. The formula used for these columns is the classic Euler formula modified by a $^{23}\!/_{12}$ safety factor. The complex, inelastic behavior of short columns was represented by the equation

$$F_a = \frac{\left[ 1 - \dfrac{\left( \dfrac{K \cdot l}{r} \right)^2}{2 C_c} \right] \cdot F_y}{\dfrac{5}{3} + \dfrac{3}{8 C_c} \cdot \left( \dfrac{K \cdot l}{r} \right) - \dfrac{1}{8 C_c} \cdot \left( \dfrac{K \cdot l}{r} \right)^3}$$

which is a parabolic formula modified by a safety factor that changes from $^5\!/_3$ when the slenderness is zero up to $^{23}\!/_{12}$ when the slenderness is $C_c$.[30] For A36 steel, $C_c$ is 126. This means that the majority of columns in any building will be short by the code definition. The practical effect of the code change was to change the formula used for *all* previous design (where a slenderness of 120 was the maximum for primary members, including all building columns) to the short-column equation and to add the long-column class to design possibility. Another provision was to decrease the safety factor for secondary members, such as bracing, as the slenderness of such members increases up to 200.

The new code provisions are still in effect, except that the 1989 ninth edition of the code eliminated the secondary-member safety-factor reduction.[31] These formulas are based on the extensive research into metal-column behavior performed during the 1940s and 1950s to improve aircraft design. The research performed since cannot drastically change these formulas, since their predicted values correspond fairly well to experimental values. It is unlikely that any radical change will be made in these equations in the future, although the new Load and Resistance Factor Design method uses slightly different formulas to achieve similar results (see "Code Changes: General," below).

The use of high-strength steel has been one reason for the growing interest in slender columns. High-strength columns are smaller in area and moment of inertia than A36 columns designed for the same load. Since slenderness is dependent on the radii of gyration, which are proportional to the moment-of-inertia to area ratio, switching to a high-strength steel does not automatically mean an increase in slenderness. The largest change comes in medium-size to small columns, where a reduction in required column size can mean a change to a smaller shape series. The change from the nominally 14-inch-deep series with nominally 14-inch-wide flanges to the 14-inch-deep by 10-inch-wide series means a minimum decrease in weak-axis radius of gyration of at least 33 percent. This is a substantial increase in column slenderness. Tube frames often use section sizes originally thought of as beam shapes, such as W24's and W36's, which have relatively narrow flanges for their depth, and thus are slenderer about their weak axes than column shapes.

## Code Changes: Inspection

One unfortunately recurring problem has been bribery or other forms of corruption in the New York Building Department. In a large city, with so much money spent on construction, this problem is perhaps inevitable, but other than investigating the people involved there was never much that could be done. The problems have declined in level, from scandals in the nineteenth century involving the entire Board of Examiners to individual cases of inspector bribery since the 1970s.[32] The methods of dealing with this problem have varied over the years, with the most recent change in an attempt to safeguard the public interest coming in the late 1970s.

Following reports in 1975 of city building inspectors soliciting bribes by making false reports of problems after inspection visits, the City Council decided to reduce the historic role of inspectors. As a temporary measure, inspectors were required to write their reports on site and immediately give copies to the architects and engineers of record.[33] The legislation proposed in early 1976 would have, in the long run, only New York State–licensed professionals inspecting the most critical aspects of construction. The original proposal was that the professionals of record for a given project be responsible for its inspections, but this was later modified to allow inspections by other licensed professionals chosen by the project owner.[34]

The regulations incorporated into the building code, referred to as "controlled inspections," include laboratory tests with long histories, such as taking concrete test cylinders from each batch of concrete and crushing them to determine the concrete ultimate compressive strength, as well as entirely new requirements such as a professional inspection of the structure for stability during all load transfers. In a new building, this last item can mean the nearly continuous presence of an inspector during the frame, floor-system, and curtain-wall erection. Before a work permit can be issued for a project, a list of required inspections must be submitted, sealed by both the architect or engineer of record and the architect or engineer for controlled inspection.

Several aspects of this requirement have been criticized, although after the passage of time it seems unlikely that much will change. Other codes, notably the Connecticut Building Code, have instituted similar requirements. When controlled inspection was first debated in the New York City Council, one perceived drawback was that the "bill's sponsors acknowledge they are relying on the integrity of the professionals and the policing of the industry by itself."[35] The inspection law was given a provision for criminal prosecution of anyone submitting false inspection reports. Professionals would risk having their licenses revoked by the state if they acted in an unethical manner. The additional expense and paperwork have been an easy target for people who dislike the requirement, but this seems less important than controlling the quality of building construction. Finally, the lack of distinction given in the law between architects and engineers has been criticized as allowing people to operate outside their professions. Inspections that require engineering knowledge, such as examining the concrete mix delivered to a site, are mixed with inspections ordinarily performed by architects, such as firestopping. Again, the individuals involved are on an honor system that they will not perform inspections for which they are unqualified. This is an imperfect solution, but it the best practical since there is no clear way to divide the process of construction into logically defined pieces—one for the architect, one for the structural engineer, one for the HVAC engineer, and so on.

While these code provisions might seem a reaction to local political conditions, there is a safety-based rationale for them. Given the increased complexity of modern construction compared to the conditions in the nineteenth century when Building Department inspections first were instituted, specialized knowledge is a requirement for inspectors. While the training required to inspect a masonry bearing wall is minimal, the training required to inspect welds thoroughly is beyond the normal standards even for licensed engineers. The current system has evolved so that inspection laboratories under the supervision of a Professional Engineer or Registered Architect send specialists, either technicians or professionals, to the site as needed for each item on the inspection list.

## Code Changes: General

Two sets of codes exist for any building project: the local building code and the national material codes. The New York City Building Code was one of dozens of local codes in use early in the twentieth century. Because it predated the AISC steel code and the ACI concrete code, originally it was forced to contain specific material information. This information was not immediately removed from the code when the national material codes became available; rather it remained in its entirety for years, and some portions still remain. This is largely an anachronism that will one day disappear. It is a throwback to the 1910s, when the perceived strength of a piece of steel varied as much as 20 percent depending on which city it was being used in.[36]

Live-load reduction is an area where national standards have never been fixed, and the local building code's provisions are therefore crucial. Since live-load reduction is based in part on probability theory and in part on empirical evidence about the degree of loading actually encountered in various building types, it is impossible to pin down. Flexibility of use requires that all office space be designed for the same live load and that the same reduction rules apply, but examples of inconsistency in actual loading are easy to find. A suite of executive offices is likely to have far more open space than an area of clerical cubbyholes, greatly reducing the probability of full loading, unless the executives decide to throw parties that heavily load the floor with people. Given such unknowable variations in loading, live-load reduction rules have changed in small ways, advancing only when the adequacy of the previous set of rules was well established. (There is an obvious fallacy in this logic: when the rules work, they are made more liberal, until they reach the point where they don't work any more, and then they stop changing. There are any number of reasons why some modern office buildings have bouncy floors and require structural reinforcement for file-storage areas, but overuse of live-load reduction is the least excusable.)

In the 1920s, live-load reduction was limited to columns in buildings taller than five floors. The roof and top-floor loads were used unreduced, and an additional 5 percent reduction could be taken at each floor, down to a maximum 50 percent reduction.[37] This was liberalized slightly in 1936. The only buildings excluded were those used for storage. Columns in other buildings used the full live load at the roof, 85 percent at the top floor, and then, as before, an additional 5 percent at each floor to a 50 percent maximum. For the first time, girder live-load reduction was introduced. Girders at levels other than roofs, supporting at least 200 square feet of floor area and not supporting columns, could have their live loads reduced 15 percent. Girders supporting columns and all footings could be sized for the reduced column live load as figured above and the full dead load.[38] In 1939, columns, piers, bearing walls, and foundations of storage buildings were allowed a 15 percent reduction in live load other than roof load.[39]

The rules that were set in place in 1939 still exist in the building code as an alternate method, which is used occasionally because it is simpler to figure than the modern reduction formulas. The modern formula gives identical rules for girders and columns. The percentage of live-load reduction allowed is based solely on the live-load-to-dead-load ratio and the tributary supported area. The 50 percent maximum still exists, along with more specific exclusions, such as all beams and girders in roofs, and areas used for storage, parking, assembly, manufacturing, or retailing. Columns, piers, and walls in the excluded areas can have their live loads reduced in the same manner as other columns, but with a maximum of 20 percent reduction.[40]

The overall effect of live-load reduction is hard to determine. Any building can be designed for the full-code live load, and probably should be if alterations or changes in use are foreseen. Flat two-way slabs rarely use reduced loads even when permitted because of the difficulty created in designing for the unreduced shear at the slab-to-column joint. This excludes most new residential construction from reduction. Retail buildings are, as a class, excluded. The buildings most often designed with reduced loads are speculative office buildings. Since the owner bears no responsibility other than to provide a building that meets the building code provisions for safety, the building is built in the least-expensive way possible. This often includes full live-load reduction on all members where it is feasible. The tenants in such buildings, if they wish to install high-density filing systems or other load-increasing furnishings, must have an engineer review the basic building structure for its actual live-load capacity. Since neither code-sponsored live-load reduction nor speculative construction is likely to disappear, this condition will remain as a caveat to all potential tenants and a source of avoidable building renovations.

The New York City building code now contains as reference standards, with minor modifications, the AISC and ACI codes. Two alternatives to the standard AISC specification are available. Plastic design, contained within the AISC "Allowable Stress Design Specification" (ASD) as an alternate since 1958 has never become popular in large buildings.[41] The AISC "Load and Resistance Factor Design Specification" (LRFD) is meant as a way to replace entirely the ASD design method, and is contained in a manual similar in format to the last four editions of the ASD manual. Most building codes long ago accepted the ASD specification in its entirety, making plastic design use possible; LRFD, after a few years of argument following its introduction in 1986, is now generally accepted. The New York City building code was one of the first to allow LRFD design, but only a handful of LRFD buildings have been built.

Both plastic design and LRFD can claim more consistent and efficient use of steel's material properties than ordinary ASD. Their lack of complete acceptance is based on several practical concerns about the way that buildings are actually designed. Plastic design is based on stressing steel members to a fixed ratio of their ultimate collapse load. This requires stability analysis for all continuous members. In simple one-bay buildings, such as typical industrial buildings, this technique is not any more time-consuming than ordinary design and can create significant reductions in the amount of steel used. The stability analysis grows oppressively complex very quickly as the frame in question gains degrees of redundancy. In order to avoid two full frame analyses, one for the moments and shears induced by external loading and one for member stability, the scope of plastic design has to be limited, thus limiting its potential savings. In the end, large buildings are rarely designed using the ASD plastic method because too much additional design and too many moment connections are required for a small material savings.

LRFD applies load factors greater than one, which represent the uncertainty of actual loading, to the design loads and resistance factors smaller than one, which represent possible material defects and construction inaccuracies, to the represented strength of various members. The basic idea is to combine these two effects into a constant safety factor for all members, designed either elastically or plastically. This is not a new idea, since the American Concrete Institute's basic code has been LRFD-based since 1971, and had been introduced even earlier.[42] Leaving aside arguments based on the design community's resistance to change, two arguments against the widespread adoption of LRFD remain. The first is that it is more complicated than ASD, and years of competition and fast-track design have already cut the amount of time for design to a minimum. While

building owners might be convinced to pay more for design if they knew that the money would be recouped in construction, there is no guarantee that LRFD will always produce a more-efficient (read "cheaper") building. This dilemma leads to the second problem in using LRFD. Modern steel design has often led to buildings and building elements that approach the maximum allowable deflections. Overly bouncy floors and buildings that sway noticeably in high winds are far more common than in the past, because there are no longer nonstructural elements such as tile partitions that help stiffen the frame and because shallow, long-span floor beams and moment frames are inherently more flexible than short beams and braced frames. While LRFD design for stress may well produce designs with lighter beams and columns, an unbiased look at steel buildings of the past twenty years raises the question of where the amount of steel that has been used can be reduced without creating even larger deflection problems. This logic does not preclude the use of LRFD, especially given the new method's more consistent safety factor, but it does remove the justification that using the method will save money.

One other issue weakens the arguments in favor of LRFD. The buildings where steel costs are pared to the minimum are typically designed with the maximum code-allowed live-load reduction. Given the greater uncertainty of live loading, the live-load factors are greater than the dead-load factors. LRFD designs therefore save more material on buildings with relatively high dead-load-to-live-load ratios. In designs where the live load has already been reduced using probabilistic logic, using LRFD to further reduce the safety factor on the grounds of there being less live load is questionable. The same probabilistic effect—that live loads are rarely as high as the full design value—is being counted twice.

The LRFD code was quickly adopted by the New York Building Department because there is little argument about its effect: it is an alternate to the already-accepted ASD code. However, an entirely new consideration for the New York code is earthquake-resistant design. The history of seismic code provisions in the United States is fairly simple. At first, there were none; then, as research into the effect of ground movement on buildings was conducted in the beginning of the twentieth century, cities and states known to be in danger of earthquakes added construction restrictions to their building codes. As each major earthquake showed up weaknesses in code seismic provisions, the codes were tightened further. Without such codes, an event such as the 1989 Loma Prieta earthquake in the San Francisco Bay area would have resulted in far greater loss of life and property destruction.

Research into the cause and effect of earthquakes has led to the unsettling conclusion that portions of the country not previously thought to be in danger may experience destructive ground movement. The current state of knowledge suggests that except for a few isolated portions of the country such as south Florida, engineers everywhere should be designing with earthquakes in mind. The New York metropolitan area is classified as "Zone II," with maximum expected ground accelerations less than one-quarter of those in Zone IV areas such as the portions of California near the San Andreas fault. Connecticut's state code has adopted seismic provisions; New Jersey uses the Building Officials Code Administration (BOCA) code, including its seismic provisions; New York City and New York State are both in the planning stages of adopting provisions similar to those in BOCA and the Uniform Building Code (UBC).

For Zone II areas, seismic design is most likely to affect detailing. The loads used to figure the effect of ground motion in Zone II are usually smaller than code wind loads for most buildings, especially high-rises. Examples of structural detailing changes likely to be introduced are minimum amounts of reinforcing in all masonry walls and energy-absorbing lateral bracing for steel frames, such as eccentric bracing.

## Present Considerations

The changes in structural steel design over the last fifty years have been fairly minor and, for the most part, well documented. In cases where original drawings are not available during renovations, simply determining the parameters of the original design is often the greatest challenge. Unlike buildings from the 1920s, where the allowable bending stress is almost certainly 16 ksi and connections are almost certainly riveted, buildings from the 1960s could contain many different steels, with bolted, riveted, or welded connections. The frames could be designed plastically, or contain beams designed compositely in one of several fashions. Beam and column loads may have been reduced by several different sets of rules, or not at all. All of these factors may come into play when an engineer attempts to determine the capacity of a "newer old building" for renovation. Fortunately, drawings for more recent buildings tend to be both more complete and more readily available than for older buildings, and copies of the sixth and later editions of the AISC manual are readily available.

# MODERN NOTATION USED IN FORMULAS AND DESCRIPTIONS

$a$ — Depth of Whitney stress block in concrete ultimate strength design

$A_b$ — Nominal area of bolt (or rivet)

$A_c$ — Area of concrete flange in composite design
Area of concrete in reinforced-concrete column

$A_s$ — Area of steel beam in composite design
Area of reinforcing in concrete beam or column

$A_w$ — Area of web

$b$ — Concrete beam width

$b_f$ — Steel beam flange width

$C_b$ — Bending coefficient based on boundary moment conditions

$C_c$ — Column slenderness ratio for steel dividing elastic from inelastic buckling

$C_m$ — Bending coefficient for columns with combined stress

$d$ — Steel beam total depth
Depth to flexural steel in concrete beams
Diameter of round columns or least dimension of rectangular columns

$E$ — Modulus of elasticity

$f_a$ — Actual compressive axial stress in steel

$F_a$ — Compressive axial stress permitted in steel

$f_b$ — Actual flexural stress in steel

$F_b$ — Flexural stress permitted in steel

$F_c$ — Compressive stress permitted in steel

$f'_c$ — 28-day compressive strength of concrete

$F'_e$ — Euler buckling stress in steel

$f_p$ — Actual bearing stress in steel

$F_p$ — Bearing stress permitted in steel

$f_t$ — Actual tensile stress in steel

$F_t$ — Tensile stress permitted in steel

$F_u$ — Ultimate stress in steel design

$F_{ult}$ — Ultimate allowable stress for a steel column

$f_v$ — Actual shear stress in steel

$F_v$ — Shear stress permitted in steel

$f_y$ — Minimum yield stress of reinforcing in concrete design

$F_y$ — Minimum yield stress in steel design

$G$ — Shear modulus

$h$ — Distance between flanges

$I_x$ — Moment of inertia about strong axis

$I_y$ — Moment of inertia about weak axis

$J$ — Polar moment of inertia

$K$ — Column effective length factor
Experimentally determined constant

$l$ — Beam or column length

$l_u$ — Unbraced length of a member

$m$ — Base-plate-edge distance in steel design

$n$ — Modular ratio in composite design
Base-plate-edge distance in steel design

$P$ — Applied axial load

$P_{ult}$ — Ultimate allowable axial load on a column

$q$ — Allowable shear on a shear connector in composite design

$r$ — Radius of gyration

$S$ — Section modulus

$t_f$ — Steel beam flange thickness

$t_w$ — Steel beam web thickness

$\gamma$ — Concrete ultimate strength design load increase factor

$\varnothing$ — Concrete ultimate strength design capacity reduction factor
Round column diameter

# BUILDING EXAMPLES

The following buildings are mentioned either because they are structural landmarks or because they represent well the typical construction of their period. With a few exceptions, such as the Statue of Liberty, these structures were not unique at the time of their construction. Wherever possible, the address, architect, engineer, and status of the buildings are given. Many of the sources are unclear or incomplete, however; missing information indicates where it has not been possible to confirm the location or status of the building.

## 1835

**Obadiah Parker house,** Parker designer, demolished. House walls were monolithic concrete, probably with natural lime cement.[1]

**Lyceum of Natural History,** Broadway near Spring Street, Andrew Jackson Davis architect, demolished. Early use of iron columns in New York, for support of facade at storefront.[2]

**Boorman, Johnston and Company store,** 119 Greenwich Street at Albany Street, demolished. Early use of iron columns in New York, for support of facade at storefront.[3]

## 1837

**G. A. Ward house,** New Brighton, Staten Island. Walls were precast concrete blocks.[4]

**Lorillard Building,** Gold Street, demolished. Iron columns for support of facade above two-story storefront.[5]

## 1841

**[Old] Merchants' Exchange,** 55 Wall Street at William Street, Isiah Rogers architect, heavily modified 1907, landmarked. Monolithic all-masonry construction.[6]

## 1846

**[Old] A. T. Stewart Department Store** [now Sun Building], 280 Broadway at Reade Street, Trench and Snook architects, Daniel Badger iron builder, landmarked. Badger provided the ground-floor storefronts and some interior structure.[7]

## 1848

**Edgar Laing stores,** 258–262 Washington Street and 97 Murray Street, James Bogardus designer, demolished. First complete cast-iron facade in New York, wood interior and brick bearing walls.[8]

**James Bogardus factory,** Centre and Duane streets, Bogardus designer, demolished 1859. Early complete cast-iron facades, two street facades, 25 feet by 100 feet, wood interior.[9]

## 1851

**Fire tower,** 33rd Street and Ninth Avenue, Bogardus designer and Hoppin Company builders, demolished. 100 feet high with six tiers of iron beams, decagonal with a column at each corner, approximately 20 feet in diameter, no bracing, fastened to bedrock with drilled iron expansion anchors.[10]

## 1853

**Fire tower,** 253 Spring Street at MacDougal Street, demolished 1885. 100 feet high with six tiers of iron beams, decagonal with a column at each corner, no bracing.[11]

**Gilsey Building** at Broadway and Cortlandt Street, demolished. Two six-story cast-iron facades.[12]

**New York Crystal Palace,** George Carstensen and Charles Gildemeister architects, burned 1858. Cast-iron columns, wrought-iron arch ribs and truss girders, arched trusses, and all-glass curtain wall, portal bracing for lateral load.[13]

## 1854

**Bogardus Building**, Centre Street, Bogardus designer, demolished. Four cast-iron facades, cast-iron girders and interior columns.[14]

**Harper & Brothers Building**, 331 Pearl Street on Franklin Square, John Corlies architect, James Bogardus engineer, James L. Jackson foundry builders, demolished 1925. Brick jack arch floor topped with concrete spanned between wrought-iron floor beams, supported by cast-iron and wrought-iron bowstring truss girders and cast-iron columns. The floor beams were the first lot of wrought-iron beams rolled in the United States, by the Trenton Iron Works. One cast-iron facade, brick side walls and rear.[15]

**United States Assay Office**, Wall Street, demolished. Used first wrought-iron beams from first Trenton Iron Works lot.[16]

## 1855

**German Winter Garden**, 45 Bowery, demolished. Cast-iron columns supporting balcony and cast-iron rib dome.[17]

**McCullough Shot and Lead Company shot tower**, near Centre and Pearl streets, demolished. Eight tiers and 175 feet high, octagon 15½ feet across at top and 25 feet at bottom, cast-iron frame supporting 12-inch-thick brick curtain wall, no bracing.[18]

**Bank of the State of New York**, demolished. Use of "members which were cold formed from 1/16- and 1/8-inch-thick steel sheets."[19]

## 1856

**Cary Building**, 105 Chambers Street, Gamaliel King and John Kellum architects, George H. Johnson facade designer, landmarked. Two cast-iron facades, among oldest remaining in New York.[20]

**Fire tower in Mount Morris Park** (now Marcus Garvey Park), Julius Kroehl engineer, landmarked. Octagonal cast-iron tower, three tiers high, with rod cross-bracing for top tier.[21]

## 1857

**254–260 Canal Street**, landmarked. Design of early building with two cast-iron facades credited to James Bogardus.[22]

## 1858

**United States Customs House**, Oswego, New York, standing. Trenton Iron Works rail beams supporting brick arch floors, wrought-iron roof trusses with cast-iron compression struts.[23]

## 1859

**Haughwout Building**, 490 Broadway at Broome Street, John P. Gaynor architect, Daniel Badger builder, landmarked. One of the oldest cast-iron facades standing in New York, five stories, large interior light court.[24]

**A. T. Stewart Department Store** [later Wanamaker's], Broadway and 9th Street, John Kellum architect, Cornell Iron Works builders, burned 1956. Two five-story, 85-foot-high cast-iron facades, wood joists on wrought-iron beams and cast-iron columns, built-up wrought-iron girders at light courts, no wind bracing.[25]

**Cooper Union Foundation Building**, Fourth Avenue and 8th Street, Frederick Peterson architect, Architectural Iron Works iron contractor, landmarked. Oldest building standing in the United States with wrought-iron floor beams, it used 7-inch I's from Trenton Iron Works's first lot, brick arch floors in part, experimental tile arches in part, cast-iron columns and brick bearing walls.[26]

## 1850s

**Colwell Lead Company shot tower**, 63–65 Centre Street, James Bogardus designer and builder, demolished 1908. Eight-tiered cast-iron column and beam tower, masonry infill in each panel.[27]

## 1860

**United States Warehousing Company grain elevator**, Brooklyn, George Johnson engineer, Architectural Iron Works builders, demolished. Cast-iron frame supporting brick curtain wall, six stories high, cross-braced in all bays by wrought-iron rods.[28]

**203 East 29th Street**, standing. Wood frame building.[29]

**Watervliet Arsenal**, Watervliet, New York, Daniel Badger designer, Architectural Iron Works builder, standing. First all-iron building, 100 feet by 196 feet, cast-iron walls and columns, wrought-iron roof trusses, cast-iron girders with wrought-iron tension rods.[30]

## 1863

**Brown Brothers Bank,** demolished. Early use of wrought-iron columns.[31]

## 1866

**324 Bedford Avenue,** Williamsburg, standing. Two four-story cast-iron facades.[32]

## 1868

**[Old] St. Patrick's Cathedral,** 260 Mulberry Street, Henry Englebert architect of rebuilding after fire, Joseph Mangin architect of original building in 1815, landmarked. Heavy timber truss roof on cast-iron columns.[33]

## 1869

**[Old] St. Ann's Church,** Clinton and Livingston Streets, Brooklyn Heights, Renwick & Sands architects, standing. Heavy timber truss roof on cast-iron columns.[34]

**1 Front Street,** Brooklyn, William Mundell architect, standing. Cast-iron facade.[35]

## 1870

**[Old] Equitable Building,** 120 Broadway, George Post architect and engineer, burned and demolished 1912. Five stories and 130 feet high, granite bearing wall, brick floor arches, and wrought-iron beams.[36]

## 1871

**Grand Central Depot,** 42nd Street and Park Avenue, John B. Snook architect, Isaac C. Buckhout engineer, Wilhelm Hildenbrand engineer for train shed, Architectural Iron Works builders, altered 1898, demolished 1913. First balloon shed in United States: wrought-iron truss barrel vault for train shed, covered with glass and galvanized iron.[37]

**William E. Ward house,** Port Chester, Robert Mook architect, Ward designer and builder, standing. First reinforced-concrete building in the United States, entire building monolithic, beams reinforced with wrought-iron I's connected for shear transfer, 3½-inch-thick floors reinforced with rods, 2½-inch-thick partitions reinforced with rods, hollow cylinder columns reinforced with hoops.[38]

## 1872

**[Old] United States General Post Office,** City Hall Park, Leonard F. Buckwith architect, demolished. First use of Kreischer's patented tile arch floors: hollow, side-construction flat arches.[39]

**287 Broadway,** John Snook architect, standing. Cast-iron facade and wrought-iron floor beams.[40]

**Bennett Building,** 93 Nassau Street, Arthur Gilman architect, standing. Built six stories with three cast-iron facades, extended upward four more stories in cast iron.[41]

## 1874

**Evening Post Building,** Broadway and Fulton Street, Charles F. Mengelson architect, demolished. Eleven stories, "one of the first large office buildings to be erected in New York."[42]

## 1875

**Western Union Building,** Broadway and Dey Street, George Post architect and engineer, demolished. Brick bearing walls, 230 feet high.[43]

**Tribune Building,** Park Row at Nassau Street, Richard Morris Hunt architect, demolished 1966. Probably highest bearing-wall building in New York at 260 feet high. First tower-type building downtown. Wrought-iron beam floors.[44]

## 1877

**1–9 Bond Street,** burned 1877, rebuilt since. Six-story, 113-foot-high cast-iron facade, 125 feet by 110 feet, wood floors.[45]

**[Old] New York Hospital,** 15th Street and Sixth Avenue, George Post architect and engineer, demolished. Entirely noncombustible construction.[46]

**Long Island Historical Society,** 128 Pierrepont Street, Brooklyn, George Post architect and engineer, landmarked. Early use of ornamental terra cotta, produced locally.[47]

## 1879

**Seventh Regiment Armory,** 643 Park Avenue, Charles Clinton architect, R. F. Hatfield "consulting architect" [apparently structural designer], landmarked. Early balloon shed with wrought-iron-truss barrel vault and glass-and-iron skylights.[48]

## 1880s

**Wucker Warehouse,** 2293 Third Avenue at 125th Street, standing empty. Heavy timber and flitch-plate frame within brick bearing walls.[49]

## 1881

**Produce Exchange,** 2 Broadway at Bowling Green, George Post architect and engineer, demolished 1957. 120-foot-high cage building, wrought-iron transfer trusses above trading rooms.[50]

## 1883

**Statue of Liberty,** Bedloe's Island, Frederic Auguste Bartholdi architectural designer, Gustave Eiffel engineer, Keystone Bridge Company and D. H. King Contracting builders, landmarked. First full-braced frame in New York, first use of steel columns in the United States, early use of concrete in foundation.[51]

**Mutual Life Insurance Building,** Nassau Street. First building with soffit tiles in tile arch floor, also first where blocks other than skewback blocks contained interior ribs.[52]

## 1884

**Chelsea Hotel,** 222 West 23rd Street, Hubert, Pirsson & Co. architects, landmarked. Twelve-story brick bearing walls.[53]

**Washington Life Insurance Building,** Broadway and Liberty Street, demolished. Nineteen stories and 273 feet high.[54]

## 1885

**Osborne Apartments,** 205 West 57th Street, James Ware architect, landmarked. Early "high" building, stone facade.[55]

**20 Henry Street,** Brooklyn, Theobold Englehart architect, standing. Factory building with brick bearing walls, heavy timber interior.[56]

## 1886

**Potter Building,** 38 Park Row at Ann Street, N. G. Starkweather architect, standing. Early extensive use of terra cotta in facade, tile arch floors, first building in New York with tile fireproofing on columns.[57]

## 1887

**Congregation Khal Adath Jeshurun,** 12–14 Eldridge Street [Eldridge St. Synagogue], Herter Brothers architects, landmarked. Heavy timber scissors roof trusses on wood columns.[58]

## 1889

**Pulitzer [World Newspaper] Building,** Park Row and Frankfort Street, George Post architect and engineer, expanded 1908, demolished. Tallest building in the city when complete, thirteen-story block with six-story tower, cage construction with exterior walls up to 9 feet thick.[59]

**Apartment houses on Ninth [Columbus] Avenue from 99th to 100th Streets,** demolished. Five stories with Guastavino fireproof vault floors.[60]

**Tower Building,** 50 Broadway, Bradford Gilbert architect, demolished. First full frame in New York, 21½ feet wide in front, 39½ feet wide in back, 108 feet deep, 129 feet high, wrought-iron floor girders across width of building supported by cast-iron columns in the side walls up to seventh floor, top six floors were wrought-iron girders supported by brick bearing walls, diagonal bracing in side walls, moment frame in short direction, received special exemption from wall-thickness provisions of code from the Building Department.[61]

## 1890

**London and Lancashire headquarters,** Cedar Street, demolished. Early complete skeleton frame.[62]

## 1892

**Havemeyer Building,** Cortland and Church streets, George Post architect and engineer, demolished. Cage building with wrought-iron frame, 193 feet high, 60 feet by 215 feet.[63]

**Jackson Building,** north side of Union Square. 28.6 feet wide by 200 feet deep, eleven stories high, demolished. "[O]ne of the earliest specimens of the skeleton construction," rectangular cast-iron columns in side walls only.[64]

**Hotel New Netherland,** 59th Street and Fifth Avenue, demolished 1926. 125 feet by 100 feet, seventeen stories, 234 feet high, tallest hotel in the world when built. Noncombustible construction.[65]

**Hotel Savoy,** Fifth Avenue and 59th Street, Ralph Townsend architect, demolished. Noncombustible steel-frame construction.[66]

**Manhattan Life Insurance Building,** 64–68 Broadway, Kimball & Thompson architects, C. O. Brown engineer, demolished. 67 feet by 119 feet, main building 254 feet high, tower 348 feet high. Tallest building in New York when built, first caisson use on a building anywhere (caissons were used in bridge and tunnel construction as early as 1850s in Europe, 1870s in the United States), fifteen caissons 55 feet below grade, 35 feet below open excavation, cantilevered built-up girders in foundations.[67]

## 1893

**Diamond Exchange Building,** 14 Maiden Lane, standing. Oversized holes in cast-iron column connections allowed out-of-plumb erection. Adjustable diagonal rods installed when thirteen-story frame was erected up to tenth floor and noticeably tilted.[68]

## 1895

**American Surety Building,** 96–100 Broadway, Bruce Price architect, standing. First complete skeleton frame in New York, twenty stories and 303 feet high, 85 feet by 85 feet, Z-bar columns, wind braced with rods, caissons to rock 72 feet below curb elevation.[69]

**Ireland Building,** 3rd Street and West Broadway [La-Guardia Place], Charles Behrens architect, John H. Parker builder, J. B. & J. M. Cornell Company iron contractor, collapsed during construction. 50 feet by 100 feet, eight stories, brick bearing walls, tile arch floors on steel beams, cast-iron columns, collapse due to overloading by plastering contractor, inadequate footings, unreliable foundation-bearing stratum, inadequate baseplates, porous iron in columns.[70]

**Manhattan Savings Bank,** Bleecker Street, destroyed by fire November 5. Fire was in Keep Building across the street, heat caused cast-iron facade of Manhattan bank to expand, interior wrought-iron beams slipped off their bearings, carrying down beams and tile arch floors.[71]

## 1896

**Gillander Building,** Nassau and Wall streets, Charles Berg architect, Charles Wills builder, Post & McCord steel erectors, demolished 1910. Sixteen stories and four-story tower, "first skyscraper to be demolished," steel frame.[72]

**Astoria Hotel,** 34th Street and Fifth Avenue, Henry Hardenbergh architect, demolished 1930. Complicated framing layout, irregular column spacing, transfer trusses over void spaces, horizontal diaphragm bracing used.[73]

**The American Tract Society Building,** 150 Nassau Street at Spruce Street, R. H. Robertson architect, standing. Twenty-three stories and 288 feet high, tallest building in Newspaper Row.[74]

## 1897

**Borax factory,** Bayonne, New Jersey, Ransome Company engineers. First building using Ransome reinforced-concrete system on the east coast.[75]

**David S. Brown Company soap factory,** 51st Street and Twelfth Avenue, Henry F. Kilburn architect, Post & McCord iron contractors, collapsed during construction. Five stories and 65 feet high, 200 feet by 150 feet, steel floor beams supported by rectangular cast-iron columns, cage construction, collapse attributed to loose bolted joints and girder connection eccentricity.[76]

## 1898

**Pabst Hotel,** Broadway, Seventh Avenue and 42nd Street, demolished 1903. Steel beams and Z-bar columns, brick curtain walls, Roebling arch floors, partitions of plaster over a single layer of wire mesh. Typical construction for its time.[77]

**Grand Central Station,** 42nd Street and Park Avenue, Samuel Huckel, Jr. architect, William Wilgus engineer, conversion of Grand Central Depot, demolished 1913. Three floors added to main building, new steel floor framing and masonry bearing walls, existing exposed cast-iron columns and wrought-iron beams on lower floors covered with tile.[78]

## 1899

**St. Paul Building,** Park Row and Ann Street, George Post architect and engineer, demolished. Steel frame supporting true curtain wall, but similar to a cage, spandrel columns are set inboard of exterior walls and carry double spandrel beams on brackets to support wall. Twenty-five stories and 310 feet high, tallest building in the United States when built.[79]

**United States Archive Building,** 641 Washington Street, W. J. Edbrooke architect, landmarked. Ten-story brick bearing walls.[80]

**Windsor Hotel,** Fifth Avenue and 46th Street, burned March 17. Built in 1873, seven stories, 200 feet by 140 feet, brick bearing wall with all-wood interior, burned in less than one hour.[81]

## 1900

**28 East 78th Street,** McKim, Mead & White architects, standing. Typical small building construction: brick bearing walls exterior and interior, five stories, built-up steel columns to support framing at elevator shaft, small wide-flange columns to support mansard roof, "E" shaped cast-iron lintels.[82]

**Parker Building,** Fourth Avenue and 19th Street, built 1900, burned January 10, 1908. Nineteen stories, considered fireproof, true curtain walls, cast-iron columns encased with 2-inch hollow terra-cotta block, flat side construction, terra-cotta arches with 8½ inches of cinder fill, steel beams were misdesigned giving allowable live loads from 5 to 35 psf, columns failed from lack of protection where electrical conduit was cut into terra cotta, already overloaded beams failed.[83]

## 1901

**Ansonia Hotel,** 2101 Broadway, Paul Duboy architect, landmarked. Seventeen-story cage with cast-iron columns.[84]

**Nassau County Courthouse,** Mineola, New York, standing. Monolithic reinforced concrete of Ransome system, imitating masonry.[85]

## 1902

**Algonquin Hotel,** 59 West 44th Street, Goldwyn Starrett architect, landmarked. Exterior cast-iron facade for bay windows, relatively late.[86]

## 1903

**Flatiron Building,** 175 Fifth Avenue, D. H. Burnham & Company architects, Corydon Purdy engineer, landmarked. Steel frame with portal bracing and knees at all wind girders, tallest building in the city when completed.[87]

**Roosevelt Building,** built 1893, demolished. Exposed cast-iron columns and terra-cotta floor arches on steel beams were damaged in a fire.[88]

**[Old] Times Building** at Park Row and Nassau Street, standing. Extended upward from fourteen to eighteen stories.[89]

**Tribune Building,** Park Row at Nassau Street, demolished. Extended upward from ten to nineteen stories, with original tower top moved upward.[90]

## 1904

**Little Singer Building,** 561 Broadway, Ernest Flagg architect, standing. Steel, glass, and terra-cotta curtain wall, over 75 percent glass.[91]

**St. Regis Hotel,** Fifth Avenue and 55th Street, landmarked. Late cage construction.[92]

**Darlington Apartments,** 59 West 46th Street, Neville & Bagge architects, Pole & Schwandtner engineers of record and builders, collapsed during construction. Thirteen-story and 148-foot-high cage building, brick exterior wall, 55 feet by 90 feet, Roebling flat slab floors on steel beams, square cast-iron columns, collapse attributed to large gravity moments in the columns, which were effectively unbraced due to oversize bolted connections.[93]

## 1905

**Tribune Building,** Park Row at Nassau Street, demolished. Extended upward again, "the addition was to be built with the assistance of partial steel framing, encased in the existing masonry work." [94]

**35 East 62nd Street** [formerly Miss Keller's Day School], George Keller architect, standing. Early reinforced-concrete frame, brick-and-marble curtain wall, 40-foot-long transfer girders above first floor.[95]

**Schirner factory,** 69 Bank Street, Turner Construction Company builders, standing. Early four-story reinforced-concrete frame.[96]

**Latteman Building,** Brooklyn. Largest foundation of reinforced concrete piles in the United States, 480 piles in 100 feet by 129 feet, Gilbreth patent piles, octagon section with semicircular cutout in center of each side, the diameter of the cutout being roughly one-third of the side length, reinforced with longitudinal bars in a circle, surrounded by hoops, 16 inches diameter at top, 11 inches at bottom, reinforced-concrete pile caps.[97]

## 1906

**Grahm & Goodman garage,** West 93rd Street, Snelling & Potter architects. Reinforced-concrete frame, three stories.[98]

**Nottingham Apartments,** 35 East 30th Street, Snelling and

Potter architects, standing. Concrete frame with brick curtain wall, nine stories high.[99]

## 1907

**Plaza Hotel,** Fifth Avenue and 59th Street, Henry Hardenburgh architect, landmarked. Cage construction with cast-iron columns, Guastavino fireproof floors, facade is partly white glazed brick.[100]

**Shepard Hall** [Main Building] at City College, near Convent Avenue, George Post architect and engineer, landmarked. Terra-cotta-block bearing walls.[101]

**Monolith Building,** 45 West 34th Street, Howells and Stokes architects, standing. The first tall, reinforced-concrete building in the city, had limestone veneer for three floors at base, but exposed concrete above to full twelve-story height, stone veneer later replaced by stucco.[102]

## 1908

**Municipal Building,** One Centre Street, McKim Mead & White architects, landmarked. Forty stories and 580 feet high, continuous solid curtain wall.[103]

**712 Fifth Avenue,** A. S. Gottlieb architect, standing altered with facade landmarked. Five stories, bearing walls, cast-iron columns, steel beams, tile arch floors.[104]

**Singer Building,** Broadway at Liberty Street, Ernest Flagg architect, O. F. Samsch engineer, demolished. Singer building tower was 65 feet square, forty-one stories and 612 feet high, tallest in the city when completed, tallest building demolished.[105]

## 1909

**[Old] New York Times Building,** 42nd Street and Broadway, Eidlitz & McKenzie architects, standing altered. First tall building in the country to be designed using live-load reduction on its columns.[106]

**55 Liberty Street,** Henry Ives Cobb architect, landmarked. Limestone ground-floor facade, entirely terra cotta for thirty-two stories above.[107]

## 1910

**40 West 20th Street,** standing. Steel-frame loft building with reinforced-concrete slabs, reinforcing consisting of small rolled channels.[108]

## 1913

**Grand Central Terminal,** Warren & Wetmore and Reed & Stem architects, landmarked. Concrete-encased steel frame, Guastavino floor arches.[109]

**Woolworth Building,** 233 Broadway, Cass Gilbert architect, Gunvald Aus Company structural engineers, landmarked. Fifty-five stories, 760 feet, 6 inches high, tallest in the city when completed, caissons to rock, with moment-resisting portal frame, all-terra-cotta facade, facade rigidly connected to steel structure, no expansion joints provided, facade restoration required in mid-1980s, designed by Ehrnkrantz Group, over 20,000 panels had to be replaced with fiberglass-reinforced polymer concrete, approximately 100,000 reanchored.[110]

## 1917

**United States Army warehouses,** Second Avenue, Brooklyn, Cass Gilbert architect, George Goethals engineer, standing. Concrete flat slabs, "largest and most heavily loaded concrete buildings in the world at the time of their construction."[111]

## 1918

**43 Exchange Place,** Trowbridge & Livingston architects, Purdy & Henderson engineers, standing. Steel skeleton frame with tile arch floors, seated moment connections at all columns, two-wythe brick curtain wall.[112]

## 1920

**Electric Welding Company of America factory,** Brooklyn, T. Leonard McBean engineer. Early use of structural welding, Brooklyn Department of Buildings required a full-scale load test before allowing construction.[113]

## 1927

**Master Printers Building.** Tallest reinforced-concrete building in the United States when completed, twenty stories, 299 feet high.[114]

## 1930

**Chrysler Building,** 405 Lexington Avenue, William Van Alen architect, landmarked. Early use of stainless steel for large facade area, tallest building in the city when completed.[115]

**137 East 57th Street,** Thompson & Churchill architects, standing. Steel frame without spandrel columns, curtain wall hung from roof.[116]

## 1931

**Starrett-Lehigh Building,** Eleventh Avenue and 26th Street, R. G. Cory, W. M. Cory, and Yasuo Matsui architects, Purdy and Henderson engineers, landmarked. Flat slab concrete floors, on concrete columns with mushroom capitals at 21 feet on center above third floor, steel columns below; slabs are cantilevered to support curtain wall, nineteen stories high.[117]

**[Old] McGraw-Hill Building,** 330 West 42nd Street, Hood, Godley & Foulihoux architects, landmarked. Entirely terra-cotta facade.[118]

**Empire State Building,** 350 Fifth Avenue, Shreve, Lamb and Harmon architects, H. G. Balcom and Associates engineers, landmarked. Eighty-five stories and 1239 feet high, tallest building in the city when completed, full moment connection wind bracing, early use of aluminum cladding for top tower (dirigible mooring mast), ornament, and spandrel panels, early use of stainless-steel cladding in window edging.[119]

## 1932

**RCA Building,** 30 Rockefeller Plaza, Hood, Godley & Foulihoux; Corbett, Harrison & MacMurray; and Reinhard & Hofmeister architects, landmarked. Curtain walls only 8 inches thick (despite building-code provision for 12-inch walls full height), finished in limestone and cast aluminum.[120]

## 1935

**Hayden Planetarium,** Trowbridge and Livingston architects, Weiskopf & Pickworth engineers, standing. Early concrete shell dome, 3 inches thick, supporting projection screen.[121]

## 1937

**United States Post Office,** Church and Canal Streets, standing. Facade of glazed terra cotta, fairly late.[122]

## 1948

**Walden Terrace Apartments,** 98th to 99th Streets, 63rd Drive to 64th Road, Rego Park, Leo Stillman architect, standing. Early exposed structural concrete.[123]

## 1950

**Secretariat Building of United Nations,** near 42nd Street and First Avenue, International Committee and Wallace Harrison architects, standing. First tall, glass curtain wall in New York.[124]

**Manhattan House Apartments,** 200 East 66th Street, Skidmore, Owings & Merrill architects, standing. Early building with entire facade of glazed brick.[125]

## 1952

**Lever House,** 390 Park Avenue, Skidmore, Owings & Merrill architects, Weiskopf & Pickworth engineers, landmarked. Second glass curtain wall in New York. Individual steel rails and mullions mounted to structural frame, steel protected by stainless-steel covers.[126]

## 1953

**99 Park Avenue,** standing. Twenty-six-story glass facade with aluminum mullions and spandrel panels.[127]

**460 Park Avenue,** Emery Roth & Sons architects, standing. Twenty-two-story glass facade with aluminum mullions and spandrel panels.[128]

## 1954

**112 West 34th Street,** Rudolph Boehler & Rene Brugnoni architects, standing. Aluminum mullion and panel facade with no form of backup.[129]

**Manufacturers Hanover Trust,** 510 Fifth Avenue, Skidmore, Owings & Merrill architects, Weiskopf & Pickworth engineers, standing. Early glass curtain wall on four-story bank, with windows up to 22 feet high by 10 feet wide.[130]

## 1955

**Mobil Building,** 150 East 42nd Street, Harrison and Abramovitz architects, Edwards & Hjorth engineers, standing. Entire 400,000-square-foot facade is stainless steel paneling with 4-inch block backup.[131]

## 1956

**425 Park Avenue,** Kahn & Jacobs architects, Charles Meyer engineer, standing. Height 375 feet, "one of the tallest to be built to date with bolted connections," 150,000 field bolts up to 1⅛ inches diameter x 7-inch grip in size; 200,000 shop rivets. Early use of two-man bolt crews.[132]

**The New York Stock Exchange Annex,** Wall Street, Kahn & Jacobs architects, Charles Meyer engineer, standing. At twenty-two stories high, largest welded frame in New York when built.[133]

## 1957

**Tishman Building,** 666 Fifth Avenue, Carson Lundin & Shaw architects, standing. At thirty-eight stories, tallest all-bolted steel frame in the world when built. All-aluminum curtain wall.[134]

**Seagram Building,** 375 Park Avenue, Ludwig Mies van der Rohe, Philip Johnson, and Kahn & Jacobs architects, Severud-Elstad-Kreuger engineers, landmarked. At thirty-eight stories and 520 feet high, tallest building using high-strength bolts when built. Shop connections riveted; unfinished bolts used for beam-to-girder connections.[135]

## 1958

**[Former] Union Carbide Building,** 270 Park Avenue, standing. At fifty-two stories and more than 700 feet high, tallest bolted frame when built.[136]

## 1959

**Kips Bay Plaza,** 30th Street to 33rd Street, First Avenue to Second Avenue, I. M. Pei & Partners and S. J. Kessler architects, August Komendant engineer, standing. Early exposed-concrete apartment houses, using load-bearing exterior walls of Vierendeel truss type.[137]

## 1960

**[Former] Pepsi Building,** 500 Park Avenue, Skidmore, Owings & Merrill architects, standing. Early glass-and-aluminum curtain wall on ten-story office building.[138]

**Western Electric Building,** Fulton Street and Broadway, Purdy & Henderson engineers, standing. At thirty-one stories, tallest steel frame with welded connections in the eastern half of the country when built.[139]

## 1961

**Chase Manhattan Building,** Cedar Street and Nassau Street, Skidmore, Owings & Merrill architects, Weiskopf & Pickworth engineers, standing. First glass curtain wall building over 800 feet high, sixty stories, largest building using solely interior bracing, steel rails and mullions mounted to structural frame.[140]

## 1963

**[Former] Pan Am Building,** 200 Park Avenue, Emery Roth & Sons, Pietro Belluschi, and Walter Gropius architects, standing. Early precast-concrete curtain wall.[141]

## 1964

**[Former] National Maritime Union Building,** now St. Vincent's Hospital, 36 Seventh Avenue, Albert Ledner & Associates architects, standing. Early use of exposed, cast-in-place, reinforced concrete, had to be covered with tile due to spalling.[142]

**New York State Pavilion,** Flushing Meadows Park, Queens, Philip Johnson and Richard Foster architects, Lev Zetlin engineer, standing empty. Early use of slip-forming to create freestanding concrete columns; roof is a bicycle-wheel cable truss.[143]

## 1965

**CBS Building,** 51 West 52nd Street, Eero Saarinen architect, standing. Early concrete tube and core structure, thirty-nine stories and 491 feet high.[144]

## 1966

**Silver Towers,** 100 & 110 Bleecker Street and 505 La-Guardia Place, I. M. Pei & Partners architects, standing. Early use of exposed, cast-in-place, reinforced concrete.[145]

**American Bible Society building,** 1865 Broadway, Skidmore, Owings & Merrill architects, Weiskopf & Pickworth engineers, standing. Exposed precast concrete.[146]

## 1967

**Ford Foundation Building,** 320 East 43rd Street, Kevin Roche John Dinkeloo Associates architects, standing. Corten steel curtain wall.[147]

## 1968

**Madison Square Garden,** Seventh Avenue and 33rd Street, Charles Luckman Associates architects, Severud Associates engineers, standing. 425-foot-diameter bicycle-wheel cable truss roof.[148]

## 1975

**Martin Luther King, Jr. High School,** 122 Amsterdam Avenue, Frost Associates architects, standing. Curtain wall of Mayari-R steel.[149]

## 1976

**World Trade Center,** near Church and Fulton streets, Minoru Yamasaki and Emery Roth architects, Worthington, Skilling, Helle and Jackson engineers, standing. Early structural tube wind frame, of Vierendeel truss type, early pressure-equalized curtain wall.[150]

**Annenberg Building,** Mount Sinai Hospital, Fifth Avenue and 99th Street, Skidmore, Owings & Merrill architects, standing. 436-foot-high Cor-ten steel facade.[151]

## 1977

**Citicorp Center,** Lexington Avenue at 53rd Street, Hugh Stubbins & Associates architects, William LeMessurier Associates engineers, standing. 900-foot-high facade of aluminum.[152]

**APPENDIX C**

# DEFINITION OF TERMS USED IN BUILDING STRUCTURE

Definitions follow their original use as closely as possible without contradictions. The terms defined are specifically those whose meanings have changed, were never properly defined, or are now taken for granted in different ways. Definitions conform to structural usage and may differ from the architectural meaning.

**American Standard.** A series of beam sections conforming to a specific standard meant to regularize the structural steel for sale by different foundries. The distinguishing characteristic is a 16⅔-percent slope to the inside faces of the flanges. Every American Standard beam can be called an I-beam, although there are "I's" that are not American Standard. The term technically also applies to the regularized sizes for channels with a 16⅔ percent flange slope.

**arch.** In building construction, most commonly a floor arch. Still in use as slang referring to any span of concrete floor slab.

**architectural terra cotta.** *See* "terra cotta."

**ashlar.** Stone masonry in blocks, as opposed to stone veneer or unit masonry.

**bar joist.** Same as an open-web joist, specifically one with round or square bars used for web members.

**bay.** One column-line-to-column-line space. In normal usage, in a rectangular building with evenly spaced columns, the rectangular area defined by four adjacent columns, or the rectangular area the width of the building between two adjacent column lines.

**beam.** An iron, steel, wood, or concrete member supporting only floor slabs or floor arches, typically connected to girders at either end.

**bearing wall.** A wall designed to carry its own weight and the weight of any other structure framed into it to a foundation or transfer girder. Made of masonry, wood stud, metal stud, or concrete. When used as the major load-carrying vertical structure in a building, it usually also serves as a shear wall.

**Bethlehem column.** Specifically, columns rolled first by the Bethlehem Steel Company with wide flanges and high weak-axis moments of inertia. In general, until wide-flange shapes became common, any steel column where the flange width was approximately the same as the shape depth. Rolling these shapes required four finishing rollers instead of two, and thus required more complex mills than American Standard shapes, channels, or angles.

**bolt.** A fastener consisting of the bolt proper, a partially threaded rod with an enlarged head at one end, and the nut, a square or hexagonal prism with a threaded circular hole in its center.

**braced frame.** The form of wind system composed of regularly located columns and girders connected by separate attached diagonal bracing members.

**brick.** "Brick are now defined as a small building unit, solid or cored not in excess of 25 percent, commonly in the form of a rectangular prism, formed from inorganic, non-metallic substances and hardened in its finished shape by heat or chemical action. The term brick, when used without a qualifying adjective, is understood to mean such a unit or a collection of such units made from clay or shale hardened by heat."[1]

**cage construction.** "In contradistinction [to the skeleton frame], the 'cage' construction is a frame work of iron or steel columns and girders which carry the floors only, and do not carry the outer walls. . . . In the cage construction the outer walls are independent walls, from the foundation to the extreme top, sustaining themselves and themselves only, and, therefore, the walls are made less in thickness than if they had to bear the floors as in ordinary buildings such walls would have to do."[2]

**cast iron.** Iron containing enough carbon to prevent it from being malleable at any temperature.

**catenary.** Specifically, the curve formed by a flexible rope or cable when suspended at its ends and loaded only by its own weight. Catenary action is the use of suspended structural elements of low flexural strength to carry vertical loads. The most familiar example is a suspension bridge, where the vertical loads on the roadway are carried to the main cables by the vertical suspender cables. The main cables carry these loads as tension to the cable anchorages.

**cement.** Material that can be molded, will adhere to surrounding materials, and hardens to a structural-range compressive strength. Natural materials first used for these properties were common lime (calcium oxide) and cement rock, which was a clayey limestone.[3]

**channel.** A wrought-iron or steel rolled section consisting of web with relatively thick, short flanges projecting on one side only.

**cinder concrete.** Concrete containing a high proportion of cinders, usually ashes from high-temperature coal boilers, as aggregate. Relatively low weight and low strength.[4]

**column.** A vertical load-carrying member in a steel or concrete frame. Depending on the form of the building's wind system, may or may not be subject to bending from lateral loads.

**combination arch.** An end-construction tile arch with side-construction-type center block.

**composite deck.** Metal deck that serves as support for the wet concrete and reinforcing for the permanent slab.

**composite design.** Structural design that combines two materials of different material properties. While wood and steel flitch plates and reinforced-concrete beams both meet this definition, the phrase is most often used to refer to a steel beam that is fastened to a concrete slab so as to allow the slab to act as an extended top flange for the beam.

**concrete.** The most basic definition that can cover all of the various materials that have used this name is a composite material made of hard, inert aggregate mixed with some form of cement to set up into a stable, relatively solid mass. Ordinarily the aggregate consists of coarse pieces up to ¾ inch across and fine pieces such as sand. The most common cements have been Rosendale and Portland, but such variations as plaster of Paris have been used.

**concrete block.** Unit masonry made of stone or lightweight aggregate concrete. Usually contains more than 50 percent voids oriented vertically.

**concrete tile.** Archaic name for concrete block.

**connection.** The joint between two or more members in a frame. Girder-to-column connections constitute the basis of the wind frame in an unbraced building.

**coupon.** Small sample of steel removed from either a member in an existing structure or a mill shipment of steel to be tested for mechanical and chemical properties.

**coupon test.** Metallurgical and mechanical property tests to determine structural properties of a metal sample.

**cross-bracing.** Frame bracing consisting of bracing members connecting diagonally opposed corners of the braced structural bays.

**curtain wall.** A nonbearing wall that encloses the perimeter of a building. Made of masonry, precast concrete panels, glass, or sheet metal. Sometimes used in cage construction to indicate the self-bearing walls at the perimeter.

**diagonal bracing.** Cross-bracing, K-bracing, or knee-bracing.

**doubly symmetric.** Members that are symmetric about both axes perpendicular to their length. Examples are rectangular tubes and "I's."

**end construction.** Terra-cotta floor arches where the voids in the tiles (and therefore the webs in the tiles as well) are parallel to the arch span (perpendicular to the floor beams).

**expanded metal.** "Sheet steel is slit on a special machine and then pulled out or expanded so as to form a diamond mesh. . . . One of the oldest forms of sheet reinforcing." Invented by John T. Golding, sizes were commonly given in inch spacing and gauges, but when it is used as concrete reinforcing, it should be compared on the basis of square inches of steel per linear foot.[5]

**falsework.** Temporary support provided during construction, most often used with timber supports for masonry arches.

**fill.** (1) Mass, nonstructural topping placed over a structural floor—e.g., lightweight concrete, typically cinder concrete placed over terra-cotta floor arches. (2) Earth placed to raise the natural grade or fill in an excavated hole. The earth used is typically inorganic soil compacted to near its maximum density.

**fire-cut.** A joist end to be embedded in a masonry wall cut so that it has full bearing on its bottom surface but slopes back so that its top does not extend into the wall. If such a joist burns through, the end can rotate freely to fall from the wall without damaging the masonry.

**fire line.** The legal limit separating the portions of the city where fireproof construction was required from those areas where it was not.

**fireproof construction.** As commonly used, building construction where the exterior walls, partitions, and floors are of noncombustible materials and those materials, such as steel, that lose strength at high temperatures are protected from fire with some kind of insulation.

**fireproofing.** Any noncombustible material applied to metal beams, girders, or columns specifically to insulate against the heat of fire.

**fire-resistant construction.** A form of construction that has the ability to withstand a fire within and still carry load. Technically, the correct term instead of "fireproof construction," since no material is truly fireproof.

**fish plate.** A small iron or steel plate used to attach a tie rod or hanger to a masonry wall. The plate sits on the opposite side of the wall as the rod or hanger shaft, and provides an anchor for tension by spreading the load over an area of masonry much larger than the rod itself.

**flange.** A projecting portion of a beam section, perpendicular to the web. In ordinary use, the beam will be arranged so that the flanges are perpendicular to the direction of the applied loads (in a floor beam, horizontal at the top and bottom of the web) and so are in position to carry the bulk of the bending stresses.

**flange tile.** *See* "soffit block."

**flat arch.** Most commonly, terra-cotta floor arches of blocks shaped so that the top and bottom surfaces of the arch are parallel to the final floor surface. The bottom surface of the arch usually forms the base for the ceiling plaster. Also used for decorative elements, such as architectural window heads in exterior walls.

**flat slab.** Typically, concrete slabs reinforced to carry load to all four edges (two-way slabs) without dropped beams. Sometimes used to indicate one-way concrete slabs.

**flexural slab.** A concrete slab that carries its load by bending, as opposed to a concrete slab reinforced with draped mesh to act as a catenary or tile floors that act as arches.

**flitch plate.** A beam consisting of an iron or steel plate sandwiched between wood. The three pieces are fastened together with through-bolts to allow them to act as a unit when loaded.

**floor arch.** The spanning structure that forms the flat floor surfaces (and, possibly, whose soffit's shape is exposed as the ceiling surface). Under ordinary circumstances, both top and bottom surfaces are covered with finish materials. The span may be a true masonry arch, a segmental brick arch, a segmental terra-cotta arch, or a flat terra-cotta arch. The supports may be iron or steel floor beams, or bearing masonry walls.

**floor plate.** The total thickness of floor construction, including nonstructural items, from the surface finish to the ceiling finish below.

**form deck.** Metal floor deck designed to serve only as a platform during construction and to support the wet concrete. The concrete is supplied with bar or mesh reinforcing to provide permanent support.

**framed connection.** A beam-to-girder or beam-to-column connection consisting of two angles connected to the supported beam's web and oriented vertically, with their outstanding legs connected to the web of a supporting girder, or the web or flange of a supporting column. Also called a double-angle connection.

**framed structure.** A building with a structure composed primarily of some form of beams and columns, as opposed to masonry bearing walls and floor arches.

**galvanizing.** The coating of iron or steel, through either physical means or electro-chemical reaction, with zinc. The zinc physically prevents oxygen from reaching the iron below, stopping rust, and if scratched, protects the uncovered steel through a sacrificial chemical reaction.

**girder.** An iron, steel, wood, or concrete member that carries floor beams, partitions, or walls.

**headed stud.** A steel rod with an enlarged end (similar in shape to a mushroom) used as a shear connector in composite beams.

**high building.** Any building taller than was practical with traditional building technology. The old limit varies with the source but is generally considered to be between 80 and 100 feet.

**high-strength bolt.** Bolts, specifically of ASTM A325 or A490, which can be used in ordinary connections, but were specifically developed for use in pre-tensioned, slip-critical connections meant to replace riveted connections.

**hung lintel.** A lintel attached to the building main structure, usually to permit the support of extensive amounts of masonry. In a curtain-wall building, the method of supporting the curtain wall (regardless of the wall material) if the wall is outboard of the spandrel beams and the edge of the floor slabs.

**I-beam.** A beam in the shape of a capital "I." Since 1890, almost always used in connection with American Standard shapes.

**jack arch.** A flat masonry arch, specifically flat terra-cotta floor arches. One explanation for the origin of this term is the resemblance of the center "keystone" arch block to the crown on the Jack in a deck of playing cards.

**joist.** A small beam of wood, light-gauge steel, open-web steel trusses, concrete, or small rolled steel sections, used as one of a series of small beams at close spacing.

**K-bracing.** Frame bracing consisting of pairs of diagonal members within each braced bay that connect two of the bay's corners on one side with the center of the opposite side. True K-bracing exists when the upper and lower corners on one side of the bay are connected to the center of the column at the opposite side. The most common type connects the two upper corners of a bay with the center of the girder at the bottom.

**kip.** "The term kip (abbreviated from kilo-pound) is extensively used in technical literature to designate one thousand pounds and is used here as being terse and convenient."[6]

**knee-bracing.** Frame bracing consisting of diagonal members within each braced bay connecting columns and beams near their ends.

**ksi.** Kips per square inch; 1000 pounds per square inch.

**leg.** One-half of a rolled angle. Not used for projecting portions of other rolled shapes.

**light-gauge steel.** Steel members of thin, cold-rolled steel sheet shaped into channels and angles. Used for structural purposes similarly to wood studs and joists in low-rise construction. Also used for framing nonstructural partitions.

**lightweight concrete.** Concrete made with aggregate lighter in weight than crushed stone. The most common types are cinder concrete and lightweight concrete made with expanded shale.

**lintel.** A nonmasonry beam embedded within a wall or partition to support the masonry over a door or window. In older construction, often wood or cast iron, now usually steel angles. An alternative, in bearing-wall buildings, to the construction of arches or flat arches over openings.

**lipped skew.** Skewback tile in floor arches that has a notch which fits around the beam lower flange. The upper lip provides support for the arch on the upper surface of the beam lower flange, and the lower lip provides fire protection for the beam soffit.

**masonry.** Cut stone, artificial stone, concrete block, terra-cotta block, gypsum block, or brick used for the construction of walls and other building elements.

**member.** Any individual, linear element of a frame: a beam, a girder, a column, a brace, and so on.

**metal deck.** Light-gauge, corrugated steel sheet used to form floors either by itself as roof deck, or in combination with concrete in composite deck and form deck.

**moment connection.** A rigid or semi-rigid connection between girders and columns.

**moment frame.** A wind system consisting of rigidly attached girders and columns. *See* "wind girder" and "moment connection."

**moment of inertia.** A fixed geometric property of a shape, describing that shape's relation to a given axis. In structures, the moment of inertia is directly related to a shape's stiffness in bending about an axis.

**natural cement.** A cement manufactured from clayey limestone.

**nonbearing wall.** A wall not intended to carry any load other than its own weight.

**noncombustible construction.** Buildings composed of material such as cast iron that do not themselves burn at temperatures encountered in ordinary fires.

**open-web joist.** A steel joist in the form of a light Warren truss, usually composed of two angles as a bottom flange, two angles as a top flange, and either bar stock or light angles as alternating diagonals connecting the two.

**partition.** A vertical enclosure that is incapable of carrying load. Depending on the era of construction, it can be built of terra-cotta tile, gypsum block, wood studs, or metal studs.

**Phoenix column.** An iron column of basically circular or polygonal cross section, composed of four, six, or eight arc sections with flanges projecting radially outward from the section center. The flanges of adjacent sections are bolted together to form the complete shape. Originally patented by the Phoenix Iron Works, later produced by other companies as well.

**pier.** (1) An isolated, vertical, load-carrying member in a masonry building. (2) A vertical masonry enclosure built around a column for fire- or waterproofing.

**portal bracing.** Frame bracing consisting of stiffened connections between columns and beams. Similar in meaning to knee-bracing.

**portal frame.** (1) A moment frame with portal bracing. (2) A frame analyzed using the portal method of approximate numeric analysis.

**Portland cement.** Artificially mixed cement, composed of primarily lime, alumina, and silica. The proportions changed over the years as the effect of the various components was studied. Named after its resemblance when hardened to English Portland stone.[7]

**quality control.** The process through which shop fabrication of building elements or field assembly and construction is checked against the design intent and code requirements. The provision in the New York City Building Code for Controlled Inspection is a quality-control requirement, forcing inspection and testing of stressed elements.

**radius of gyration.** A geometric property of any shape about a specific axis, consisting of the square root of the ratio of the moment of inertia about the axis over the shape area.

**rebar.** Reinforcing rods for concrete similar in form to modern reinforcing: round (or, in older forms, occasionally square) steel rods with deformed surfaces.

**reinforcing.** In concrete, any steel used to provide tensile strength, including modern rebar, wire mesh, and older patented systems.

**reinforced concrete.** Concrete with steel embedded within it before hardening. The steel is needed primarily to carry tensile stresses in flexural members and to provide ductility in columns. Forces are transferred between the steel and the concrete through the chemical adhesion of the cement and the geometric interlocking shapes of the steel surface.

**rivet.** A fastener consisting of a steel or iron rod with two shaped heads that are tight to the steel members being joined on either side. Ordinarily, one head was shop fabricated, the other created through hammering on the red-hot rivet in place. There is some clamping action created by the shrinkage of the rivet during cooling.

**rolled steel.** Ordinary structural steel members, produced by rolling a hot bar of steel to force it into the required cross section.

**Rosendale cement.** Naturally occurring cement, composed of clayey limestone. The proportions could vary widely depending on where the parent rock was mined. Named after the town in Ulster County, New York, that supplied cement to much of the eastern seaboard.[8]

**segmental arch.** Most commonly, terra-cotta arches of blocks shaped so that the top and bottom surfaces of the arch form a circular arc. Fill was provided to create a flat final floor surface. On occasion, a flat ceiling was hung below.

**shear tab.** A steel beam-to-column or beam-to-girder connection consisting of a single vertical plate welded to the supporting member and bolted to the web of the supported beam.

**shear wall.** A wall designed to resist lateral forces applied to a building. The name comes from the dominance of shear stresses in the relatively deep walls when compared to flexural stresses.

**shoe tile.** A terra-cotta block designed to protect only the bottom flange of a downward projecting girder. In section, a square with a notch in one side, which in use fits around the girder flange below the main arch soffit. Also called a "shoe block."

**side construction.** Terra-cotta floor arches where the voids in the tiles (and therefore the webs in the tiles as well) are perpendicular to the arch span (parallel to the floor beams).

**sidewalk vault.** An extension of the cellar space of a building that extends past the building line underneath the adjacent sidewalk or street. The sidewalk or street above is supported by a portion of the building frame.

**skeleton construction.** "What is understood by 'skeleton construction' is a frame work of iron or steel columns and girders which carry the weight of the outer inclosing brick walls, together with the floors, down to the foundations at initial points. . . . In the skeleton construction the outer walls are in panels, each panel extending horizontally from column to column, and vertically from girder to girder, acting as curtain walls, sustaining nothing, and being carried, each panel, on a girder."[9]

**skewback blocks.** In terra-cotta tile flat arches, the end blocks that rest on the bottom flanges of the support beams. These blocks usually had a notched rectangle cross section, to allow the bulk of the block and the rest of the arch to sit with their soffits approximately 2 inches below the beam flange underside. Some had a notched side inside which the flange sat. They provided support for soffit blocks, where used. Also called "skewback tiles."

**skyscraper.** A tall building that is notably high and notably slender for the time under consideration.

**slenderness.** In buildings, the height divided by the smallest base dimension. In columns, the unbraced length divided by the smallest radius of gyration. An archaic usage for columns was the unbraced length divided by the smallest lateral dimension.

**soffit blocks.** Long, narrow, terra-cotta tiles that cover the bottom flange of the supporting beams. Supported by the skewback blocks on either side.

**spandrel.** At the exterior edge of a building (e.g., spandrel beam, spandrel column, etc.).

**spiral.** Length of steel rod bent into a helix, used as a composite-beam shear connector.

**steel.** An alloy of iron, containing relatively low amounts of carbon and, depending on the particular alloy, small amounts of silicon, copper, chromium, manganese, molybdenum, nickel, and vanadium. Sulphur and phosphorus are limited to extremely small amounts, as they considerably degrade the metal's properties.

**stem.** The web of a structural tee.

**stone concrete.** Concrete made with gravel or crushed stone coarse aggregate; normal-weight concrete.

**strike plate.** *See* "shear tab."

**structural clay tile.** Hollow burned clay masonry units with parallel internal cells. *See* "terra cotta."

**structural tee.** (1) Before World War II, one of a series of structural shapes rolled in "T" section, usually with equal stem and flange thicknesses. (2) A "T" section created by splitting an "I" beam or wide-flange longitudinally at its midheight.

**stud.** (1) Small repetitive compression member used in constructing bearing walls and partitions, either wood or light-gauge steel. (2) Shear connector type for composite beams. *See* "headed stud."

**substructure.** The portion of the building below grade; the foundation.

**superstructure.** The portion of the building above grade.

**tall building.** Usually, a high building that is relatively slender.

**terra cotta.** Literally, in Italian, baked clay. A form of hollow block more properly designated by its use: architectural terra cotta for curtain walls, terra-cotta lumber for interior finish work, and structural clay tile for floor arches. Architectural terra cotta is usually glazed.

**terra-cotta lumber.** Light, porous terra-cotta blocks used for partitions and furring. In general, softer and weaker than structural terra cotta, often containing up to 50 percent sawdust. Called lumber because it could be nailed and worked with ordinary carpentry tools, and thus provided a noncombustible substitute for wood.

**tier building.** Specifically, a building with a frame arranged in levels; generally, a skeleton-frame building.

**tile.** A generic term usually referring to terra-cotta construction materials, specifically referring to the form of terra-cotta units, which are almost always hollow blocks with thin walls.

**tile arch.** Floor arches made of terra cotta. Usually flat, but occasionally segmental, either end or side bearing in action.

**tube frame.** A wind frame consisting of a rigid exterior tube composed of the spandrel members. For buildings without setbacks, this frame is more efficient in resisting lateral loads than frames with interior bracing. The tube can be a moment frame or braced.

**unbraced frame.** A moment frame.

**unit masonry.** Brick or block.

**vault.** (1) An extension of the cellar space of a building beyond the building line. The roof of the vault is a framed slab that can be exposed or buried. *See* "sidewalk vault." (2) A floor arch supported on masonry bearing walls.

**veneer wall.** An archaic name for a masonry curtain wall.

**waffle slab.** A concrete slab with joists running parallel to at least two different axes.

**wall.** A vertical enclosure more substantial than a partition. Depending on its construction, may be load bearing, self-supporting, or supported.

**web.** The central portion of a beam section, to which flanges are attached. In ordinary use, the beam will be arranged so that the web is parallel to the direction of the applied loads (in a floor beam, vertical) and so in position to carry the bulk of the shearing stresses.

**weld.** Joining wrought iron, steel, or aluminum by heating the metal until molten. Usually, the heating is performed by electric current, and the electrode passed over the weld is a matching metal that joins with the metal of the individual pieces joined.

**wide flange.** Technically, an "I" section rolled shape whose flange is relatively wider than that of an American Standard shape. Almost always used in connection with shapes with parallel flange surfaces. Used now for shapes that were previously not considered to be wide flanges because of their relative lightness (e.g., a W12x14, which is 12 inches deep and has a 4-inch flange width, would have been classified as a junior beam in the 1940s, but is now called a wide flange).

**wind bracing.** In a frame building, members or connections specifically designed to prevent lateral movement of the main frame members.

**wind girder.** A girder that serves as part of the wind system of the building.

**wind system.** The portion of a building either designed to resist lateral forces or resisting those forces by default. Common forms are shear walls, cross- or portal bracing, and moment-resisting girder-to-column connections.

**wire fabric (wire mesh).** Reinforcing for concrete consisting of two layers of parallel wires at right angles to one another. When used for reinforcing slabs on grade or mass concrete, the wires in the different directions are usually the same size and spacing. When used in framed one-way slabs, the wires in the bearing direction are typically larger and spaced closer.

**wrought iron.** Almost pure iron, containing slag fibers, malleable, and not capable of strengthening through tempering.

**wythe.** One thickness of masonry in use. Common brick is 4 inches thick; a 12-inch brick wall is a "three-wythe" wall.

**x-bracing.** *See* "cross-bracing."

**yard.** In concrete work, slang for a cubic yard of concrete. Used as a measure during placing and transporting operations.

**zee.** A rolled structural shape similar in form to two angles joined leg tip to leg tip. Phased out of use after World War II because as individual shapes, zees were not particularly useful. Their shape made them vulnerable to torsional loading in almost any situation. Most often used as part of combination members serving as lintels, hung lintels, and columns.

**zee column.** A built-up column composed of four zees and a short web plate. The column provided a relatively large moment of inertia about both principal axes for its weight with only two lines of rivets needed for assembly. Limited in capacity by the relatively small size of rolled zees.

# MATERIAL PROPERTIES FOR CAST IRON, WROUGHT IRON, AND STEEL*

| Material | Year | Property |
|---|---|---|
| cast iron[1] | 1840s | tensile mean strength = 16.5 ksi<br>crushing compressive strength = 80 to 130 ksi |
| cast iron[2] | 1854 | $F_t$ = 12 ksi |
| wrought iron[3] | 1854 | $F_t$ = 27 ksi; $E_{WI}/E_{CI}$ = 2.25 |
| wrought iron[4] | 1873 | $F_b$ = 14 ksi, factor of safety = 3 |
| wrought iron[5] | 1874 | $F_b$ = 12 ksi |
| wrought iron[6] | 1877 | $F_u$ = 36 ksi |
| cast iron[7] | 1877 | $F_u$ = 80 ksi |
| cast iron[8] | 1880 | $F_u$ = 80 ksi<br>$E$ = 15,960 ksi |
| wrought iron[9] | 1880 | $F_u$ = 36 ksi<br>$E$ = 24,000 ksi |
| steel[10] | 1880 | $F_u$ = 114 ksi<br>$E$ = 31,000 ksi |
| wrought iron[11] | 1881–84 | $F_b$ = 10 to 12 ksi |
| wrought iron[12] | 1883 | $E$ = 23,000 ksi<br>$F_u$ = 52 to 58 ksi |
| cast iron[13] | 1883 | $E$ = 12,000 ksi in compression |
| steel[14] | 1884 | $E_{avg}$ = 29,000 ksi |
| wrought iron[15] | 1885–87 | $F_b$ = 12 ksi |
| wrought iron[16] | 1886 | $E$ = 27,000 ksi<br>$F_u$ = 42 ksi |
| cast iron[17] | 1886 | $E$ = 16,000 ksi<br>$F_u$ = 80 ksi |
| mild steel (C = .12%)[18] | 1886 | $E$ = 27,000 ksi<br>$F_u$ = 52.5 ksi |
| hard steel (C = .36%)[19] | 1886 | $E$ = 27,000 ksi<br>$F_u$ = 80 ksi |

*Note: Data sources are the New York City Building Code, the AISC code, the ACI code, or other sources as available.

| Material | Year | Property |
|---|---|---|
| cast iron[20] | 1887 | safety factor for "columns made of the best iron, perfectly molded and with both ends turned" = 6<br>safety factor for "columns [with] imperfections in casting: such as airholes, unequal thickness of metal, etc. . . ., deviation of pressure from axis of columns, and the effect of lateral forces accidentally applied" = 10<br>safety factor = 8 in bending |
| wrought iron[21] | 1887 | $F_b$ = 12 ksi<br>$F_b$ = 10 ksi for riveted plate girders<br>$F_v$ = 9 ksi on beam webs |
| steel[22] | 1887 | $F_b$ = 15.6 ksi |
| steel[23] | 1887 | $F_b$ = 16 ksi |
| cast iron[24] | 1889 | $F_u$ = 15 to 30 ksi in tension |
| wrought iron[25] | 1889 | $F_b$ = 12 ksi |
| cast iron[26] | 1889 | density = 450 pcf<br>$E$ = 12,000 to 23,000 ksi, average 17,500 ksi<br>$F_y$ = 4.5 to 8.0 ksi, average 6.25 ksi |
| wrought iron[27] | 1889 | density = 485 pcf<br>$E$ = 18,000 to 40,000 ksi, average 29,000 ksi<br>$F_y$ = 20 to 40 ksi, average 30 ksi |
| steel[28] | 1889 | density = 490 pcf<br>$E$ = 29,000 to 42,000 ksi, average 35,500 ksi<br>$F_y$ = 34 to 44 ksi, average 39 ksi |
| steel[29] | 1889–93 | $F_b$ = 16 ksi |
| unspecified iron[30] | 1891 | safety factor = 3 in bending<br>safety factor = 6 in compression or tension |
| cast iron[31] | 1892 | safety factor for building use = 6 |
| wrought iron[32] | 1892 | safety factor for building use = 4 |
| iron rivets[33] | 1892 | $F_v$ = 9 ksi |
| steel[34] | 1893–1908 | $F_b$ = 12.5 to 16.0 ksi |
| steel[35] | 1896–1919 | $F_b$ = 16 ksi |
| rivet steel[36] | 1896 | $F_u$ = 48 to 58 ksi, $F_y \geq \frac{1}{2} F_u$ |
| soft steel[37] | 1896 | $F_u$ = 52 to 62 ksi, $F_y \geq \frac{1}{2} F_u$ |
| medium steel[38] | 1896 | $F_u$ = 60 to 70 ksi, $F_y \geq \frac{1}{2} F_u$ |
| ASTM A9 Rivet Steel[39] | 1900 | $F_u$ = 50 to 60 ksi<br>$F_y$ = 30 ksi |
| ASTM A9 Medium Steel[40] | 1900 | $F_u$ = 60 to 70 ksi<br>$F_y$ = 35 ksi |
| steel[41] | 1900–1903 | $F_b$ = 16 ksi |

| Material | Year | Property |
|---|---|---|
| wrought iron[42] | 1901 | $F_u$ = 48 ksi in tension<br>$F_y$ = 24 ksi in tension<br>$F_c$ = 12 ksi<br>$F_t$ = 12 ksi<br>$F_b$ = 12 ksi<br>$F_b$ = 10 ksi on net section of riveted plate girders<br>$F_v$ = 6 ksi |
| steel[43] | 1901 | $F_u$ = 54 to 64 ksi in tension<br>$F_y$ = 32 ksi in tension<br>$F_c$ = 16 ksi<br>$F_t$ = 16 ksi<br>$F_b$ = 16 ksi<br>$F_b$ = 14 ksi on net section of riveted plate girders<br>$F_v$ = 9 ksi |
| steel rivets[44] | 1901 | $F_u$ = 50 to 58 ksi in tension |
| cast iron[45] | 1901 | $F_u$ = 16 ksi in tension<br>$F_c$ = 16 ksi<br>$F_t$ = 3 ksi<br>$F_b$ = 16 ksi in compression<br>$F_b$ = 3 ksi in tension<br>$F_v$ = 3 ksi |
| all metals[46] | 1901 | factors of safety = 4 |
| steel rivets and pins[47] | 1901 | $F_c$ = 20 ksi |
| steel shop rivets and pins[48] | 1901 | $F_v$ = 10 ksi |
| steel field rivets[49] | 1901 | $F_v$ = 8 ksi |
| steel field bolts[50] | 1901 | $F_v$ = 7 ksi |
| steel pins, rivets, and bolts[51] | 1901 | $F_b$ = 20 ksi |
| wrought-iron rivets and pins[52] | 1901 | $F_c$ = 15 ksi |
| wrought-iron pins and shop rivets[53] | 1901 | $F_v$ = 7.5 ksi |
| wrought-iron field rivets[54] | 1901 | $F_v$ = 6 ksi |
| wrought-iron field bolts[55] | 1901 | $F_v$ = 5.5 ksi |
| wrought-iron pins, bolts, and rivets[56] | 1901 | $F_b$ = 15 ksi |
| ASTM A9 Rivet Steel[57] | 1901–8 | $F_u$ = 50 to 60 ksi<br>$F_y = \frac{1}{2} F_u$ |
| ASTM A9 Medium Steel[58] | 1901–8 | $F_u$ = 60 to 70 ksi<br>$F_y = \frac{1}{2} F_u$ |
| cast iron[59] | 1904 | safety factor for steady stress = 6<br>safety factor for varying stress = 15<br>$E$ = 15,000 ksi<br>$F_y$ = 6 ksi in tension<br>$F_u$ = 20 ksi in tension<br>$F_u$ = 90 ksi in compression<br>ultimate shear strength = 20 ksi<br>ultimate flexural strength = 36 ksi |

| Material | Year | Property |
|---|---|---|
| wrought iron[60] | 1904 | safety factor for steady stress = 4<br>safety factor for varying stress = 6<br>$E$ = 25,000 ksi<br>$F_y$ = 25 ksi in tension<br>$F_u$ = 50 ksi<br>ultimate shear strength = 47 ksi<br>ultimate flexural strength = 50 ksi |
| steel[61] | 1904 | safety factor for steady stress = 5<br>safety factor for varying stress = 7<br>$E$ = 30,000 ksi<br>$F_y$ = 50 ksi in tension<br>$F_u$ = 100 ksi in tension<br>$F_u$ = 150 ksi in compression<br>ultimate shear strength = 70 ksi<br>ultimate flexural strength = 120 ksi |
| steel[62] | 1907–11 | $F_b$ = 16 ksi |
| ASTM A9 Rivet Steel[63] | 1909–12 | $F_u$ = 48 to 58 ksi<br>$F_y = \frac{1}{2} F_u$ |
| ASTM A9 Structural Steel[64] | 1909–12 | $F_u$ = 55 to 65 ksi<br>$F_y = \frac{1}{2} F_u$ |
| wrought iron[65] | 1912 | $E$ = 24,000 to 29,000 ksi; average 25,000 ksi<br>$G$ = 10,000 ksi |
| structural steel[66] | 1912 | $E$ = 27,000 to 32,000 ksi; average 30,000 ksi<br>$G$ = 11,000 to 13,000 ksi; average 12,000 ksi |
| nickel steel[67] | 1912 | $E$ = 27,000 to 33,000 ksi; average 30,000 ksi<br>$G$ = 12,000 ksi |
| cast iron[68] | 1912 | $E$ = 13,000 to 22,000 ksi; average 15,000 ksi in tension<br>$E$ = 13,000 to 25,000 ksi; average 15,000 ksi in compression<br>$G$ = 5,500 to 9,000 ksi; average 6,000 ksi |
| steel[69] | 1912 | $F_t$ = 16 ksi on net section<br>$F_c$ = 16 ksi<br>$F_b$ = 16 ksi<br>$F_v$ = 10 ksi<br>$(f_a + f_b) / F_a \leq 1$<br>$(f_a + f_b) / F_b \leq 1$ |
| steel pins[70] | 1912 | $F_b$ = 24 ksi |
| steel shop rivets and pins[71] | 1912 | $F_v$ = 12 ksi<br>$F_p$ = 24 ksi |
| steel field rivets and bolts[72] | 1912 | $F_v$ = 10 ksi<br>$F_p$ = 20 ksi |
| steel expansion rollers[73] | 1912 | $F_p$ = 600 × roller diameter (in inches) in psi |
| ASTM A9 Rivet Steel[74] | 1913 | $F_u$ = 48 to 58 ksi<br>$F_y = \frac{1}{2} F_u$ |

| Material | Year | Property |
|---|---|---|
| ASTM A9 Structural Steel[75] | 1913 | $F_u$ = 55 to 65 ksi<br>$F_y = \frac{1}{2} F_u$ |
| ASTM A9 Rivet Steel[76] | 1914–23 | $F_u$ = 46 to 56 ksi<br>$F_y = \frac{1}{2} F_u$ |
| ASTM A9 Structural Steel[77] | 1914–23 | $F_u$ = 55 to 65 ksi<br>$F_y = \frac{1}{2} F_u$ |
| steel[78] | 1915 | $F_b$ = 16 ksi |
| mild steel[79] | 1917 | $F_y$ = 35 to 40 ksi<br>$F_u$ = 60 to 70 ksi |
| high steel[80] | 1917 | $F_y$ = 50 to 60 ksi<br>$F_u$ = 80 to 100 ksi |
| cast iron[81] | 1920 | modulus of rupture = 40 ksi<br>$E$ = 15,000 ksi<br>$G$ = 6,000 ksi<br>$F_u$ = 80 ksi in compression<br>$F_u$ = 20 ksi in tension<br>ultimate shear stress is determined by tensile failure, $F_u$ |
| wrought iron[82] | 1920 | $E$ = 27,000 ksi<br>$G$ = 10,000 ksi<br>$F_y$ = 30 ksi<br>shear proportional limit = 18 ksi<br>$F_u$ = 50 ksi in tension<br>$F_u$ = 31 ksi in compression<br>ultimate shear stress = 40 ksi |
| medium steel[83] | 1920 | $E$ = 30,000 ksi<br>$G$ = 12,000 ksi<br>$F_y$ = 35 ksi<br>shear proportional limit = 21 ksi<br>$F_u$ = 60 ksi in tension<br>$F_u$ = 40 ksi in compression<br>ultimate shear stress = 48 ksi |
| 3.5% nickel steel[84] | 1920 | $E$ = 30,000 ksi<br>$G$ = 12,000 ksi<br>$F_y$ = 35 ksi<br>shear proportional limit = 25 ksi<br>$F_u$ = 85 ksi<br>$F_u$ = 48 ksi in compression<br>ultimate shear stress = 68 ksi |
| cast iron[85] | 1921 | $F_t$ = 3 ksi<br>$F_v$ = 3 ksi |
| wrought iron[86] | 1921 | $F_t$ = 12 ksi<br>$F_v$ = 7.5 ksi |
| medium steel[87] | 1921 | $F_t$ = 16 ksi<br>$F_v$ = 12 ksi |

| Material | Year | Property |
| --- | --- | --- |
| steel[88] | 1922 | $F_t = 16$ ksi on net section<br>$F_c = 16$ ksi<br>$F_b = 16$ ksi<br>$F_b = 16$ ksi on net section of riveted plate girders<br>$F_v = 9$ ksi<br>$F_v = 10$ ksi on net section of plate girder webs |
| steel pins[89] | 1922 | $F_b = 20$ ksi |
| steel pins and shop rivets[90] | 1922 | $F_p = 24$ ksi<br>$F_v = 12$ ksi |
| steel field rivets[91] | 1922 | $F_p = 16$ ksi<br>$F_v = 8$ ksi |
| steel bolts[92] | 1922 | $F_p = 12$ ksi<br>$F_v = 7$ ksi |
| steel[93] | 1923 | $F_t = 18$ ksi<br>$F_c = 18$ ksi<br>$F_b = 18$ ksi when $l_u < 15b_f$<br>$F_b = 20$ ksi $/ [1 + l_u^2 / (2000b_f^2)]$ when $l_u \geq 15b_f$ $(l_{umax} = 40b_f)$<br>$F_v = 12$ ksi on gross area of beam webs when $h \leq 60t_w$<br>$F_v = 18$ ksi $/ [1 + h^2 / (7200t^2)]$ on gross area of beam webs, when $h > 60t_w$<br>$(f_a + f_b) / F_a \leq 1$;<br>$(f_a + f_b) / F_b \leq 1$<br>$\frac{1}{3}$ increase in allowable stresses when wind load is combined with dead and live loads |
| steel pins[94] | 1923 | $F_b = 27$ ksi |
| steel pins, power-driven rivets, and turned bolts in reamed holes with a clearance of not more than $\frac{1}{50}$ inch[95] | 1923 | $F_v = 13.5$ ksi<br>$F_p = 30$ ksi in double shear<br>$F_p = 24$ ksi in single shear |
| steel hand-driven rivets and unfinished bolts[96] | 1923 | $F_v = 10$ ksi<br>$F_p = 20$ ksi in double shear<br>$F_p = 16$ ksi in single shear |
| steel expansion rollers[97] | 1923 | $F_p = 600 \times$ roller diameter (in inches) in psi |
| ASTM A9 Rivet Steel[98] | 1924–31 | $F_u = 46$ to $56$ ksi<br>$F_y = \frac{1}{2} F_u$; $F_{ymin} = 25$ ksi |
| ASTM A9 Structural Steel[99] | 1924–31 | $F_u = 55$ to $65$ ksi<br>$F_y = \frac{1}{2} F_u$; $F_{ymin} = 30$ ksi |
| all metals[100] | 1926 | when allowable stresses not specified, factor of safety = 4 |
| steel[101] | 1926 | $F_c = 16$ ksi<br>$F_t = 16$ ksi<br>$F_b = 16$ ksi<br>$F_b = 16$ ksi on net section of riveted plate girders<br>$F_v = 10$ ksi |
| cast steel[102] | 1926 | $F_c = 16$ ksi<br>$F_t = 16$ ksi |

| Material | Year | Property |
|---|---|---|
| cast iron[103] | 1926 | $F_c$ = 16 ksi<br>$F_t$ = 8 ksi<br>$F_v$ = 8 ksi<br>$F_b$ = 16 ksi in compression<br>$F_b$ = 8 ksi in tension |
| steel pins[104] | 1926 | $F_p$ = 24 ksi<br>$F_v$ = 12 ksi<br>$F_b$ = 20 ksi |
| steel rivets, power or shop driven[105] | 1926 | $F_p$ = 24 ksi<br>$F_v$ = 12 ksi<br>$F_b$ = 20 ksi |
| steel field rivets, hand driven[106] | 1926 | $F_p$ = 16 ksi<br>$F_v$ = 8 ksi<br>$F_b$ = 20 ksi |
| steel field bolts[107] | 1926 | $F_p$ = 18 ksi<br>$F_v$ = 7 ksi<br>$F_b$ = 20 ksi |
| ASTM A140-32T shapes[108] | 1932 | $F_u$ = 60 to 72 ksi<br>$F_y = \frac{1}{2} F_u$; $F_{ymin}$ = 33 ksi |
| ASTM A140-32T eyebar flats unannealed[109] | 1932 | $F_u$ = 67 to 82 ksi<br>$F_y = \frac{1}{2} F_u$; $F_{ymin}$ = 36 ksi |
| ASTM A141-32T Rivet Steel[110] | 1932 | $F_u$ = 52 to 62 ksi<br>$F_y = \frac{1}{2} F_u$; $F_{ymin}$ = 28 ksi |
| ASTM A9 Steel[111] | 1933 | $F_u$ = 55 to 65 ksi<br>$F_y = \frac{1}{2} F_u$; $F_{ymin}$ = 30 ksi |
| ASTM A9 Steel[112] | 1934–38 | $F_u$ = 60 to 72 ksi<br>$F_y = \frac{1}{2} F_u$; $F_{ymin}$ = 33 ksi |
| ASTM A141 Rivet Steel[113] | 1934–48 | $F_u$ = 52 to 62 ksi<br>$F_y = \frac{1}{2} F_u$; $F_{ymin}$ = 28 ksi |
| steel[114] | 1934 | $F_t$ = 18 ksi on net section<br>$F_c$ = 18 ksi<br>$F_b$ = 18 ksi on net section where lateral deflection is prevented;<br>$F_b$ = 20 ksi / $[1 + l_u^2 / (2000b_f^2)]$, when $l_u > 15b_f$; $l_{umax} = 40b_f$<br>$F_v$ = 12 ksi when $h \leq 60t_w$;<br>$F_v$ = 18 ksi / $[1 + h^2 / (7200t_w^2)]$ when $h > 60t_w$<br>$(f_a + f_b) / F_a \leq 1$;<br>$(f_a + f_b) / F_b \leq 1$<br>$\frac{1}{3}$ increase in allowable stresses when wind load is combined with dead and live loads |
| steel pins[115] | 1934 | $F_b$ = 27 ksi |
| steel rivets[116] | 1934 | $F_t$ = 13.5 ksi |
| steel pins, power-driven rivets, and turned bolts in reamed holes with a clearance of not more than $\frac{1}{50}$ inch[117] | 1934 | $F_v$ = 13.5 ksi<br>$F_p$ = 30 ksi in double shear<br>$F_p$ = 24 ksi in single shear |

| Material | Year | Property |
|---|---|---|
| hand-driven steel rivets and unfinished bolts[118] | 1934 | $F_v = 10$ ksi<br>$F_p = 20$ ksi in double shear<br>$F_p = 16$ ksi in single shear |
| steel expansion rollers[119] | 1934 | $F_p = 600 \times$ roller diameter |
| steel welds[120] | 1934 | $F_v = 11.3$ ksi on effective throat<br>$F_t = 13$ ksi on effective throat<br>$F_c = 15$ ksi on effective throat |
| ASTM A9 Steel[121] | 1937 | $F_t = 20$ ksi on net section<br>$F_c = 20$ ksi on gross section of plate girder stiffeners<br>$F_c = 24$ ksi on rolled section webs (crippling)<br>$F_{bmax} = 20$ ksi<br>$F_b = 22.5$ ksi $/ [1 + l_u{}^2 / (1800 b_f{}^2)]$; $l_u / b_f \leq 40$<br>$F_v = 13$ ksi<br>$F_p = 30$ ksi on contact areas of milled stiffeners and milled surfaces<br>$F_p = 27$ ksi on contact areas of fitted stiffeners<br>$(f_a + f_b) / F_a \leq 1$; $(f_a + f_b) / F_b \leq 1$<br>$\frac{1}{3}$ increase in allowable stresses when wind load is combined with dead and live loads. |
| steel pins[122] | 1937 | $F_b = 30$ ksi<br>$F_p = 32$ ksi |
| ASTM A141 Rivets[123] | 1937 | $F_t = 15$ ksi<br>$(f_v + f_t) / F_t \leq 1$ |
| ASTM A141 Rivets, pins, and turned bolts in drilled or reamed holes[124] | 1937 | $F_v = 15$ ksi<br>$F_p = 40$ ksi in double shear<br>$F_p = 32$ ksi in single shear |
| unfinished steel bolts[125] | 1937 | $F_v = 10$ ksi<br>$F_p = 25$ ksi in double shear<br>$F_p = 20$ ksi in single shear |
| steel expansion rollers[126] | 1937 | $F_p = 600 \times$ roller diameter |
| ASTM A9 beam composite with slab through mesh-reinforced encasement of poured "stone or gravel concrete with Portland Cement" haunch with 2-inch cover on side and bottom and $1\frac{1}{2}$-inch cover on top and mesh on sides and bottom. Steel beam carries construction dead load; steel and concrete jointly carry loads applied later.[127] | 1937 | $F_b = 20$ ksi for steel |
| ASTM A7 Steel[128] | 1939–48 | $F_u = 60$ to $72$ ksi<br>$F_y = \frac{1}{2} F_u$; $F_{ymin} = 33$ ksi |

| Material | Year | Property |
|---|---|---|
| ASTM A7 Steel[129] | 1941 | $F_t$ = 20 ksi on net section<br>$F_c$ = 20 ksi on gross section of plate girder stiffeners<br>$F_c$ = 24 ksi on rolled section webs (crippling)<br>$F_{bmax}$ = 20 ksi<br>$F_b$ = 22.5 ksi / $[1 + l_u^2 / (1800 b_f^2)]$; $l_u / b_f \leq 40$<br>$F_v$ = 13 ksi on gross area<br>$F_p$ = 30 ksi on contact areas of milled stiffeners and milled surfaces<br>$F_p$ = 27 ksi on contact areas of fitted stiffeners<br>$(f_a / F_a) + (f_b / F_b) \leq 1$<br>$\frac{1}{3}$ increase in allowable stresses when wind load is combined with dead and live loads |
| steel pins[130] | 1941 | $F_b$ = 30 ksi<br>$F_p$ = 32 ksi |
| ASTM A141 Rivets[131] | 1941 | $F_t$ = 15 ksi<br>$(f_v + f_t) / F_t \leq 1$ |
| bolts and threaded rods[132] | 1941 | $F_t$ = 12 ksi<br>$F_v$ = 10 ksi<br>$F_p$ = 25 ksi in double shear<br>$F_p$ = 20 ksi in single shear |
| ASTM A141 Rivets, pins, and turned bolts in drilled or reamed holes[133] | 1941 | $F_v$ = 15 ksi<br>$F_p$ = 40 ksi in double shear<br>$F_p$ = 32 ksi in single shear |
| steel expansion rollers[134] | 1941 | $F_p$ = 600 × roller diameter |
| ASTM A7 beam composite with slab through mesh-reinforced encasement of poured "stone or gravel concrete with Portland Cement" haunch with 2-inch cover on side and bottom and $1\frac{1}{2}$ inch cover on top and mesh on sides and bottom. Alternate 1: steel beam carries construction dead load; steel and concrete jointly carry loads applied later. Alternate 2: temporary support provided and steel and concrete split all loads.[135] | 1941 | $F_b$ = 20 ksi for steel |
| all steel and connections[136] | 1941 | forces increased for impact: 100% for elevator supports, 25% for traveling crane supports, 20% for light machinery (shaft or motor driven) supports, 50% for reciprocating machinery |
| cast iron[137] | 1942 | $F_t$ = 3 ksi<br>$F_v$ = 3 ksi<br>$F_b$ = 3 ksi in tension<br>$F_b$ = 16 ksi in compression |

| Material | Year | Property |
|---|---|---|
| steel[138] | 1942 | $F_t = 18$ ksi<br>$F_c = 18$ ksi<br>$F_b = 18$ ksi<br>$F_v = 12$ ksi<br>$\frac{1}{3}$ increase in allowable stresses when wind load is combined with dead and live loads |
| steel rivets[139] | 1942 | $F_t = 13.5$ ksi |
| fully encased steel beams[140] | 1942 | $F_b = 20$ ksi |
| steel pins[141] | 1942 | $F_b = 27$ ksi |
| steel pins, power-driven rivets, and turned bolts in reamed holes with $\frac{1}{50}$-inch clearance[142] | 1942 | $F_v = 13.5$ ksi<br>$F_p = 30$ ksi in double shear<br>$F_p = 24$ ksi in single shear |
| hand-driven rivets and unfinished bolts[143] | 1942 | $F_v = 10$ ksi<br>$F_p = 20$ ksi in double shear<br>$F_p = 16$ ksi in single shear |
| welds[144] | 1942 | $F_v = 11.3$ ksi on effective throat<br>$F_t = 13$ ksi on effective throat<br>$F_c = 15$ ksi on effective throat |
| all steel and connections[145] | 1947 | forces increased for impact: 100% for elevator supports, 25% for traveling crane supports, 20% for light machinery (shaft or motor driven) supports, 50% for reciprocating machinery, 33% for hangers supporting floors and balconies |
| ASTM A7 beam composite with slab through mesh-reinforced encasement of poured "stone or gravel concrete with Portland Cement" haunch with 2-inch cover on side and bottom and $1\frac{1}{2}$-inch cover on top and mesh on sides and bottom. Alternate 1: steel beam carries construction dead load; steel and concrete jointly carry loads applied later. Alternate 2: temporary support provided and steel and concrete split all loads.[146] | 1947 | $F_b = 20$ ksi for steel |
| ASTM A7 steel[147] | 1947 | $(f_a / F_a) + (f_b / F_b) \leq 1$<br>$F_t = 20$ ksi on net section<br>$F_c = 20$ ksi on gross section of plate girder stiffeners<br>$F_c = 24$ ksi on rolled section webs (crippling)<br>$F_b = 20$ ksi in tension<br>$F_b = 20$ ksi in compression when $l_u d / b_f t_f \leq 600$;<br>$F_b = 12000$ ksi / $(l_u d / b_f t_f)$ when $l_u d / b_f t_f > 600$<br>$F_v = 13$ ksi<br>$F_p = 30$ ksi on contact areas of milled stiffeners and other milled surfaces<br>$F_p = 27$ ksi on contact areas of fitted stiffeners<br>$\frac{1}{3}$ increase in allowable stresses when wind load is combined with dead and live loads |

| Material | Year | Property |
|---|---|---|
| steel butt welds made with E6010, E6011, E6012, E6013, E6020, or E6030 electrodes[148] | 1947 | $F_t$ = 20 ksi on effective throat<br>$F_c$ = 20 ksi on effective throat<br>$F_b$ = $F_t$ or $F_c$ as indicated<br>$F_v$ = 13 ksi on effective throat |
| steel fillet welds made with E6010, E6011, E6012, E6013, E6020, or E6030 electrodes[149] | 1947 | $F_v$ = 13.6 ksi on effective throat |
| steel plug or slot welds made with E6010, E6011, E6012, E6013, E6020, or E6030 electrodes[150] | 1947 | $F_v$ = 13.6 ksi on faying surface |
| ASTM A141 Rivets, bolts, and threaded fasteners[151] | 1947 | $F_t$ = 20 ksi on nominal area |
| steel pins[152] | 1947 | $F_b$ = 30 ksi<br>$F_p$ = 32 ksi |
| ASTM A141 Rivets, pins, and turned bolts in reamed or drilled holes[153] | 1947 | $F_v$ = 15 ksi on nominal area<br>$F_p$ = 40 ksi in double shear<br>$F_p$ = 32 ksi in single shear |
| unfinished bolts[154] | 1947 | $F_v$ = 10 ksi<br>$F_p$ = 25 ksi in double shear<br>$F_p$ = 20 ksi in single shear |
| expansion rollers and rockers[155] | 1947 | $F_p$ = 600 × diameter |
| cast steel[156] | 1947 | $F_c$ = $F_c$ for ASTM A7 Steel<br>$F_p$ = $F_p$ for ASTM A7 Steel<br>all other stresses = 75% of values for ASTM A7 Steel |
| ASTM A7 Steel tension members[157] | 1947 | $l_u / r \leq 240$ for main tension members; $l_u / r \leq 300$ for secondary tension members |
| ASTM A7 Steel[158] | 1949 | $F_u$ = 60 to 72 ksi<br>$F_y$ = $\frac{1}{2} F_u$; $F_{ymin}$ = 33 ksi |
| ASTM A141 Rivet Steel[159] | 1949 | $F_u$ = 52 to 62 ksi<br>$F_y$ = $\frac{1}{2} F_u$; $F_{ymin}$ = 28 ksi |
| butt welds[160] | 1950 | $F_c$ = 20 ksi<br>$F_t$ = 20 ksi<br>$F_v$ = 13 ksi |
| fillet welds[161] | 1950 | $F_v$ = 13.6 ksi |

# IRON AND STEEL COLUMN FORMULAS FOR ALLOWABLE STRESS*

| Material | Year | End Condition | Formula |
|---|---|---|---|
| round cast iron[1] | 1848 | 2 round | $P_{ult} = 29.8 \text{ kips} \cdot \dfrac{d^{3.6}}{l^{1.7}}$ ; $l > 15d$ |
| | | 2 flat | $P_{ult} = 88.32 \text{ kips} \cdot \dfrac{d^{3.6}}{l^{1.7}}$ ; $l > 15d$ |
| | | not specified | $P_{ult} = 98.922 \text{ kips} \cdot \dfrac{d^{3.6}}{l^{1.7}}$ ; $l < 15d$ |
| hollow, symmetric cast iron[2] | 1877 | 2 fixed | $F_{ult} = \dfrac{80 \text{ ksi}}{1 + \left(\dfrac{1}{800}\right) \cdot \left(\dfrac{l}{d}\right)^2}$ |
| | | 2 pinned | $F_{ult} = \dfrac{80 \text{ ksi}}{1 + \left(\dfrac{4}{800}\right) \cdot \left(\dfrac{l}{d}\right)^2}$ |
| | | 1 pinned and 1 fixed | $F_{ult} = \dfrac{80 \text{ ksi}}{1 + \dfrac{\left(\dfrac{16}{9}\right)}{800} \cdot \left(\dfrac{l}{d}\right)^2}$ |
| hollow, symmetric wrought iron[3] | 1877 | 2 fixed | $F_{ult} = \dfrac{45 \text{ ksi}}{1 + \left(\dfrac{1}{3000}\right) \cdot \left(\dfrac{l}{d}\right)^2}$ |
| | | 2 pinned | $F_{ult} = \dfrac{45 \text{ ksi}}{1 + \left(\dfrac{4}{3000}\right) \cdot \left(\dfrac{l}{d}\right)^2}$ |
| | | 1 pinned and 1 fixed | $F_{ult} = \dfrac{45 \text{ ksi}}{1 + \dfrac{\left(\dfrac{16}{9}\right)}{3000} \cdot \left(\dfrac{l}{d}\right)^2}$ |

*Note: 1) The end condition was recognized early on as affecting the maximum load that a given column could carry. Nineteenth-century experiments tended to consider three cases: a flat end (cut flat at 90 degrees to the column axis) which was usually considered theoretically to be similar to full fixity, a theoretically pinned end, and a round end with partial fixity. For cast-iron columns, the first case corresponded roughly to the bottom shaft of a column when anchored to a large, stiffened base plate and the third case corresponded to the typical shaft with flat, but loosely bolted splices at its top and bottom. The modern "K" factor has replaced physical end-condition qualifications with restraint qualifications based on effective length. The common case in steel unbraced frames, where K is larger than 1.0, could not apply to cast-iron columns, which were always used in braced frames or cage frames with masonry shear walls.

2) Equations are given in modern form, not in the original presentation.

3) Equations from different sources conflict severely.

| Material | Year | End Condition | Formula |
|---|---|---|---|
| wrought iron other than hollow, symmetric[4] | 1877 | not specified | $F_{ult} = \dfrac{36 \text{ ksi}}{1 + \dfrac{(1.5 \cdot l)^2}{36000 \cdot r^2}}$ |
| solid cylindrical cast iron[5] | 1880 | not specified | $F_a = 9.89 \text{ ksi} \cdot \dfrac{d^{3.6}}{l^{1.7}}$ |
| | | | $F_a = \dfrac{80 \text{ ksi}}{6 \cdot \left[ 1 + \dfrac{1}{266} \cdot \left( \dfrac{l}{d} \right)^2 \right]} ; \; F_a = F_{ult} / 6$ |
| hollow cylindrical cast iron[6] | 1880 | not specified | $F_a = 9.923 \text{ ksi} \cdot \dfrac{d_o^{3.6} - d_i^{3.6}}{l^{1.7}}$ |
| | | | $F_a = \dfrac{80 \text{ ksi}}{6 \cdot \left[ 1 + \dfrac{1}{400} \cdot \left( \dfrac{l}{d} \right)^2 \right]} ; \; F_a = F_{ult} / 6$ |
| rectangular cast iron[7] | 1880 | not specified | $F_a = \dfrac{80 \text{ ksi}}{6 \cdot \left[ 1 + \dfrac{1}{500} \cdot \left( \dfrac{l}{d} \right)^2 \right]} ; \; F_a = F_{ult} / 6$ |
| solid cylindrical wrought iron[8] | 1880 | not specified | $F_a = \dfrac{36 \text{ ksi}}{4 \cdot \left[ 1 + \dfrac{1}{3000} \cdot \left( \dfrac{l}{d} \right)^2 \right]}$ |
| hollow cylindrical wrought iron[9] | 1880 | not specified | $F_a = \dfrac{36 \text{ ksi}}{4 \cdot \left[ 1 + \dfrac{1}{4500} \cdot \left( \dfrac{l}{d} \right)^2 \right]}$ |
| solid rectangular wrought iron[10] | 1880 | not specified | $F_a = \dfrac{36 \text{ ksi}}{4 \cdot \left[ 1 + \dfrac{1}{3000} \cdot \left( \dfrac{l}{d} \right)^2 \right]}$ |
| hollow rectangular wrought iron[11] | 1880 | not specified | $F_a = \dfrac{36 \text{ ksi}}{4 \cdot \left[ 1 + \dfrac{1}{6000} \cdot \left( \dfrac{l}{d} \right)^2 \right]}$ |
| "H" section wrought iron[12] | 1880 | not specified | $F_a = \dfrac{36 \text{ ksi}}{4 \cdot \left[ 1 + \dfrac{1}{3000} \cdot \left( \dfrac{l}{b_f} \right)^2 \cdot \dfrac{A_{tot}}{A_{tot} + A_{web}} \right]}$ |

| Material | Year | End Condition | Formula |
|---|---|---|---|
| wrought iron Phoenix columns[13] | 1881 | not specified | $F_{ult} = \dfrac{36 \text{ ksi}}{\left(1 + \dfrac{l^2}{36000 \cdot r^2}\right)}$ |
| wrought iron[14] | 1886 | 2 flat | $F_{ult} = 42 \text{ ksi} - .128 \text{ ksi} \cdot \left(\dfrac{l}{r}\right); \ l/r < 218.1$ |
| | | | $F_{ult} = \dfrac{666090 \text{ ksi}}{\left(\dfrac{l}{r}\right)^2}; \ l/r > 218.1$ |
| | | 2 pinned | $F_{ult} = 42 \text{ ksi} - .157 \text{ ksi} \cdot \left(\dfrac{l}{r}\right); \ l/r < 178.1$ |
| | | | $F_{ult} = \dfrac{444150 \text{ ksi}}{\left(\dfrac{l}{r}\right)^2}; l/r > 178.1$ |
| | | 2 round | $F_{ult} = 42 \text{ ksi} - .203 \text{ ksi} \cdot \left(\dfrac{l}{r}\right); \ l/r < 138.0$ |
| | | | $F_{ult} = \dfrac{266490 \text{ ksi}}{\left(\dfrac{l}{r}\right)^2}; l/r > 138.0$ |
| cast iron[15] | 1886 | 2 flat | $F_{ult} = 80 \text{ ksi} - .438 \text{ ksi} \cdot \left(\dfrac{l}{r}\right); \ l/r < 121.6$ |
| | | | $F_{ult} = \dfrac{394720 \text{ ksi}}{\left(\dfrac{l}{r}\right)^2}; l/r > 121.6$ |
| | | 2 pinned | $F_{ult} = 80 \text{ ksi} - .537 \text{ ksi} \cdot \left(\dfrac{l}{r}\right); \ l/r < 99.3$ |
| | | | $F_{ult} = \dfrac{263200 \text{ ksi}}{\left(\dfrac{l}{r}\right)^2}; l/r > 99.3$ |
| | | 2 round | $F_{ult} = 80 \text{ ksi} - .693 \text{ ksi} \cdot \left(\dfrac{l}{r}\right); \ l/r < 77.0$ |
| | | | $F_{ult} = \dfrac{157920 \text{ ksi}}{\left(\dfrac{l}{r}\right)^2}; l/r > 77.0$ |

| Material | Year | End Condition | Formula |
|---|---|---|---|
| mild steel (C = .12%)[16] | 1886 | 2 flat | $F_{ult} = 52.5 \text{ ksi} - .179 \text{ ksi} \cdot \left(\dfrac{l}{r}\right);\ l/r < 195.1$ |
| | | | $F_{ult} = \dfrac{666090 \text{ ksi}}{\left(\dfrac{l}{r}\right)^2};\ l/r > 195.1$ |
| | | 2 pinned | $F_{ult} = 52.5 \text{ ksi} - .200 \text{ ksi} \cdot \left(\dfrac{l}{r}\right);\ l/r < 159.3$ |
| | | | $F_{ult} = \dfrac{444150 \text{ ksi}}{\left(\dfrac{l}{r}\right)^2};\ l/r > 159.3$ |
| | | 2 round | $F_{ult} = 52.5 \text{ ksi} - .284 \text{ ksi} \cdot \left(\dfrac{l}{r}\right);\ l/r < 123.3$ |
| | | | $F_{ult} = \dfrac{266490 \text{ ksi}}{\left(\dfrac{l}{r}\right)^2};\ l/r > 123.3$ |
| hard steel (C = .36%)[17] | 1886 | 2 flat | $F_{ult} = 80 \text{ ksi} - .337 \text{ ksi} \cdot \left(\dfrac{l}{r}\right);\ l/r < 158.0$ |
| | | | $F_{ult} = \dfrac{666090 \text{ ksi}}{\left(\dfrac{l}{r}\right)^2};\ l/r > 158.0$ |
| | | 2 pinned | $F_{ult} = 80 \text{ ksi} - .414 \text{ ksi} \cdot \left(\dfrac{l}{r}\right);\ l/r < 129.0$ |
| | | | $F_{ult} = \dfrac{444150 \text{ ksi}}{\left(\dfrac{l}{r}\right)^2};\ l/r > 129.0$ |
| | | 2 round | $F_{ult} = 80 \text{ ksi} - .534 \text{ ksi} \cdot \left(\dfrac{l}{r}\right);\ l/r < 99.9$ |
| | | | $F_{ult} = \dfrac{266490 \text{ ksi}}{\left(\dfrac{l}{r}\right)^2};\ l/r > 99.9$ |
| round cast iron[18] | 1887 | not specified | $F_{ult} = \dfrac{80 \text{ ksi}}{1 + \dfrac{1}{800} \cdot \left(\dfrac{l}{d}\right)^2}$ |

| Material | Year | End Condition | Formula |
|---|---|---|---|
| rectangular cast iron[19] | 1887 | not specified | $F_{ult} = \dfrac{80 \text{ ksi}}{1 + \dfrac{1}{1066} \cdot \left(\dfrac{l}{d}\right)^2}$ |
| wrought-iron Z-bar[20] | 1888 | not specified | $F_{ult} = 46 \text{ ksi} - .125 \text{ ksi} \cdot \left(\dfrac{l}{r}\right); l/r > 90$<br><br>$F_{ult} = 35 \text{ ksi}; l/r < 90$ |
| cast iron[21] | 1889 | 2 flat or 2 fixed | $F_{ult} = \dfrac{80 \text{ ksi}}{1 + \dfrac{1}{a} \cdot \left(\dfrac{l}{r}\right)^2}$ ; where $a = 4500$ to $5000$ |
| | | 2 pinned | $F_{ult} = \dfrac{80 \text{ ksi}}{1 + \dfrac{1}{a} \cdot \left(\dfrac{l}{r}\right)^2}$ ; where $a = 2250$ to $2500$ |
| | | 1 pinned and 1 fixed or flat | $F_{ult} = \dfrac{80 \text{ ksi}}{1 + \dfrac{1}{a} \cdot \left(\dfrac{l}{r}\right)^2}$ ; where $a = 3000$ to $3750$ |
| wrought iron[22] | 1889 | 2 flat or 2 fixed | $F_{ult} = \dfrac{40 \text{ ksi}}{1 + \dfrac{1}{a} \cdot \left(\dfrac{l}{r}\right)^2}$ ; where $a = 36000$ to $40000$ |
| | | 2 pinned | $F_{ult} = \dfrac{40 \text{ ksi}}{1 + \dfrac{1}{a} \cdot \left(\dfrac{l}{r}\right)^2}$ ; where $a = 18000$ to $20000$ |
| | | 1 pinned and 1 fixed or flat | $F_{ult} = \dfrac{40 \text{ ksi}}{1 + \dfrac{1}{a} \cdot \left(\dfrac{l}{r}\right)^2}$ ; where $a = 24000$ to $30000$ |
| mild steel (C = .15%)[23] | 1889 | 2 flat or 2 fixed | $F_{ult} = \dfrac{52 \text{ ksi}}{1 + \dfrac{1}{a} \cdot \left(\dfrac{l}{r}\right)^2}$ ; where $a = 36000$ to $40000$ |
| | | 2 pinned | $F_{ult} = \dfrac{52 \text{ ksi}}{1 + \dfrac{1}{a} \cdot \left(\dfrac{l}{r}\right)^2}$ ; where $a = 18000$ to $20000$ |
| | | 1 pinned and 1 flat or fixed | $F_{ult} = \dfrac{52 \text{ ksi}}{1 + \dfrac{1}{a} \cdot \left(\dfrac{l}{r}\right)^2}$ ; where $a = 24000$ to $30000$ |

| Material | Year | End Condition | Formula |
|---|---|---|---|
| hard steel (C = .36%)[24] | 1889 | 2 flat or 2 fixed | $F_{ult} = \dfrac{83 \text{ ksi}}{1 + \dfrac{1}{a} \cdot \left(\dfrac{l}{r}\right)^2}$ ; where $a$ = 36000 to 40000 |
| | | 2 pinned | $F_{ult} = \dfrac{83 \text{ ksi}}{1 + \dfrac{1}{a} \cdot \left(\dfrac{l}{r}\right)^2}$ ; where $a$ = 18000 to 20000 |
| | | 1 pinned and 1 flat or fixed | $F_{ult} = \dfrac{83 \text{ ksi}}{1 + \dfrac{1}{a} \cdot \left(\dfrac{l}{r}\right)^2}$ ; where $a$ = 24000 to 30000 |
| cast iron[25] | 1893 | not specified | $F_{ult} = \dfrac{80 \text{ ksi}}{1 + \dfrac{1}{800} \cdot \left(\dfrac{l}{r}\right)^2}$ |
| wrought iron[26] | 1893 | not specified | $F_{ult} = \dfrac{40 \text{ ksi}}{1 + \dfrac{1}{3000} \cdot \left(\dfrac{l}{r}\right)^2}$ |
| steel[27] | 1893 | not specified | $F_{ult} = \dfrac{40 \text{ ksi}}{1 + \dfrac{1}{3000} \cdot \left(\dfrac{l}{r}\right)^2}$ |
| steel and wrought iron[28] | 1893 | not specified | $F_a = F_{ult} / 4$ |
| cast iron[29] | 1898 | not specified | $F_{ult} = \dfrac{80 \text{ ksi}}{1 + \dfrac{1}{800} \cdot \left(\dfrac{l}{r}\right)^2}$ |
| wrought iron[30] | 1898 | not specified | $F_{ult} = \dfrac{40 \text{ ksi}}{1 + \dfrac{1}{3000} \cdot \left(\dfrac{l}{r}\right)^2}$ |
| steel[31] | 1898 | not specified | $F_{ult} = \dfrac{48 \text{ ksi}}{1 + \dfrac{1}{3000} \cdot \left(\dfrac{l}{r}\right)^2}$ |
| all metals[32] | 1901 | not applicable | *Live-load reduction* Columns over 5 stories are allowed reduction: full live load used at roof and top floor, 5% reduction additional for each floor load down, to a minimum live load of 50% at lower floors. |

| Material | Year | End Condition | Formula |
|---|---|---|---|
| cast iron[33] | 1901 | not specified | $F_a = 11.3 \text{ ksi} - .030 \text{ ksi} \cdot \dfrac{l}{r}$; $l/r \leq 70$ |
| wrought iron[34] | 1901 | not specified | $F_a = 14 \text{ ksi} - .080 \text{ ksi} \cdot \dfrac{l}{r}$; $l/r \leq 120$ |
| steel[35] | 1901 | not specified | $F_a = 15.2 \text{ ksi} - .058 \text{ ksi} \cdot \dfrac{l}{r}$; $l/r \leq 120$ |
| built up steel Z-bar and channel boxes[36] | 1903 | not specified | $F_a = 12 \text{ ksi}$; $l/r < 90$ <br> $F_a = 17.1 \text{ ksi} - .057 \text{ ksi} \cdot \dfrac{l}{r}$; $l/r > 90$ <br> $F_a = F_{ult}/4$ |
| round cast iron[37] | 1903 | 2 pinned | $F_{ult} = \dfrac{80 \text{ ksi}}{1 + \left(\dfrac{12}{400}\right) \cdot \left(\dfrac{l^2}{d^2}\right)}$ |
| | | 2 flat | $F_{ult} = \dfrac{80 \text{ ksi}}{1 + \left(\dfrac{12}{800}\right) \cdot \left(\dfrac{l^2}{d^2}\right)}$ |
| | | 1 pinned and 1 flat | $F_{ult} = \dfrac{80 \text{ ksi}}{1 + \left(\dfrac{36}{1600}\right) \cdot \left(\dfrac{l^2}{d^2}\right)}$ |
| rectangular cast iron[38] | 1903 | 2 pinned | $F_{ult} = \dfrac{80 \text{ ksi}}{1 + \left(\dfrac{36}{1600}\right) \cdot \left(\dfrac{l^2}{d^2}\right)}$ |
| | | 2 flat | $F_{ult} = \dfrac{80 \text{ ksi}}{1 + \left(\dfrac{36}{3200}\right) \cdot \left(\dfrac{l^2}{d^2}\right)}$ |
| | | 1 pinned and 1 flat | $F_{ult} = \dfrac{80 \text{ ksi}}{1 + \left(\dfrac{108}{6400}\right) \cdot \left(\dfrac{l^2}{d^2}\right)}$ |
| cast iron[39] | 1907 | not specified | $F_a = F_{ult}/8$ |
| cylindrical cast iron[40] | 1907 | 2 pinned | $F_{ult} = \dfrac{80 \text{ ksi}}{1 + \left(\dfrac{12}{400}\right) \cdot \left(\dfrac{l^2}{d^2}\right)}$ |
| | | 2 flat | $F_{ult} = \dfrac{80 \text{ ksi}}{1 + \left(\dfrac{12}{800}\right) \cdot \left(\dfrac{l^2}{d^2}\right)}$ |
| | | 1 flat and 1 pinned | $F_{ult} = \dfrac{80 \text{ ksi}}{1 + \left(\dfrac{36}{1600}\right) \cdot \left(\dfrac{l^2}{d^2}\right)}$ |

| Material | Year | End Condition | Formula |
|---|---|---|---|
| rectangular cast iron[41] | 1907 | 2 pinned | $F_{ult} = \dfrac{80 \text{ ksi}}{1 + \left(\dfrac{36}{1600}\right) \cdot \left(\dfrac{l^2}{d^2}\right)}$ |
| | | 2 flat | $F_{ult} = \dfrac{80 \text{ ksi}}{1 + \left(\dfrac{36}{3200}\right) \cdot \left(\dfrac{l^2}{d^2}\right)}$ |
| | | 1 flat and 1 pinned | $F_{ult} = \dfrac{80 \text{ ksi}}{1 + \left(\dfrac{108}{6400}\right) \cdot \left(\dfrac{l^2}{d^2}\right)}$ |
| cast iron[42] | 1909 | 2 flat | $F_{ult} = 34 \text{ ksi} - .088 \text{ ksi} \cdot \left(\dfrac{l}{r}\right); \, l/r < 120$ |
| wrought iron[43] | 1909 | 2 flat | $F_{ult} = 34 \text{ ksi} - .00043 \text{ ksi} \cdot \left(\dfrac{l}{r}\right)^2; \, l/r < 210$ |
| | | | $F_{ult} = \dfrac{675000 \text{ ksi}}{\left(\dfrac{l}{r}\right)^2}; \, l/r \geq 210$ |
| mild steel[44] | 1909 | 2 flat | $F_{ult} = 42 \text{ ksi} - .00062 \text{ ksi} \cdot \left(\dfrac{l}{r}\right)^2; \, l/r < 190$ |
| | | | $F_{ult} = \dfrac{712000 \text{ ksi}}{\left(\dfrac{l}{r}\right)^2}; \, l/r \geq 190$ |
| | | 2 pinned | $F_{ult} = 42 \text{ ksi} - .00097 \text{ ksi} \cdot \left(\dfrac{l}{r}\right)^2; \, l/r < 150$ |
| | | | $F_{ult} = \dfrac{456000 \text{ ksi}}{\left(\dfrac{l}{r}\right)^2}; \, l/r \geq 150$ |
| steel[45] | 1912 | not specified | $F_a = 19 \text{ ksi} - .100 \text{ ksi} \cdot \left(\dfrac{l}{r}\right); \, l/r \leq 120$ |
| | | | $F_a = 7 \text{ ksi} - .050 \text{ ksi} \cdot \left(\dfrac{l}{r}\right); \, 120 < l/r \leq 200$ |
| | | | $F_{amax} = 13 \text{ ksi}$ |

| Material | Year | End Condition | Formula |
|---|---|---|---|
| steel[46] | 1916 | not specified | $F_a = 16 \text{ ksi} - .070 \text{ ksi} \cdot \left( \dfrac{l}{r} \right)$; $l/r \leq 120$ |
| cast iron[47] | 1916 | not specified | $F_a = 9 \text{ ksi} - .040 \text{ ksi} \cdot \left( \dfrac{l}{r} \right)$; $l/r \leq 70$ |
| all metals[48] | 1916 | not applicable | *Live-load reduction* Columns over 5 stories are allowed reduction: full live load used at roof and top floor, 5% reduction additional for each floor load down, to a minimum live load of 50% at lower floors. |
| steel[49] | 1917 | not specified | $F_a = 16 \text{ ksi} - .070 \text{ ksi} \cdot \left( \dfrac{l}{r} \right)$; $l/r \leq 120$ |
| cast iron[50] | 1917 | not specified | $F_a = 9 \text{ ksi} - .040 \text{ ksi} \cdot \left( \dfrac{l}{r} \right)$; $l/r \leq 70$ |
| cast iron[51] | 1920 | 2 fixed | $F_a = \dfrac{F_c}{1 + \left( \dfrac{1}{5000} \right) \cdot \left( \dfrac{l}{r} \right)^2}$ |
| | | 2 pinned | $F_a = \dfrac{F_c}{1 + \left( \dfrac{4}{5000} \right) \cdot \left( \dfrac{l}{r} \right)^2}$ |
| | | 1 fixed and 1 pinned | $F_a = \dfrac{F_c}{1 + \left( \dfrac{1.95}{5000} \right) \cdot \left( \dfrac{l}{r} \right)^2}$ |
| wrought iron[52] | 1920 | 2 fixed | $F_a = \dfrac{F_c}{1 + \left( \dfrac{1}{36000} \right) \cdot \left( \dfrac{l}{r} \right)^2}$ |
| | | 2 pinned | $F_a = \dfrac{F_c}{1 + \left( \dfrac{4}{36000} \right) \cdot \left( \dfrac{l}{r} \right)^2}$ |
| | | 1 fixed and 1 pinned | $F_a = \dfrac{F_c}{1 + \left( \dfrac{1.95}{36000} \right) \cdot \left( \dfrac{l}{r} \right)^2}$ |

| Material | Year | End Condition | Formula |
|---|---|---|---|
| steel[53] | 1920 | 2 fixed | $$F_a = \frac{F_c}{1 + \left(\dfrac{1}{25000}\right) \cdot \left(\dfrac{l}{r}\right)^2}$$ |
| | | 2 pinned | $$F_a = \frac{F_c}{1 + \left(\dfrac{4}{25000}\right) \cdot \left(\dfrac{l}{r}\right)^2}$$ |
| | | 1 fixed and 1 pinned | $$F_a = \frac{F_c}{1 + \left(\dfrac{1.95}{25000}\right) \cdot \left(\dfrac{l}{r}\right)^2}$$ |
| steel[54] | 1923 | not specified | $$F_a = \frac{18 \text{ ksi}}{1 + \dfrac{1}{18000} \cdot \left(\dfrac{l}{r}\right)^2}$$ |
| steel[55] | 1926 | not specified | $F_a = 16 \text{ ksi} - .070 \text{ ksi} \cdot \left(\dfrac{l}{r}\right); l/r \leq 120$ |
| cast iron[56] | 1926 | not specified | $F_a = 9 \text{ ksi} - .040 \text{ ksi} \cdot \left(\dfrac{l}{r}\right); l/r \leq 70$ |
| all metals[57] | 1926 | not applicable | *Live-load reduction* In buildings over five stories, the second floor from the top can be reduced 5%, 5% additional for each floor below, up to a limit of 50%. |
| steel[58] | 1934 | not specified | $$F_a = \frac{18 \text{ ksi}}{1 + \dfrac{1}{18000} \cdot \left(\dfrac{l}{r}\right)^2}$$ $F_{amax} = 15 \text{ ksi}$ <br> $l/r \leq 120$ for main compression members <br> $l/r \leq 200$ for bracing and secondary members |
| all metals[59] | 1936 | not applicable | *Live-load reduction* is not allowed in buildings designed for storage. Otherwise, 0% reduction at roof, 15% at the top floor, 5% additional each floor but not more than 50%. Girders not at roofs and not supporting columns can have 15% reduction if tributary area is > 200 sf. Girders supporting columns and footings can be sized for the full dead load plus the reduced column live load. |

| Material | Year | End Condition | Formula |
|---|---|---|---|
| steel[60] | 1937 | not specified | $F_a = 17 \text{ ksi} - .000485 \text{ ksi} \cdot \left(\dfrac{l}{r}\right)^2$<br><br>for axially loaded columns with $l/r \leq 120$<br><br>$F_a = \dfrac{18 \text{ ksi}}{1 + \dfrac{1}{18000} \cdot \left(\dfrac{l}{r}\right)^2}$<br><br>for axially loaded columns with $l/r > 120$<br><br>$l/r$ maximum of 120 for main compression members and $l/r$ maximum of 200 for bracing and secondary members |
| all metals[61] | 1939 | not applicable | *Live-load reduction* 15% live-load reduction is allowed on columns, piers, walls and foundations of buildings intended for storage. In other buildings, 0% reduction at roof, 15% at top floor, 5% additional at each successive floor with a maximum of 50%. Girders not at the roof, not supporting columns, and with a tributary area of > 200 sf can have 15% reduction. Girders supporting columns and footings can be designed for full dead load and the reduced column live load. |
| steel[62] | 1941 | not specified | Straight Line formula<br><br>$F_a = 15 \text{ ksi} - .050 \text{ ksi} \cdot \left(\dfrac{l}{r}\right)$, $F_{amax} = 12.5 \text{ ksi}$<br><br>Parabolic formula  $F_a = 15 \text{ ksi} - .0004 \text{ ksi} \cdot \left(\dfrac{l}{r}\right)^2$<br><br>Rankine-Gordon formula  $F_a = \dfrac{18 \text{ ksi}}{1 + \left(\dfrac{1}{18000}\right) \cdot \left(\dfrac{l}{r}\right)^2}$<br><br>Secant formula  $F_a = \dfrac{\dfrac{33 \text{ ksi}}{1.76}}{1 + \dfrac{\sec\left(\dfrac{3}{8} \cdot \dfrac{l}{r} \cdot \sqrt{\dfrac{1.76 f_a}{E}}\right)}{4}}$; $F_a = F_{ult}/1.76$<br><br>Euler formula  $F_a = \dfrac{16E}{2.2 \cdot \left(\dfrac{l}{r}\right)^2}$ |

| Material | Year | End Condition | Formula |
|---|---|---|---|
| steel[63] | 1941 | not specified | $F_a = 17 \text{ ksi} - .000485 \text{ ksi} \cdot \left(\dfrac{l}{r}\right)^2$ ; $l/r \leq 120$. <br><br> $F_a = \dfrac{18 \text{ ksi}}{1 + \dfrac{1}{18000} \cdot \left(\dfrac{l}{r}\right)^2}$ ; $l/r > 120$ <br><br> $l/r$ maximum of 120 for main compression members and $l/r$ maximum of 200 for bracing and secondary members. |
| cast iron[64] | 1942 | not specified | $F_a = 9 \text{ ksi} - .040 \text{ ksi} \cdot \left(\dfrac{l}{r}\right)$; $l/r \leq 70$ |
| steel[65] | 1942 | not specified | $F_a = \dfrac{18 \text{ ksi}}{1 + \dfrac{1}{18000} \cdot \left(\dfrac{l}{r}\right)^2}$; $F_{amax} = 15 \text{ ksi}$; $l/r \leq 120$; <br><br> $l/r \leq 200$ for braces and secondary framing |
| all metals[66] | 1942 | not applicable | *Live-load reduction* Storage-building columns, piers, walls, and foundations can be designed for 85% of live load. Columns in all other buildings: full live load used at roof, 10% reduction additional for each floor load down, to a minimum live load of 50% at lower floors. Girders not in roof and with < 200 sf tributary area, 85% live load. Column transfer trusses or girders and column footings: live load as described above. |
| steel[67] | 1947 | not specified | $F_a = 17 \text{ ksi} - .000485 \text{ ksi} \cdot \left(\dfrac{l}{r}\right)^2$ ; $l/r < 120$ <br><br> $F_a = \left(1.6 - \dfrac{1}{200r}\right) \cdot \left[17 \text{ ksi} - .00485 \text{ ksi} \cdot \left(\dfrac{l}{r}\right)^2\right]$; $l/r > 120$ <br><br> $F_a = \dfrac{18 \text{ ksi}}{1 + \dfrac{1}{18000} \cdot \left(\dfrac{l}{r}\right)^2}$; secondary columns with $120 < l/r < 200$ |
| ASTM A53 pipe lally columns, $1:1\frac{1}{2}:3$ concrete filling, A7 steel connections[68] | 1953 | not specified | $P_a = \left(A_c + 12A_s\right) \cdot \left(1600 - 24 \cdot \dfrac{l}{d}\right)$; $l/d \leq 40$ |

| Material | Year | End Condition | Formula |
|---|---|---|---|
| steel[69] | 1963 | K factor used | $$F_a = \dfrac{\left[1 - \dfrac{\left(\dfrac{K \cdot l}{r}\right)^2}{2C_c}\right] \cdot F_y}{\dfrac{5}{3} + \dfrac{3}{8C_c} \cdot \left(\dfrac{K \cdot l}{r}\right) - \dfrac{1}{8C_c} \cdot \left(\dfrac{K \cdot l}{r}\right)^3};$$ $$Kl/r < C_c = \sqrt{\dfrac{2 \cdot \pi^2 \cdot E}{F_y}}$$ $$F_a = \dfrac{149000 \text{ ksi}}{\left(\dfrac{K \cdot l}{r}\right)^2}; Kl/r > C_c$$ $$F_{as} = \dfrac{F_a}{1.6 - \dfrac{l}{200r}} \text{ for bracing and secondary}$$ members with $l/r > 120$ ($K$ assumed to be 1) |
| steel[70] | 1970 | K factor used | $$F_a = \dfrac{\left[1 - \dfrac{\left(\dfrac{K \cdot l}{r}\right)^2}{2C_c}\right] \cdot F_y}{\dfrac{5}{3} + \dfrac{3}{8C_c} \cdot \left(\dfrac{K \cdot l}{r}\right) - \dfrac{1}{8C_c} \cdot \left(\dfrac{K \cdot l}{r}\right)^3};$$ $$Kl/r > C_c = \sqrt{\dfrac{2 \cdot \pi^2 \cdot E}{F_y}}$$ $$F_a = \dfrac{12 \cdot \pi^2 \cdot E}{23 \cdot \left(\dfrac{K \cdot l}{r}\right)^2}; Kl/r > C_c$$ $$F_{as} = \dfrac{F_a}{1.6 - \dfrac{l}{200r}} \text{ for bracing and secondary}$$ members with $l/r > 120$ ($K$ assumed to be 1) |

| Material | Year | End Condition | Formula |
|---|---|---|---|
| steel[71] | 1980 | $K$ factor used | |

$$F_a = \frac{\left[1 - \dfrac{\left(\dfrac{K \cdot l}{r}\right)^2}{2C_c}\right] \cdot F_y}{\dfrac{5}{3} + \dfrac{3}{8C_c} \cdot \left(\dfrac{K \cdot l}{r}\right) - \dfrac{1}{8C_c} \cdot \left(\dfrac{K \cdot l}{r}\right)^3};$$

$$Kl/r < C_c = \sqrt{\frac{2 \cdot \pi^2 \cdot E}{F_y}}$$

$$F_a = \frac{12 \cdot \pi^2 \cdot E}{23 \cdot \left(\dfrac{K \cdot l}{r}\right)^2} ; Kl/r > C_c$$

$$F_{as} = \frac{F_a}{1.6 - \dfrac{1}{200r}} \quad \text{for bracing and secondary}$$

members with $l/r > 120$ ($K$ assumed to be 1)

# NEW YORK CITY FIRE LINE LOCATION

1849    One hundred feet north of 32nd Street, Hudson River to East River.[1]

1860    "Followed 52nd Street," river to river.[2]

1866    "Followed 86th Street," river to river.[3]
        "Between 86th and 87th streets."[4]

1887    155th Street, river to river.[5]

1891    Manhattan, 155th Street centerline from Hudson to Eleventh Avenue, Eleventh Avenue centerline, and the midpoint of the blocks between 155th and 156th streets to Harlem River; Bronx, 138th Street centerline from Harlem River to Cypress Avenue, Cypress Avenue centerline from 138th Street to East River.[6]

1898    Manhattan, 190th Street.[7]
        Bronx, 149th Street.[8]

# DESIGN LIVE LOADS

| Occupancy | Year | Live Load |
|---|---|---|
| "greatest weight of people"[1] | 1880 | 85 psf |
| "people in crowds"[2] | 1880 | 70 psf |
| minimum for all buildings[3] | 1892 | 75 psf |
| public assembly[4] | 1892 | 120 psf |
| stores, factories, warehouses, manufacturing, and commercial[5] | 1892 | 150 psf and up |
| dwellings, hotels, tenements, apartments[6] | 1892 | 70 psf |
| office use[7] | 1892 | 100 psf |
| roofs[8] | 1892 | 50 psf |
| office buildings[9] | 1893 | 75 psf for upper floors<br>150 psf for first floors and basements |
| dwellings and hotels[10] | 1898 | 70 psf |
| offices[11] | 1898 | 100 psf |
| public buildings[12] | 1898 | 120 psf |
| factories, warehouses, and stores[13] | 1898 | 150 psf and up |
| dwellings, including hotels[14] | 1901 | 60 psf |
| office buildings[15] | 1901 | 75 psf for upper floors<br>150 psf for first floors and basements |
| schools[16] | 1901 | 75 psf |
| stables or carriage houses[17] | 1901 | 75 psf |
| public assembly[18] | 1901 | 90 psf |
| stores, storage, light manufacturing[19] | 1901 | 120 psf |
| heavy stores, factories, warehouses, manufacturing, commercial[20] | 1901 | 150 psf |
| roofs[21] | 1901 | 50 psf with slopes < 20°<br>30 psf with slopes > 20° |
| sidewalks[22] | 1901 | 300 psf |

| Occupancy | Year | Live Load |
|---|---|---|
| wind load (note that combined wind load and gravity load allows a 50% increase in allowable stresses)[23] | 1901 | 0 psf on buildings shorter than 100 feet with height-to-base width ratio < 4<br>30 psf on all other buildings |
| apartments, dwellings, hotels, and lodging houses[24] | 1917 | 40 psf |
| roofs[25] | 1917 | 40 psf with slopes < 20°<br>30 psf with slopes > 20° |
| office buildings[26] | 1917 | 60 psf |
| schools and colleges[27] | 1917 | 75 psf |
| hospitals, jails, municipal buildings, churches, libraries, museums, theaters, auditoriums, public assembly, corridors and hallways, stairs and fire escapes[28] | 1917 | 100 psf |
| light manufacturing exclusive of weight and impact load of machinery[29] | 1917 | 120 psf |
| stores, garages, and stables[30] | 1917 | 120 psf |
| sidewalks[31] | 1917 | 300 psf |
| wind[32] | 1917 | 0 psf on buildings shorter than 100 feet with height-to-base width ratio < 4<br>30 psf on all other buildings |
| wind load[33] | 1922 | 30 psf |
| apartments[34] | 1922 | 60 psf |
| auditoriums[35] | 1922 | 100 psf |
| churches[36] | 1922 | 100 psf |
| dance halls[37] | 1922 | 150 psf |
| drill halls[38] | 1922 | 150 psf |
| factories[39] | 1922 | 120 psf |
| garages[40] | 1922 | 75 psf |
| hotels[41] | 1922 | 60 psf |
| office buildings[42] | 1922 | 150 psf at first floor<br>60 psf above first floor |
| retail stores[43] | 1922 | 120 psf |
| schools[44] | 1922 | 75 psf |
| storehouses[45] | 1922 | 120 psf |
| theaters[46] | 1922 | 100 psf |
| warehouses[47] | 1922 | 150 psf |
| residences[48] | 1926 | 40 psf |
| assembly "or public purpose" except for classrooms[49] | 1926 | 100 psf |

| Occupancy | Year | Live Load |
|---|---|---|
| classrooms[50] | 1926 | 75 psf |
| offices[51] | 1926 | 60 psf |
| any purpose not listed[52] | 1926 | 120 psf |
| roofs[53] | 1926 | 40 psf with slopes < 20° <br> 30 psf with slopes > 20° |
| sidewalks between curbs and building lines[54] | 1926 | 300 psf |
| yards and courts inside the building line[55] | 1926 | 120 psf |
| steel beams in commercial or industrial buildings[56] | 1926 | 4 kip concentrated live load at midspan |
| wind[57] | 1926 | 0 psf on buildings shorter than 150 feet from grade with height-to-base width ratio < 4 <br> 30 psf on all other buildings |
| commercial with incidental factory use taking up less than 25% of floor area[58] | 1936 | 75 psf |
| factories, storage, library stacks[59] | 1936 | 120 psf |
| stables and car-only garages[60] | 1936 | 75 psf |
| all-vehicle garages[61] | 1936 | 175 psf for floor construction <br> 120 psf for beams, girders, and columns <br> 12 kip concentrated load anywhere on beams and girders |
| trucking space within a building[62] | 1936 | 300 psf <br> 24 kip concentrated load anywhere on beams and girders |
| roofs[63] | 1936 | 40 psf with slopes < 20° <br> 30 psf with slopes > 20° |
| sidewalks[64] | 1936 | 300 psf |
| courts and yards[65] | 1936 | 120 psf |
| wind pressure[66] | 1936 | 0 psf on buildings shorter than 100 feet with height-to-base width ratio < $2\frac{1}{2}$ <br> For other buildings: 20 psf on exposed surface from top of building down to 100 feet above grade or upper 50% of height for buildings shorter than 100 feet |
| stables and car-only garages[67] | 1939 | 75 psf |
| car-only garages[68] | 1939 | 2 kip concentrated load located anywhere if the garage can hold two or more cars |
| all-vehicle garages[69] | 1939 | 175 psf <br> 6 kip or actual axle load concentrated load anywhere |
| truck spaces and driveways within a building[70] | 1939 | 175 psf <br> 24 kip or actual axle load concentrated load anywhere |

| Occupancy | Year | Live Load |
|-----------|------|-----------|
| sidewalks[71] | 1939 | 300 psf<br>12 kip or actual axle load concentrated load anywhere |
| roofs[72] | 1939 | 40 psf with slope < 3 / 12<br>30 psf with 3 / 12 < slope < 12 / 12<br>0 snow load, but 20 psf wind load normal to roof with slope > 12 / 12 |
| wind pressure[73] | 1939 | 0 psf on buildings less than 100 feet high with height-to-base width ratio < $2\frac{1}{2}$<br>For other buildings: 20 psf on exposed surface from top of building down to 100 feet above grade or upper 50% of height for buildings shorter than 100 feet |
| residences and sleeping quarters[74] | 1939 | 40 psf |
| offices including corridors[75] | 1939 | 50 psf |
| theaters and assembly halls[76] | 1939 | 75 psf |
| classrooms, churches, reading rooms, etc. with fixed seating[77] | 1939 | 60 psf |
| corridors, halls, lobbies, public places in hotels and public structures, assembly halls without fixed seats, museums, art galleries, grandstands, dance halls, stairways, and other places where crowds of people are likely to assemble[78] | 1939 | 100 psf |
| display and sale of light merchandise[79] | 1939 | 75 psf |
| factories, wholesale stores, storage, and stock rooms in libraries[80] | 1939 | 120 psf |
| partitions not located on drawings[81] | 1942 | 20 psf |
| partitions not located on drawings in residential, non-fireproof buildings[82] | 1942 | 12 psf |
| cinder fill[83] | 1942 | 60 pcf |
| residences[84] | 1942 | 40 psf |
| offices[85] | 1942 | 50 psf minimum |
| classrooms, churches, etc. with fixed seating[86] | 1942 | 60 psf minimum |
| theaters, assembly halls with fixed seating[87] | 1942 | 75 psf minimum |
| assembly spaces, including corridors[88] | 1942 | 100 psf |
| light merchandise sales, incidental factory space (< 25% of factory floor area)[89] | 1942 | 75 psf |
| factory, storage, stacks in libraries[90] | 1942 | 120 psf |
| stables and private passenger-car garages[91] | 1942 | 75 psf |

| Occupancy | Year | Live Load |
|---|---|---|
| other garages[92] | 1942 | 175 psf for floor construction<br>120 psf for beams, columns, and girders<br>6 kip concentrated load anywhere |
| trucking areas within buildings[93] | 1942 | 175 psf<br>12 kip concentrated load or heaviest known concentrated load anywhere |
| sidewalks[94] | 1942 | 300 psf<br>12 kip concentrated load anywhere |
| roofs[95] | 1942 | 40 psf with slope $\leq 3 / 12$<br>30 psf with $3 / 12 <$ slope $\leq 12 / 12$<br>0 psf live load, but 20 psf wind load normal to roof with slope $\leq 12 / 12$ |
| wind load[96] | 1942 | "in general, wind pressure in such structures may be neglected" for buildings shorter than 100 feet<br>20 psf from top to 100 feet above grade for buildings taller than 100 feet<br>20 psf applied to top 50% of height for buildings shorter than 200 feet with height-to-base width ratio $> 2\frac{1}{2}$,<br>and mill buildings, shops, auditoriums, or drill shed roofs<br>30 psf on tank towers, stacks, other similar structures on top of buildings |

## APPENDIX H

# DESIGN DATA FOR FLOOR SYSTEMS

## 1896

Common formulas for floor arches[1]
$H = Ql/8r$ where $H$ = thrust, $l$ = arch span, $r$ = arch rise, $Q$ = distributed load
$H = Pl/4r$ where $P$ = concentrated load at arch center

## 1897

Elastic concrete beam theory[2]
$M_o = (f_s)[(by_2/3e)(E_c/E_s) + ae]$
$y_2 = \sqrt{[(2ae/b)(E_s/E_c)]}$
$f_c = (f_s)[(E_c/E_s)(y_2/e)]$ where $M_o$ = resisting moment, $a$ = reinforcing steel area, $y_2$ = distance to neutral axis from compression face, $e$ = distance to neutral axis from reinforcing steel centroid, $b$ = beam width, $E_c$ = concrete elastic modulus, $f_c$ = concrete stress, $E_s$ = steel elastic modulus

## 1911

New York building code concrete elastic design[3]
Limited to $f_c$ = 650 psi, $f_s$ = 16 ksi, $n$ = 15

## 1913

Joint Committee Report flexural concrete design[4]
$M = wl^2/12$ at center of span and at supports for continuous floor slabs
$M = wl^2/12$ at center of span and at supports for interior spans for continuous beams
$M = wl^2/10$ for end span center and first interior support for continuous beams
$M = wl^2/8$ for beams and slabs continuous over two spans only at the interior support
$M = wl^2/10$ for beams and slabs continuous over two spans only near the span centers
$k = [\sqrt{(2pn + [pn]^2)}] - pn$
$j = 1 - k/3$
$f_s = M/Ajd = M/pjbd^2$
$f_c = 2M/jkbd^2 = 2pf_s/k$

$\rho_{balanced} = 1 / \{(2f_s / f_c)[(f_s / nf_c) + 1]\}$ where $z$ = depth of resultant compression below beam top, $j$ = ratio of lever arm of resisting couple to depth $d$, $jd = d - z$ = arm of resisting couple, $\rho$ = steel ratio, $kd$ = location of neutral axis below beam top
$F_s$ = 16 ksi
$n$ = 15 $f'_c \leq$ 2200 psi
$n$ = 12 when 2900 psi > $f'_c$ > 2200 psi
$n$ = 10 when $f'_c \geq$ 2900 psi
$n$ = 8 for deflection calculations
$f'_c$ = 3300 for granite or trap rock aggregate, 1:1:2 mixture
$f'_c$ = 2800 for granite or trap rock aggregate, 1:1$\frac{1}{2}$:3 mixture
$f'_c$ = 2200 for granite or trap rock aggregate, 1:2:4 mixture
$f'_c$ = 1800 for granite or trap rock aggregate, 1:2$\frac{1}{2}$:5 mixture
$f'_c$ = 1400 for granite or trap rock aggregate, 1:3:6 mixture
$f'_c$ = 3000 for gravel, hard limestone or hard sandstone aggregate, 1:1:2 mixture
$f'_c$ = 2500 for gravel, hard limestone or hard sandstone aggregate, 1:1$\frac{1}{2}$:3 mixture
$f'_c$ = 2000 for gravel, hard limestone or hard sandstone aggregate, 1:2:4 mixture
$f'_c$ = 1600 for gravel, hard limestone or hard sandstone aggregate, 1:2$\frac{1}{2}$:5 mixture
$f'_c$ = 1300 for gravel, hard limestone or hard sandstone aggregate, 1:3:6 mixture
$f'_c$ = 2200 for soft limestone or soft sandstone aggregate, 1:1:2 mixture
$f'_c$ = 1800 for soft limestone or soft sandstone aggregate, 1:1$\frac{1}{2}$:3 mixture
$f'_c$ = 1500 for soft limestone or soft sandstone aggregate, 1:2:4 mixture
$f'_c$ = 1200 for soft limestone or soft sandstone aggregate, 1:2$\frac{1}{2}$:5 mixture
$f'_c$ = 1000 for soft limestone or soft sandstone aggregate, 1:3:6 mixture
$f'_c$ = 800 for cinder aggregate, 1:1:2 mixture
$f'_c$ = 700 for cinder aggregate, 1:1$\frac{1}{2}$:3 mixture
$f'_c$ = 600 for cinder aggregate, 1:2:4 mixture

$f'_c$ = 500 for cinder aggregate, 1:2$\frac{1}{2}$:5 mixture
$f'_c$ = 400 for cinder aggregate, .1:3:6 mixture
$F_v$ = .02$f'_c$ for beams without web reinforcing
$F_v$ = .06$f'_c$ for beams with web reinforcing
$F_v$ = .03$f'_c$ for beams with bent up longitudinal bars
Allowable bond stress, plain bars: .04$f'_c$
Allowable bond stress, cold drawn wire: .02$f'_c$
Floor slabs may be used as part of the beam for T-beam analysis, $b_{eff} \le span / 4$
$(b_{eff} - b_{web})/2 \le 4 \times$ slab thickness

## 1915

Draped-mesh design rules proposed by the New York Building Department[5]
$W_t$ = 3,500,000$da/l^2$ for fully restrained mesh slabs
$W_t$ = 2,500,000$da/l^2$ for slabs with longitudinal bars hooked over beams where $W_t$ = total allowable distributed load, average cinder concrete slab weight = 108 pcf, $f'_c$ = 800 to 1000 psi, $d$ = slab depth, $a$ = reinforcing steel area, $l$ = slab span, 1 inch cover assumed

Elastic concrete design rules proposed by the New York Building Department[6]
$W_t$ = 550,000$d^2a/l^2$ for simple slabs, unrestrained, without anchored reinforcing, where average slab weight = 108 pcf, $f'_c$ = 800 to 1000 psi, 400 for poor material, bending moment coefficient = $\frac{1}{20}$ for continuous slabs, $\frac{1}{8}$ for simply supported slabs, $F_b$ = 300 psi for concrete, $F_b$ = 16 ksi for steel, $n$ = 30 for slabs with continuous mesh, cover = 1 inch

## 1919

Design formulas for tile arches[7]
$p = 3wL^2 / 2R$
$P = 3wL^2L_b / 2R$
$A = 3wL^2L_b / 2fR = wL^2L_b / 10667R$
$L_s = 2faR / 3wL^2 = 10667aR / wL^2$
$M_{1-1} = (12L_b)(\frac{1}{2}wLL_b) / 8 = 3wLL_b^2/4$
$M_{2-2} = (12L_s)(pL_s) / 12 = pL_s^2$ [for a continuous beam supported by tie rods]
$f = M_{1-1}/S_{1-1} + M_{2-2}/S_{2-2}$ where
$w$ = distributed load on arch in psf, $L$ = span of arch in feet, $L_b$ = length of floor beam supporting arch in feet, $R$ = effective rise of arch in inches [vertical distance from highest point of underside to spring line for segmental arch; arch depth − 2.4 inches for flat arches], $p$ = thrust of arch per lineal foot in pounds, $P$ = total thrust of arch per panel in pounds, $A$ = total net area of tie rods per panel in square inches, $a$ = net area of one

tie rod in square inches, $L_s$ = spacing of tie rods, center to center in feet, $f$ = allowable combined steel stress not to exceed 16000 psi, $S_{1-1}$ = section modulus of beam, axis 1-1 [major axis], $S_{2-2}$ = section modulus of beam, axis 2-2, $M_{1-1}$ = bending moment due to vertical loading in inch pounds, $M_{2-2}$ = bending moment due to arch thrust in inch pounds

## 1923

Building code rules for flexural concrete[8]
$F_b$ = .325$f'_c$
$F_b$ = .375$f'_c$ adjacent to supports of continuous or fixed beams or slabs or rigid frames
$F_v$ = .02$f'_c$ on beams without web reinforcing and without special anchorage of longitudinal steel
$F_v$ = .075$f'_c$ on beams with web reinforcing and with special anchorage of longitudinal steel (web steel design for full shear)
Allowable bond stress for plain bars: .04$f'_c$
Allowable bond stress for deformed bars: .05$f'_c$
$F_p$ = .25$f'_c$ on full area
Concrete reinforcing allowable working stresses:
$F_s$ = 16 ksi for structural grade billet steel
$F_s$ = 16 ksi for intermediate grade billet steel
$F_s$ = 16 ksi for hard grade steel
$F_s$ = 20 ksi for cold drawn steel wire

ACI code rules for flexural concrete[9]
$F_b$ = .375$f'_c$
$F_b$ = .41$f'_c$ adjacent to supports of continuous or fixed beams or slabs or rigid frames
$F_v$ = .02$f'_c$ on beams without web reinforcing and without special anchorage of longitudinal steel
$F_v$ = .03$f'_c$ on beams without web reinforcing and with special anchorage of longitudinal steel
$F_v$ = .12$f'_c$ on beams with web reinforcing and with special anchorage of longitudinal steel (web steel design for full shear)
Allowable bond stress for plain bars: .04$f'_c$
Allowable bond stress for deformed bars: .05$f'_c$
$F_p$ = .25$f'_c$ on full area
Concrete reinforcing allowable working stresses:
$F_s$ = 16 ksi for structural grade billet steel
$F_s$ = 16 ksi for intermediate grade billet steel
$F_s$ = 18 ksi for hard grade steel
$F_s$ = 18 ksi for cold drawn steel wire

## 1938

Draped mesh design formulas[10]
$w = 3CA_s / L^2$, where $w$ is the gross uniform load on the floor in pounds, $A_s$ is the area of steel reinforcing in square inches per foot of slab, $L$ is the clear span

between steel beam flanges (but limited to 10 feet ordinarily and to 8 feet if $w$ is greater than 200 psf), and $C$ is a coefficient equal to 20000 for cinder concrete when reinforcing is continuous, 14000 for cinder concrete when reinforcing is hooked or attached to supports, 23000 for stone concrete when reinforcing is continuous, and 15000 for stone concrete when reinforcing is hooked or attached to supports. Concrete is required to have $f_c' \geq 700$ psi and a maximum mixture ratio of 1:2:5, and steel is required to have $F_u \geq 55000$ psi.

The depth of the slab was limited by the following formula:

$t = L/2 + (w - 75)/200$ where $t$ is the thickness of the structural slab in inches (minimum 4 inches), with the exception that 4-inch-thick slabs could be used if $L < 8$ feet and $w < 200\#/ft^2$. The minimum steel sectional area allowed was .054 square inches in a 4-inch slab.

## 1942

Building code rules for flexural concrete[11]
Controlled concrete allowable working stresses, where $f_c' \leq 3500$ psi:
$F_b = .40f_c'$
$F_b = .45f_c'$ adjacent to supports of continuous or fixed beams or slabs or rigid frames
$F_v = .02f_c'$ on beams without web reinforcing and without special anchorage of longitudinal steel
$F_v = .03f_c'$ on beams without web reinforcing and with special anchorage of longitudinal steel
$F_v = .06f_c'$ on beams with web reinforcing and without special anchorage of longitudinal steel
$F_v = .09f_c'$ on beams with web reinforcing and with special anchorage of longitudinal steel
Allowable bond stress in slabs, with plain bars or rolled structural steel: $.04f_c'$
Allowable bond stress in slabs, with deformed bars: $.05f_c'$
Allowable bond stress in slabs, with plain bars or rolled structural steel specially anchored: $.06f_c'$
Allowable bond stress in slabs, with deformed bars specially anchored: $.075f_c'$
$F_p = .25f_c'$ (400 psi maximum) on full area
$F_p = .375f_c'$ on $\frac{1}{3}$ of full support area
Bearing on more than $\frac{1}{3}$ of full support area, but less than full support area: interpolate two values above
$n = E_s / E_c = 30000 / f_c'$
Concrete allowable working stresses, $1:5\frac{1}{2}:5\frac{1}{2}$ design mix, $7\frac{1}{2}$ gallons of water per sack of cement:
$F_b = 650$ psi
$F_b = 750$ psi adjacent to supports of continuous or fixed beams or slabs or rigid frames
$F_v = 40$ on beams without web reinforcing and without special anchorage of longitudinal steel

$F_v = 60$ on beams without web reinforcing and with special anchorage of longitudinal steel
$F_v = 100$ on beams with web reinforcing and without special anchorage of longitudinal steel
$F_v = 150$ on beams with web reinforcing and with special anchorage of longitudinal steel
Allowable bond stress in slabs, deformed bars: 100 psi
$F_p = 500$ psi
$n = 15$
Concrete allowable working stresses, $1:4\frac{1}{2}:4\frac{1}{2}$, $6\frac{3}{4}$ gallons of water per sack of cement:
$F_b = 800$ psi
$F_b = 920$ psi adjacent to supports of continuous or fixed beams or slabs or rigid frames
$F_v = 50$ on beams without web reinforcing and without special anchorage of longitudinal steel
$F_v = 75$ on beams without web reinforcing and with special anchorage of longitudinal steel
$F_v = 125$ on beams with web reinforcing and without special anchorage of longitudinal steel
$F_v = 190$ on beams with web reinforcing and with special anchorage of longitudinal steel
Allowable bond stress in slabs, with plain bars or rolled structural steel: 100 psi
Allowable bond stress in slabs, with deformed bars: 125 psi
$F_p = 625$ psi
$n = 12$

## 1944

Draped mesh design formulas[12]
$W = 30da/L^2$ for stone concrete
$W = 26da/L^2$ for cinder concrete, where $W$ is the total load in kips per square foot, $d$ is the depth to steel in inches, $a$ is the area of steel in square inches per foot, and $L$ is the span in feet

## 1946

Design of standard flat and segmental arches[13]
$p = 3wD^2/2R$
$P = pL$
$A = 3wD^2L / (2F_bR) = P/F_b$
$T = 2aF_bR/(3wD^2) = aF_b/p$
$M_x = 12L(DLw/2)/8 = 3wDL^2/4$
$M_y = 12T(pT)/12 = pT^2$
$f_b = M_x / S_x + M_y / S_y$ where $w$ = unit load on arch in psf, $D$ = arch span in feet, $L$ = length of beam supporting arch in feet, $R$ = effective rise of arch in inches, $p$ = thrust of arch per lineal foot in pounds, $P$ = total thrust per panel in pounds, $A$ = net area of tie rods per panel in square inches, $a$ = net area of one tie rod in square inches, $T$ = spacing of tie rods in feet

# ENDNOTES

CHAPTER ONE
**INTRODUCTION**

[1]"The Development of Structural Ironwork Design," pp. 417–18.

CHAPTER TWO
**TRADITIONAL CONSTRUCTION**

[1]Badger, *Illustrations of Iron Architecture*, p. x.

[2]Jensen, "Board and Batten Siding and the Balloon Frame."

[3]White and Willensky, *AIA Guide to New York City*, p. 123; Gray, "Status for an 1866 House."

[4]White and Willensky, *AIA Guide to New York City*, p. 390.

[5]Goldberger, *The City Observed*, p. 6; White and Willensky, *AIA Guide to New York City*, p. 53.

[6]Salmon and Johnson, *Steel Structures*, p. 26.

[7]Freyer, *History of Real Estate*, p. 82.

[8]Hool and Kinne, *Steel and Timber Structures*, p. 94; Hool and Kinne, *Structural Members and Connections*, p. 356.

[9]*Building Laws*, 1891.

[10]Hool and Kinne, *Structural Members and Connections*, p. 425.

[11]Ibid., p. 425; Sturgis, *Illustrated Dictionary*, vol. I, col. 369, and vol. III, col. 623.

[12]Hool and Kinne, *Steel and Timber Structures*, p. 96.

[13]Feld, *Construction Failure*, p. 201.

[14]*New York Times*, March 18, 1899; March 19, 1899; March 26, 1899.

[15]*Engineering News*, March 23, 1899.

[16]Freyer, *History of Real Estate*, p. 395.

[17]Stokes, *The Iconography of Manhattan*, vol. 5, p. 1581.

[18]Albion, *The Rise of New York Port*, p. 261.

[19]Liscombe, "A 'New Era in My Life': Ithiel Town Abroad."

[20]*History of Architecture and the Building Trades*, I, p. 31.

[21]Liscombe, "A 'New Era in My Life': Ithiel Town Abroad."

[22]*History of Architecture and the Building Trades*, II, p. 289.

[23]Sante, *Low Life*, p. 32.

[24]Condit, *American Building*, pp. 115–16.

[25]Birkmire, *Planning and Construction of High Office Buildings*, pp. 39–40.

[26]Condit, *American Building*, p. 116.

[27]van Leeuwen, *The Skyward Trend of Thought*, p. 94.

[28]White and Willensky, *AIA Guide to New York City*, p. 105.

[29]Diamonstein, *The Landmarks of New York*, p. 215.

[30]Freyer, *History of Real Estate*, p. 53.

[31]Ibid., p. 82.

[32]Author's observations at Fifth Avenue Presbyterian Church and Eldridge Street Synagogue.

[33]*Building Laws*, 1891, p. 28.

[34]Ibid., p. 29.

[35]Ibid., p. 30.

[36]Ibid., p. 29.

[37]Ibid., p. 25.

[38]Ibid., p. 31.

[39]Crowly, "The Use of Terra Cotta in the United States."

[40]Ibid.; Meadows, *Historic Building Facades*, p. 23.

[41]van Leeuwen, *The Skyward Trend of Thought*, p. 89; King, *King's Handbook of New York City*, p. 824.

[42]*The City Record*, February 29, 1980.

CHAPTER THREE
**CAST-IRON FACADES**

[1]Condit, *American Building*, p. 81.

[2]Condit, "James Bogardus,"*Macmillan*, vol. I, pp. 233–34; Badger, *Illustrations of Iron Architecture*, p. 5.

[3]Condit, "James Bogardus," *Macmillan*, vol. I, pp. 233–34; Waite and Gayle, *The Maintenance and Repair of Architectural Cast Iron*, p. 2; Freyer, *History of Real Estate*, p. 457.

[4]Badger, *Illustrations of Iron Architecture*, p. v.

[5]Ibid., p. vii.

[6]Condit, *American Building*, p. 84.

[7]Badger, *Illustrations of Iron Architecture*, p. viii and numerous references.

[8]Ibid., p. vi.

[9]Sturgis, *Dictionary of Architecture*, vol. II, col. 506.

[10]Badger, *Illustrations of Iron Architecture*, p. vi.

[11]Aitchison, "On Iron as a Building Material," pp. 97–107.

[12]Diamonstein, *The Landmarks of New York*, p. 121; Condit, *American Building*, p. 84; Badger, *Illustrations of Iron Architecture*, p. x.

[13]Starrett, *Skyscrapers*, p. 22; Condit, *American Building*, p. 83; Condit, "James Bogardus," *Macmillan*, vol. I, p. 234; Badger, *Illustrations of Iron Architecture*, p. v; Fitch, *American Building*, p. 123.

[14]Freyer, *History of Real Estate*, p. 463.

[15]White and Willensky, *AIA Guide to New York City*, pp. 392, 572; Badger, *Illustrations of Iron Architecture*, p. vii; Goldberger, *The City Observed*, p. 14.

[16]White and Willensky, *AIA Guide to New York City*, pp. 45–49.

[17]Gayle and Gillon, *Cast-Iron Architecture in New York*, p. vi.

[18]*Manual of the Bouton Foundry*, pp. 11, 14.

[19]"Iron Castings," p. 197.

[20]Condit, *American Building*, pp. 85–86; Fleming, "Whence the Skyscraper," pp. 505–9.

[21]Aitchison, "On Iron as a Building Material," pp. 97–107.

[22]Condit, *American Building Art: the 19th Century*, p. 28; MacKay, *The Building of Manhattan*, p. 23; Spann, *The New Metropolis*, p. 101.

[23]Albion, *The Rise of New York Port*, p. 261.

[24]Badger, *Illustrations of Iron Architecture*, p. vi.

[25]Hitchcock, "American Influence Abroad," p. 22.

[26]Badger, *Illustrations of Iron Architecture*, p. xii.

[27]Ibid., p. viii.

28Gayle and Gillon, *Cast-Iron Architecture in New York*, pp. 88–89; *New York Times*, account of the fire at 1–5 Bond Street, March 1877.

29Badger, *Illustrations of Iron Architecture*, p. vi.

30Condit, *American Building Art: The 19th Century*, p. 44.

31Freitag, *Fireproofing*, p. 17.

32Birkmire, *Skeleton Construction in Buildings*, pp. 26–28.

33Badger, *Illustrations of Iron Architecture*, p. vi.

34Freyer, *History of Real Estate*, pp. 353, 363; Federal Writers Project, *New York Panorama*, p. 212.

35Freitag, *Fireproofing*, p. 25.

36Freitag, *Fire Prevention*, p. 129.

37Birkmire, *Skeleton Construction in Buildings*, p. 17.

38McCullough, *The Great Bridge*, p. 326; Hoerr, *And the Wolf Finally Came*, p. 6.

39Merriman and Jacoby, *Roofs and Bridges, Part III*, p. 3.

40*Manual of the Bouton Foundry*, pp. 11, 14.

41*Building Laws*, 1891, p. 25.

42Ibid., p. 43.

43Ibid., pp. 35, 36.

44Birkmire, *Architectural Iron and Steel*, pp. 30–31; *Manual of the Bouton Foundry*, p. 65.

CHAPTER FOUR
STEEL SKELETON FRAME

1Birkmire, *Skeleton Construction in Buildings*, p. 1.

2van Leeuwen, *The Skyward Trend of Thought*, pp. 94, 96; Condit, *American Building*, p. 116.

3Diamonstein, *The Landmarks of New York*, p. 215.

4Freitag, *Fireproofing*, p. 19; Freyer, *History of Real Estate*, p. 465.

5Baier, "Wind Bracing in the St. Louis Tornado," pp. 276–77.

6*Pocket Companion*, 1903, pp. 121, 122; Sturgis, *Dictionary of Architecture*, vol. II, col. 35.

7Birkmire, *Skeleton Construction in Buildings*, pp. 18–19.

8Birkmire, *Architectural Iron and Steel*, pp. 57–58.

9Parsons, "Collapse of a Building During Construction," pp. 454–56; "The Collapse of the Darlington," pp. 217–19; "Further Notes on the Collapse of the Darlington," pp. 250–51.

10"Concerning the Fall of the Darlington Building."

11Starrett, *Skyscrapers*, p. 41.

12"Structural Engineering in Apartment House Construction," pp. 353–54.

13Starrett, *Skyscrapers*, p. 41.

14Freyer, *History of Real Estate*, pp. 477–78.

15*History of Architecture and the Building Trades*, vol. II, p. 257.

16*Engineering News*, editorial, November 5, 1896; *History of Architecture and the Building Trades*, vol. II, p. 254; *New York Times*, accounts of Ireland Building collapse, August 1895.

17"Fall of a Building," pp. 111–12; "Some Details of the Ireland Building," pp. 127–28.

18*Engineering News*, editorial, August 15, 1895.

19"Fatal Collapse of a Part of a Cast-Iron Building in New York," pp. 33–35; *History of Architecture and the Building Trades*, vol. II, p. 256; *New York Times*, account of the collapse in June 1897.

20"The Collapse of a Portion of a New Building in New York City," p. 23.

21*Pocket Companion*, 1903, pp. 121, 122.

22Condit, *Port of New York*, vol. I, p. 109, quoted with permission of The University of Chicago Press; Condit, *American Building*, p. 117; Federal Writers Project, *New York Panorama*, p. 216.

23Birkmire, *Skeleton Construction in Buildings*, pp. 115–16.

24*Engineering News*, editorial on high buildings, October 25, 1894.

25Birkmire, *Skeleton Construction in Buildings*, p. 2; *Building Laws*, 1891, pp. 28, 30.

26Freyer, *History of Real Estate*, pp. 465–66.

27Condit, *American Building*, pp. 117–18, quoted with permission of The University of Chicago Press.

28Fleming, "A Half-Century of the Skyscraper," pp. 634–38; *Building Laws*, 1891, p. 28.

29Freyer, *History of Real Estate*, pp. 467–68; Condit, *The Port of New York*, vol. I, p. 109.

30Condit, *American Building*, p. 118

31van Leeuwen, *The Skyward Trend of Thought*, pp. 103–6; Kirby, et al., *Engineering in History*, p. 322; Stern, Gilmartin, and Massengale, *New York 1900*, p. 148.

32Freyer, *History of Real Estate*, pps. 465–66.

33Condit, *American Building*, p. 119.

34Ibid., p. 83; Weisman, "A New View of Skyscraper History," p. 131; Silver, *Lost New York*, p. 93; Condit, "James Bogardus," *Macmillan*, vol. I, p. 234.

35Goldberger, *The City Observed*, pp. 293–94; Silver, *Lost New York*, p. 92.

36Condit, *American Building*, pp. 81, 85; Silver, *Lost New York*, p. 183.

37Condit, *American Building*, p. 84.

CHAPTER FIVE
THE FIREPROOF BUILDING (I)

1MacKay, *The Building of Manhattan*, p. 23; Spann, *The New Metropolis*, p. 101.

2Gray, "Status for an 1866 House," p. 6; Freyer, *History of Real Estate*, p. 288.

3Freyer, *History of Real Estate*, pp. 289–93; *Building Laws*, 1891, pp. 46–47.

4Albion, *The Rise of New York Port*, p. 261.

5Freyer, *History of Real Estate*, p. 384.

6Badger, *Illustrations of Iron Architecture*, pp. v, viii, x, and numerous references to illustrations; Condit, "James Bogardus," *Macmillan*, vol. I, pp. 233–34; White and Willensky, *AIA Guide to New York City*, p. 563; Sturgis, *Dictionary of Architecture*, vol. II, col. 506; Waite and Gayle, *The Maintenance and Repair of Architectural Cast Iron*, p. 2; Waite, *Iron Architecture in New York City*, pp. 8, 9, 37.

7Badger, *Illustrations of Iron Architecture*, p. vi.

8"Iron Castings," p. 197.

9Federal Writers Project, *New York Panorama*, p. 212; Freyer, *History of Real Estate*, pp. 353, 379; Badger, *Illustrations of Iron Architecture*, p. vi.

10Freyer, *History of Real Estate*, p. 363.

11*History of Architecture and the Building Trades*, vol. II, p. 314.

12Lowe, ed., *The Great Chicgo Fire*, pp. 9, 34, 35, and numerous illustrations.

13van Leeuwen, *The Skyward Trend of Thought*, p. 96; Condit, *American Building*, p. 116.

14"The New-York Sketch Book of Architecture," edited by H. H. Richardson, as quoted in van Leeuwen, *The Skyward Trend of Thought*, p. 94.

15Birkmire, *Skeleton Construction in Buildings*, p. 18.

16Freitag, *Fireproofing*, p. 25.

17White and Willensky, *AIA Guide to New York City*, pp. 45–49.

18Badger, *Illustrations of Iron Architecture*, p. v; Condit, *American Building*, p. 83; White and Willensky, *AIA Guide to New York City*, p. 101; Sturgis, *Dictionary of Architecture*, vol. II, column 35; Waite, *Iron Architecture in New York City*, pp. 54, 59, 60, 62, 66–70; Condit, "James Bogardus," *Macmillan*, vol. I, pp. 233–34; Birkmire, *Architectural Iron and Steel*, pp. 30–31; *Manual of the Bouton Foundry*, p. 65; Fleming, "Whence the Skyscraper," pp. 505–9; Freyer, *History of Real Estate*, p. 461; Starrett, *Skyscrapers*, p. 22.

19Freyer, *History of Real Estate*, p. 463.

20"Early Bridge Building Pointed Way," pp. 83–84.

21Fleming, "Whence the Skyscraper," pp. 505–9.

22Condit, *American Building*, pp. 115–16, quoted with permission of The University of Chicago Press; van Leeuwen, *The Skyward Trend of Thought*, p. 4.

23Merriman and Jacoby, *Roofs and Bridges, Part III*, p. 3; McCullough, *The Great Bridge*, p. 326; Hoerr, *And the Wolf Finally Came*, p. 6.

[24]McCullough, *The Great Bridge*, p. 477.

[25]Fleming, "Whence the Skyscraper," pp. 505–9.

[26]Condit, *American Building*, p. 116.

[27]Condit, *American Building*, p. 116; Fleming, "Whence the Skyscraper," pp. 505–9; Starrett, *Skyscrapers*, p. 24; Sturgis, *Dictionary of Architecture*, vol. II, col. 35.

[28]Freitag, *Fireproofing*, p. 11; Fleming, "Whence the Skyscraper," pp. 505–9; Freyer, *History of Real Estate*, p. 406.

[29]Freyer, *History of Real Estate*, p. 476.

[30]Sturgis, *Dictionary of Architecture*, vol. II, col. 35; Freitag, *Fireproofing*, p. 13.

[31]Wight, "Hollow Tile Fire Proof Floor Construction"; Fleming, "Whence the Skyscraper," pp. 505–9.

[32]*Pocket Companion 1923*, p. 268.

[33]Condit, *American Building*, p. 116.

[34]Freitag, *Fireproofing*, pp. 17–18.

[35]Freyer, *History of Real Estate*, p. 115; van Leeuwen, *The Skyward Trend of Thought*, p. 89; White and Willensky, *AIA Guide to New York City*, p. 31.

[36]Freitag, *Fireproofing*, p. 17; Birkmire, *Skeleton Construction in Building*, pp. 13–16, 26–28.

[37]Birkmire, *Skeleton Construction in Buildings*, p. 17.

[38]Freyer, *History of Real Estate*, pp. 477–78.

**CHAPTER SIX**
**STANDARDIZATION OF STEEL FRAMING**

[1]Birkmire, *Architectural Iron and Steel*, pp. 57–58; Birkmire, *Skeleton Construction in Buildings*, pp. 50–51.

[2]*Building Laws*, 1891, p. 43.

[3]Gillette, *AISC: The First 60 Years*, p. 10.

[4]Condit, *American Building*, p. 83.

[5]Hodgkinson, *Experimental Researches*, pp. 334–35; Fairbairn, *On the Application*, p. v.

[6]Peterson, "Inventing the I-beam," p. 89.

[7]Merriman and Jacoby, *Roofs and Bridges, Part III*, p. 3.

[8]Ferris, *AISC Iron and Steel Beams*, p. 1.

[9]"Standard Specifications for Steel," p. 130.

[10]Blakely, *Dimensions, Weights and Properties*, p. 6.

[11]Condit, *American Building*, pp. 85–86; Burnham, "Last Look at a Structural Landmark," pp. 273–79; Salmon and Johnson, *Steel Structures*, p. 567.

[12]Author's observations, [Old] U.S. Customs House, One Bowling Green, New York.

[13]*Manual of the Bouton Foundry*, pp. 149, 150.

[14]Fisher, *The Epic of Steel*, p. 104; Gillette, *AISC: The First 60 Years*, p. 10; *Arbed-Rolled Wide Flange Beams*, p. 4.

[15]Hool and Kinne, *Steel and Timber Structures*, p. 23; Starrett, *Skyscrapers*, p. 172.

[16]Blakely, *Dimensions, Weights and Properties*, pp. 6, 8–10, 12.

[17]Hool and Kinne, *Steel and Timber Structures*, p. 14.

[18]Birkmire, *Planning and Construcuion of High Office Buildings*, p. 192.

[19]Birkmire, *Skeleton Construction in Buildings*, p. 22.

[20]Ibid., pp. 23–25, 57.

[21]Dencer, *Detailing and Fabricating Structural Steel*, pp. 121–22.

[22]Birkmire, *Skeleton Construction in Buildings*, p. 22.

[23]Dencer, *Detailing and Fabricating Structural Steel*, pp. 121–22.

[24]Urquhart, ed., *Civil Engineering Handbook*, 3d ed., pp. 272–74.

[25]Hool and Kinne, *Structural Members and Connections*, pp. 132–36; Urquhart, *Civil Engineering Handbook*, 3d ed., pp. 272–74; Grinter, *Design of Modern Steel Structures*, p. 184.

[26]Hool and Kinne, *Structural Members and Connections*, pp. 132–36; Urquhart, *Civil Engineering Handbook*, 3d ed., pp. 272–74; Grinter, *Design of Modern Steel Structures*, p. 184.

[27]Hool and Kinne, *Structural Members and Connections*, pp. 132–36; Urquhart, *Civil Engineering Handbook*, 3d ed., pp. 272–74; Grinter, *Design of Modern Steel Structures*, p. 184.

[28]Grinter, *Design of Modern Steel Structures*, p. 184.

[29]Birkmire, *Skeleton Construction in Buildings*, pp. 26–28.

[30]Birkmire, *Planning and Construction of High Office Buildings*, p. 198.

[31]*Building Code*, 1901, p. 112.

[32]*Pocket Companion*, 1919, pp. 294, 320.

[33]*New York City Building Code*, 1926, pp. 15–21

[34]"Specification for the Design, Fabrication, and Erection of Structural Steel for Buildings," *Steel Construction Manual*, 1st ed., pp. 8–16; 2d ed., pp. 227–34; 3d ed., pp. 261–78; 4th ed., pp. 271–92; 5th ed., pp. 285–86; 6th ed., pp. 5-16–21.

[35]*Building Laws*, 1891, p. 43.

[36]*Building Code*, 1901, p. 112.

[37]Cowan, *Science and Building*, p. 91; Mujica, *History of the Skyscraper*, p. 29.

[38]*Pocket Companion*, 1919, p. 325; *New York City Building Code*, 1926, pp. 15–21.

[39]Winterton, "5000 B.C.–1962 A.D."

[40]Dencer, *Detailing and Fabricating Structural Steel*, pp. 325–29; Hool and Kinne, *Structural Members and Connections*, p. 320.

[41]Birkmire, *Skeleton Construction in Buildings*, pp. 73–74.

[42]Ibid., p. 75.

[43]Birkmire, *Skeleton Construction in Buildings*, pp. 115–16; Condit, "The Wind Bracing of Buildings," pp. 92–105.

[44]Birkmire, *Planning and Construction of High Office Buildings*, pp. 189–90.

[45]Sturgis, *Dictionary of Architecture*, vol. II, col. 515–19.

[46]Hool and Kinne, *Steel and Timber Structures*, pp. 32, 33.

[47]Winterton, "5000 B.C.–1962 A.D."

[48]Condit, *American Building Art: The 20th Century*, p. 30; Condit, *American Building*, p. 192.

[49]Fisher, *The Epic of Steel*, pp. 103–4.

[50]"Experiments with a Flitch Plate Girder," p. 325.

[51]Gayle, Look, and Waite, *Metals in America's Historic Buildings*, vol. I, p. 15.

[52]Williams and Harris, *Structural Design in Metals*, p. 348.

[53]Hool and Kinne, *Steel and Timber Structures*, p. 623; Williams and Harris, *Structural Design in Metals*, p. 348.

[54]Hool and Kinne, *Steel and Timber Structures*, p. 623; Ramsey and Sleeper, *Architectural Graphic Standards*, 1st ed., pp. 80–81; Walker, *The Building Estimator's Reference Book*, p. 1558.

[55]Williams and Harris, *Structural Design in Metals*, p. 570.

[56]Ramsey and Sleeper, *Architectural Graphic Standards*, 1st ed., pp. 80–81; Sprauge, *50-Year Steel Joist Digest*, p. 9.

[57]Walker, *The Building Estimator's Reference Book*, p. 1558.

[58]Badger, *Illustrations of Iron Architecture*, p. vii and numerous illustrations.

[59]Birkmire, *Architectural Iron and Steel*, pp. 30–31.

[60]Hool and Kinne, *Structural Members and Connections*, pp. 230–31.

[61]*Building Laws*, 1891, pp. 35, 36.

[62]Dencer, *Detailing and Fabricating Structural Steel*, p. 128.

[63]Hool and Kinne, *Steel and Timber Structures*, p. 35; Hool and Kinne, *Structural Members and Connections*, pp. 230–31; Dencer, *Detailing and Fabricating Structural Steel*, pp. 127–28.

[64]Dencer, *Detailing and Fabricating Structural Steel*, p. 127.

[65]Hool and Kinne, *Steel and Timber Structures*, p. 35; Dencer, *Detailing and Fabricating Structural Steel*, p. 128.

[66]Birkmire, *Planning and Construction of High Office Buildings*, p. 198.

[67]*Building Code*, 1901, pp. 118–21; *Pocket Companion*, 1919, pp. 158–64; Hool and Kinne, *Steel and Timber Structures*, pp. 6, 21;

"Standard Specification for Structural Steel for Buildings," *Steel Construction Manual*, 1st ed., pp. 8–16; *Building Code*, 1926, pp. 15–21.

[68]Dencer, *Detailing and Fabricating Structural Steel*, p. 65.

CHAPTER SEVEN
**FLOOR SYSTEMS**

[1]Birkmire, *Skeleton Construction in Buildings*, pp. 77–78.

[2]Ibid., p. 80.

[3]Birkmire, *Planning and Construction of High Office Buildings*, p. 121; *Building Laws*, 1891, p. 44; *Building Code*, 1901, pp. 77–83.

[4]Perrine and Strehan, "Cinder Concrete Floor Construction," pp. 523–621.

[5]Freitag, *Fireproofing*, pp. 58–60.

[6]"Comparative Standard Fireproof Floor Tests of the New York Building Department"; Freitag, *Fireproofing*, pp. 58–60.

[7]Perrine and Strehan, "Cinder Concrete Floor Construction," pp. 523–621.

[8]"Fireproof Floor Construction in New York City," pp. 72–75.

[9]Aus, "Tests of Long-Span Terra-Cotta Floor Arches," p. 314.

[10]Freitag, *Fireproofing*, pp. 61–75.

[11]"Comparative Standard Fireproof Floor Tests of the New York Building Department."

[12]Aus, "Tests of Long-Span Terra-Cotta Floor Arches," p. 314.

[13]Collins, "Guastavino y Moreno, Rafael and Guastavino y Esposito, Rafael", *Macmillan*, vol. II, pp. 279–80; "A New System of Fireproof Floor Construction"; "Comparative Standard Fireproof Floor Tests"; Freitag, *Fire Prevention*, p. 328.

[14]"Comparative Standard Fireproof Floor Tests"; Freitag, *Fireproofing*, pp. 61–75; Birkmire, *Planning and Construction of High Office Buildings*, p. 153.

[15]Birkmire, *Planning and Construction of High Office Buildings*, pp. 157–58.

[16]"Comparative Standard Fireproof Floor Tests"; Freitag, *Fireproofing*, pp. 61–75; Birkmire, *Planning and Construction of High Office Buildings*, p. 154.

[17]"Comparative Standard Fireproof Floor Tests"; Freitag, *Fireproofing*, pp. 61–75.

[18]"Comparative Standard Fireproof Floor Tests"; Birkmire, *Planning and Construction of High Office Buildings*, p. 138; Hool, *Reinforced Concrete Construction*, vol. II, pp. 137–40.

[19]"Comparative Standard Fireproof Floor Tests."

[20]Hool and Johnson, *Concrete Engineers' Handbook*, pp. 55–57.

[21]"Comparative Standard Fireproof Floor Tests"; Freitag, *Fireproofing*, pp. 61–75.

[22]"Comparative Standard Fireproof Floor Tests"; Freitag, *Fireproofing*, pp. 61–75.

[23]Birkmire, *Planning and Construction of High Office Buildings*, p. 158.

[24]"Comparative Standard Fireproof Floor Tests of the New York Building Department"; Freitag, *Fireproofing*, pp. 61–75.

[25]"Comparative Standard Fireproof Floor Tests"; Freitag, *Fireproofing*, pp. 61–75.

[26]Hool, *Reinforced Concrete Construction*, vol. II, pp. 137–40; Freitag, *Fireproofing*, pp. 61–75.

[27]"Comparative Standard Fireproof Floor Tests"; Birkmire, *Planning and Construction of High Office Buildings*, p. 131; Hool, *Reinforced Concrete Construction*, vol. II, pp. 137–40.

[28]Birkmire, *Planning and Construction of High Office Buildings*, pp. 135–36.

[29]"Comparative Standard Fireproof Floor Tests"; Freitag, *Fireproofing*, pp. 61–75; Birkmire, *Planning and Construction of High Office Buildings*, pp. 155–56; "The Metropolitan Concrete and Wire Floor," p. 333.

[30]Ketchum, *Structural Engineer's Handbook*, p. 34.

[31]Birkmire, *Skeleton Construction in Buildings*, pp. 81–82.

[32]Waite, "Cinder Concrete Floors," pp. 1805–6; Perrine and Strehan, "Cinder Concrete Floor Construction," pp. 523–621.

[33]Norton, "The Protection of Steel from Corrosion," p. 29.

[34]Ketchum, *Structural Engineer's Handbook*, p. 69.

[35]Ibid., p. 513.

[36]Perrine and Strehan, "Cinder Concrete Floor Construction," pp. 523–621.

[37]Waite, "Cinder Concrete Floors," pp. 1774, 1803.

[38]"Second Report of the Joint Committee on Concrete and Reinforced Concrete," in Hool, *Reinforced Concrete Construction*, vol. II, pp. 615–59.

[39]Waite, "Cinder Concrete Floors," pp. 1778–812.

[40]"Second Report of the Joint Committee on Concrete and Reinforced Concrete," in Hool, *Reinforced Concrete Construction*, vol. II, p. 648; Ketchum, *Structural Engineer's Handbook*, pp. 522–23; Hool, *Reinforced Concrete Construction*, vol. I, p. 2; Waite, "Cinder Concrete Floors," p. 1800.

[41]*Pocket Companion*, 1919, p. 329.

[42]*American Welded Wire Fabric*, pp. 46, 48, 49.

[43]Waite, "Cinder Concrete Floors," p. 1776.

[44]"The Removal of a Steel Frame Building," pp. 113–14.

[45]"The Corrosion of Reinforcing Metal in Cinder Concrete Floors."

[46]Fox, "The Corrosion of Steel in Reinforced Cinder Concrete," pp. 5669–70.

[47]Feld, *Construction Failure*, p. 260.

[48]Freitag, *Fire Prevention*, p. 324.

[49]Feld, *Construction Failure*, p. 7.

[50]Collins, *Concrete: The Vision of a New Architecture*, p. 87.

[51]Wight, "The Pioneer Concrete Residence in America," pp. 359–63; Collins, *Concrete: The Vision of a New Architecture*, pp. 57–58.

[52]Bruce, "The Transverse Strength of Concrete," pp. 367–68.

[53]Hool, *Reinforced Concrete Construction*, vol. II, p. 67.

[54]Wang and Salmon, *Reinforced Concrete Structures*, p. 4; Collins, *Concrete: The Vision of a New Architecture*, p. 62; Kurtz, "A Review of Concrete Metal Construction," pp. 61–62.

[55]Collins, *Concrete: The Vision of a New Architecture*, p. 62; "A Concrete Court House," p.87.

[56]Collins, *Concrete: The Vision of a New Architecture*, pp. 88–89.

[57]Ibid., p. 63; "Concrete," pp. 97–100.

[58]Collins, *Concrete: The Vision of a New Architecture*, p. 63; *Evaluation of Reinforcing Steel Systems*, p. 3.

[59]Winter and Nilson, *Design of Concrete Structures*, p. 107.

[60]Birkmire, *Planning and Construction of High Office Buildings*, pp. 135–36.

[61]Waite, "Cinder Concrete Floors," pp. 1774–75.

[62]Hool and Kinne, *Structural Members and Connections*, p. 436.

[63]Hool and Kinne, *Reinforced Concrete and Masonry*, p. 3; "History and Properties of Light-Weight Aggregates," pp. 802–4.

[64]*Evaluation of Reinforcing Steel Systems*, p. 3.

[65]Author's observations at 40 West 20th Street.

[66]Hool and Johnson, *Concrete Engineers' Handbook*, p. 36.

[67]Merriman, ed. *American Civil Engineers' Handbook*, p. 530.

[68]*Pocket Companion*, 1919, pp. 124-31; Merriman, ed., *American Civil Engineers' Handbook*, p. 532.

[69]*Pocket Companion*, 1919, pp. 124-31; Merriman, ed., *American Civil Engineers' Handbook*, p. 529.

[70]*Pocket Companion*, 1919, pp. 124-31; Merriman, ed., *American Civil Engineers' Handbook*, p. 532; Hool and Johnson, *Concrete Engineers' Handbook*, pp. 43-45.

[71]*Evaluation of Reinforcing Steel Systems*, p. 3; *Pocket Companion*, 1923, p. 68.

[72]Hool and Johnson, *Concrete Engineers' Handbook*, pp. 46–53; Hool, *Reinforced Concrete Construction*, vol. II, pp. 163–71.

[73]Hool, *Reinforced Concrete Construction*, vol. II, pp. 137–40.

[74]Plummer and Wanner, *Principles of Tile Engineering*, pp. 286–88; Hool and Kinne, *Reinforced Concrete and Masonry Structures*, p. 162.

[75]Plummer and Wanner, *Principles of Tile Engi-

*neering*, pp. 280–81.

[76]Plummer and Wanner, *Principles of Tile Engineering*, pp. 280–81; Merriman, ed., *American Civil Engineers' Handbook*, pp. 542–43; Hool and Kinne, *Reinforced Concrete and Masonry Structures*, p. 162.

[77]Plummer and Wanner, *Principles of Tile Engineering*, pp. 285–86.

[78]Hool, *Reinforced Concrete Construction*, vol. II, pp. 171–78.

[79]Plummer and Wanner, *Principles of Tile Engineering*, pp. 279–80.

[80]Ibid., pp. 283–85.

[81]Hool, *Reinforced Concrete Construction*, vol. II, pp. 171–78.

[82]Plummer and Wanner, *Principles of Tile Engineering*, pp. 288–89.

[83]Ibid., p. 289.

[84]Ibid., pp. 279–80, 290–96.

[85]Ibid., p. 298.

## CHAPTER EIGHT
## CURTAIN WALL SYSTEMS

[1]Freyer, "Skeleton Construction," pp. 228–35.

[2]Birkmire, *Skeleton Construction in Buildings*, p. 2.

[3]Freyer, "The New York Building Laws," pp. 69–82.

[4]*Building Code*, 1901, pp. 34–35.

[5]Birkmire, *Skeleton Construction in Buildings*, pp. 13–16.

[6]*Building Code*, 1901, pp. 34–35.

[7]Fleming, "A Half Century of the Skyscraper," pp. 634–38.

[8]Birkmire, *Planning and Construction of High Office Buildings*, p. 101.

[9]Ibid., p. 103.

[10]Ibid., p. 107.

[11]"Tall Buildings of New York," pp. 400–403.

[12]Stockbridge, "The Interaction Between Exterior Walls and Building Frames," pp. 257–65.

[13]Ibid.

[14]Meadows, *Historic Building Facades*, p. 24.

[15]White and Willensky, *AIA Guide to New York City*, p. 23; Condit, *American Building*, pp. 186–87.

[16]Meadows, *Historic Building Facades*, pp. 31, 65.

[17]Freitag, *Fireproofing*, p. 17; Birkmire, *Skeleton Construction in Buildings*, pp. 13–16, 26–28.

[18]van Leeuwen, *The Skyward Trend of Thought*, p. 89; King, *King's Handbook of New York City*, p. 824; Gayle and Gillon, *Cast-Iron Architecture in New York*, p. 7.

[19]Goldberger, *The City Observed*, p. 144.

[20]Birkmire, *Planning and Construction of High Office Buildings*, pp. 189–90.

[21]Gray, "A Case of Terra Cotta Failure," p. 7.

[22]Bixby, "Wind Pressure in Engineering Construction," pp. 178–84.

[23]Baier, "Wind Bracing in the St. Louis Tornado," p. 221.

[24]Ibid., pp. 251, 273, 285.

[25]*Building Laws*, 1891, p. 31; *Building Code*, 1901, p. 38.

[26]Ramsey and Sleeper, *Architectural Graphic Standards*, 1st ed., p. 16.

[27]Belle, Hoke, and Kliment, eds.,*Traditional Details*, p. 43; Plummer and Wanner, *Principles of Tile Engineering*, p. 176.

[28]Nolan, *Kidder's Architects' and Builders' Handbook*, p. 234.

[29]White and Willensky, *AIA Guide to New York City*, p. 170.

[30]Comer, *New York City Building Control*, p. 124.

[31]White and Willensky, *AIA Guide to New York City*, p. 48.

[32]Dencer, *Detailing and Fabricating Structural Steel*, pp. 127–28.

[33]For example, *Building Laws*, 1891, p. 30.

[34]Condit, *The Port of New York*, vol. I, p. 92; Condit, *American Building*, p. 134.

[35]McCullough, *The Great Bridge*, p. 146.

[36]Marinelli, "Architectural Glass," pp. 34–43.

[37]Stern, Gilmartin, and Mellins, *New York 1930*, p. 537.

[38]Prudon, "Repairing Early Curtain Walls," pp. 123–26.

[39]Wang, Sakamoto, and Bassler, eds., *Cladding*, p. xiii.

[40]Goldberger, *The City Observed*, p. 141; Goldberger, *On the Rise*, p. 253.

[41]Prudon, "Repairing Early Curtain Walls," pp. 123–26; Clausen, "Belluschi and the Equitable Building in History," pp. 109–29.

[42]Clausen, "Beluschi and the Equitable Building in History," pp. 109–29; Goldberger, *The Skyscraper*, p. 107; Goldberger, *The City Observed*, p. 141; White and Willensky, *AIA Guide to New York City*, p. 152; Schaal, *Curtain Walls*, p. 59.

[43]Condit, "Wind Bracing of Tall Buildings," pp. 92–105.

[44]"Skyscraper Gets Aluminum Skin"; "Held by Rails"; "A Second Aluminum Faced Skyscraper."

[45]Trimble, "Brick Institute Recommendations," pp. 117–18; Wang, Sakamoto, and Bassler, eds., *Cladding*, p. 13.

[46]"Technics Topics; SS/BV Walls: Discussion," *Progressive Architecture,* pp. 113–16.

[47]Wang, Sakamoto, and Bassler, eds., *Cladding*, p. 14.

[48]Clausen, "Belluschi and the Equitable Building in History," pp. 109–29.

[49]Gayle, Look, and Waite, *Metals in America's Historic Buildings*, vol. I, p. 85; White and Willensky, *AIA Guide to New York City*, p. 170.

[50]Goldberger, *The Skyscraper*, p. 109.

[51]Goldberger, *The City Observed*, pp. 158–59.

[52]White and Willensky, *AIA Guide to New York City*, p. 156.

[53]"Miracle at Grand Central."

[54]Gayle, Look, and Waite, *Metals in America's Historic Buildings*, vol. I, p. 85; White and Willensky, *AIA Guide to New York City*, p. 156.

[55]Goldberger, *The City Observed*, pp. 134, 200, 261; White and Willensky, *AIA Guide to New York City*, p. 181.

[56]White and Willensky, *AIA Guide to New York City*, pp. 148, 174.

[57]Schaal, *Curtain Walls*, p. 8.

[58]White and Willensky, *AIA Guide to New York City*, p. 247.

[59]Meadows, *Historic Building Facades*, p. 22.

[60]Talk with Pamela McMullen at MJM Studios, Inc., May 1991.

[61]Condit, *American Building*, pp. 186–87; Mujica, *History of the Skyscraper*, p. 58; Condit, "Wind Bracing of Buildings," pp. 92–105; Stockbridge, "The Interaction Between Exterior Walls and Building Frames," pp. 258–62; MacKay, *The Building of Manhattan*, pp. 38–39; Shopsin, *Restoring Old Buildings for Contemporary Uses*, pp. 110, 114–15; "Historic Terra Cotta-Clad Tower Gets $9-Million Face-Lift."

[62]"Falling Masonry Fatally Injures Barnard Student," p. B3-6; Gupte, "City Is Studying Why Lintel Fell, Killing Student," p. B1-4.

[63]"Marble Falls from Skyscraper Facade"; "Spandrel Repairs Avert Collapse."

[64]"Local Laws of the City of New York," *The City Record*, February 9, 1980.

[65]"Chicken Little, the Sky Is Falling," p. 82.

## CHAPTER NINE
## THE FIREPROOF BUILDING (II)

[1]Hinton, *Catalogue of the National Fireproofing Company*, p. 6; King, *King's Handbook of New York City*, p. 220.

[2]Connor, *Labor Costs of Construction*, p. 60.

[3]Friberg, "Combined Form and Reinforcement for Concrete Slabs."

[4]Badger, *Illustrations of Iron Architecture*, pp. 104, 117.

[5]Salmon and Johnson, *Steel Structures*, p. 911.

[6]Walker, *The Building Estimator's Reference Book*, pp. 672–74.

[7]Salmon and Johnson, *Steel Structures*, p. 911.

[8]"Office Building Assigned Triple Role"; "Why the Finest New Buildings Have Q-Floor."

[9]"Spray Fireproofing on Steel Frame."

[10]Sabnis, ed., *Handbook of Composite Construction Engineering*, p. 81.

[11]Friberg, "Combined Form and Reinforcement for Concrete Slabs."

[12]Sabnis, ed., *Handbook of Composite Construction Engineering*, p. 82.

[13]Ibid.

[14]"Lightweight Fireproofing for Steel Framing."

[15]Troup, "Fire-Resistant Design of Interior Structural Steel," p. 348.

[16]"Metal Deck Moves Into Bay Area."

[17]"Spray Fireproofing on Steel Floors."

[18]Rains, "A New Era in Fre Protective Coatings for Steel," pp. 80–83; Wildt, "Fire Protection of Structural Steel Framing."

[19]Rains, "A New Era in Fire Protective Coatings for Steel," pp. 80–83.

[20]Ibid.

[21]Ibid.

CHAPTER TEN
"MODERN" STEEL CONSTRUCTION

[1]Singleton, *Fireproofing Structural Steel*, p. 4.

[2]Jenny, "Structural Lightweight Concrete for Composite Design," pp. 122–24.

[3]Leabu, "Composite Design with Lightweight Aggregate," pp. 135–44.

[4]Cook, *Composite Construction Methods*, p. 2.

[5]"Composite Beam Patentee Forgoes Royalties."

[6]Toprac, "Strength of Three New Types of Composite Beams," pp. 21–30; Page, "New Type of Shear Connector."

[7]Cook, *Composite Construction Methods*, p. 2.

[8]Toprac, "Strength of Three New Types of Composite Beams."

[9]"Metal Deck, Connectors Pass Composite-Beam Test."

[10]Ferris, ed., *AISC Iron and Steel Beams*, p. 6.

[11]Toprac, "Strength of Three New Types of Composite Beams."

[12]Kulak, Fisher, and Struik, *Guide to Design Criteria for Bolted and Riveted Joints*, p. 2; Gillette, *AISC: The First 60 Years*, p. 60.

[13]"High Tensile Bolts Speed Erection."

[14]Salmon and Johnson, *Steel Structures*, pp. 95–96.

[15]"Metal Diamonds Adorn Skyscraper"; "Contract is Awarded for Largest Bolted Skyscraper"; "Bolting is New, TV Too—Both Meet"; "A Skyscraper Crammed with Innovation."

[16]Stetina, "Choosing the Best Structural Fastener."

[17]Salmon and Johnson, *Steel Structures*, p. 567.

[18]"Girder Will Carry 23 Floors."

[19]"Tallest Welded Building in East Topped Out."

[20]*Pocket Companion, 1923.*

[21]*Bethlehem Structural Shapes*, 1938, pp. 65ff.

[22]*AISC Steel Construction Manual,* 5th ed., pp. 12, 18; *AISC Steel Construction Manual,* 6th ed., 1st printing, pp. 1–12.

[23]*AISC Steel Construction Manual,* 7th ed., pp. 1–15; *AISC Steel Construction Manual,* 8th ed., pp. 1–20.

[24]*Arbed-Rolled Wide Flange Beams.*

[25]*AISC Steel Construction Manual ,* 9th ed., pp. 1-10–1-24.

[26]Fisher and Pense, "Experience with the Use of Heavy W Shapes in Tension," pp. 63–77.

[27]*AISC Steel Construction Manual: ASD,* 9th ed., pp. 1–6; "AISC Develops New Rules for Heavy Shapes," p. 21.

[28]*AISC Steel Construction Manual,* 9th ed., pp. 5-26, 5-27, 5-63, 5-64, 5-69, 5-87.

[29]Ibid., p. 5–69.

[30]*AISC Steel Construction Manual,* 6th ed.

[31]*AISC Steel Construction Manual,* 7th ed., pp. 5-3 to 5-166; *AISC Steel Construction Manual,* 8th ed., pp. 5-3–5-168; *AISC Steel Construction Manual: ASD,* 9th ed., p. 5-42.

[32]"Investigating the New York Building Department," pp. 285–86; "Corruption Clean-up," p. 69.

[33]"Corruption Clean-up," p. 69.

[34]"Self Inspection for N.Y.C. Builders?" p. 94; "Inspection Proposal Deserves Support"; "Bill to Allow Building Inspection by A-E's Debated," p. 14.

[35]"Self Inspection for N.Y.C. Builders?" p. 94.

[36]Shryock, "Steel Struts or Columns."

[37]*Building Code of the City of New York*, 1926, pp. 15–21.

[38]*Building Code of the City of New York,* 1936, p. 191.

[39]*Building Code of the City of New York,* 1939, p. 250.

[40]*Building Code of the City of New York,* 1991–92, p. 117.

[41]Salmon and Johnson, *Steel Structures*, p. 26.

[42]Wang and Salmon, *Reinforced Concrete Structures*, p. 31.

APPENDIX B

[1]Condit, *American Building*, pp. 156–57.

[2]Ibid., p. 81.

[3]McKay, *South Street*, pp. 193–94.

[4]Collins, *Concrete: The Vision of a New Architecture*, p. 56; Condit, *American Building*, p. 157.

[5]Condit, *American Building*, p. 81.

[6]*History of Architecture and the Building Trades,* vol. I, p. 31.

[7]Badger, *Illustrations of Iron Architecture*, p. vii.

[8]White and Willensky, *AIA Guide to New York City*, p. 563; Condit, "James Bogardus," *Macmillan*, vol. I, pp. 233–34; Waite and Gayle, *The Maintenance and Repair of Architectural Cast Iron*, p. 2; Waite, ed. *Iron Architecture in New York City*, pp. 8, 9, 37.

[9]Condit, "James Bogardus," *Macmillan*, vol. I, pp. 233–34; Badger, *Illustrations of Iron Architecture*, p. v; Freyer, ed., *History of Real Estate*, p. 457.

[10]Kahn, "Bogardus, Fire, and the Iron Tower," pp. 186–203; Bannister, "Bogardus Revisited"; Silver, *Lost New York*, p. 92.

[11]Kahn, "Bogardus, Fire, and the Iron Tower," pp. 186–203; Bannister, "Bogardus Revisited."

[12]Badger, Illustrations *of Iron Architecture*, p. vii.

[13]*History of Architecture and the Building Trades,* vol. II, p. 300; Condit, *American Building,* pp. 81, 85; Silver, *Lost New York,* p. 183.

[14]Sturgis, *Dictionary of Architecture,* vol. II, col. 506.

[15]Huxtable, "Harper & Brothers Building," pp. 153–54; Fitch, *American Building*, p. 123; Condit, *American Building*, p. 83; Starrett, *Skyscrapers*, p. 22; Condit, "James Bogardus," *Macmillan*, vol. I, p. 234; Badger, *Illustrations of Iron Architecture*, p.v.

[16]Jewett, "Solving the Puzzle of the First American Structural Rail-Beam"; Condit, "James Bogardus," *Macmillan*, vol. I, p. 234; Condit, *American Building*, p. 83; Peterson, "Inventing the I-Beam," p. 78.

[17]Silver, *Lost New York*, p. 48.

[18]Bannister, "Bogardus Revisited"; Weisman, "A New View of Skyscraper History," p. 131; Condit, *American Building Art: The 19th Century*, p. 37; Condit, *American Building*, p. 83.

[19]Williams and Harris, *Structural Design in Metals*, p. 348.

[20]Badger, Illustrations of Iron Architecture, pp. vi, xi.

[21]Goldberger, *The City Observed*, pp. 293–94; Kahn, "Bogardus, Fire, and the Iron Tower," pp. 186–203.

[22]Diamonstein, *The Landmarks of New York*, p. 121.

[23]Interview with Martin Weaver, August 1992.

[24]Condit, *American Building*, p. 84; Badger, *Illustrations of Iron Architecture*, p. x.

[25]Burnham, "Last Look at a Structural Landmark," pp. 273–79; Fleming, "Whence the Skyscraper," pp. 505–9; Condit, *American Building,* pp.85–86; Badger, *Illustrations of Iron Architecture*, p. vi.

[26]Sturgis, *Dictionary of Architecture,* vol. II, col. 35; White and Willensky, *AIA Guide to New York City,* p. 101; Fleming, "Whence the Skyscraper," pp. 505–9; Waite, ed., *Iron Architecture in New York City*, pp. 54, 59, 60, 62, 66–70; Freyer, ed., *History of Real Estate*, p. 461; Wight, "Origin and History of Hollow Tile."

27Silver, *Lost New York*, p. 93.

28Condit, *American Building Art: The l9th Century*, p. 31; Condit, *American Building*, p. 84.

29White and Willensky, *AIA Guide to New York City*, p. 123.

30Badger, *Illustrations of Iron Architecture*, p. xii.

31Condit, "The Two Centuries of Technical Evolution," pp. 11–24.

32White and Willensky, *AIA Guide to New York City*, p. 572.

33White and Willensky, *AIA Guide to New York City*, p. 50; Goldberger, *The City Observed*, p. 49.

34Gray, "An 1869 Work with a Shaky Future," p. 6.

35White and Willensky, *AIA Guide to New York City*, p. 392.

36Condit, *American Building*, pp. 115–16; van Leeuwen, *The Skyward Trend of Thought*, p. 4; Fleming, "Whence the Skyscraper," pp. 505–9.

37Condit, *American Building*, p. 134; Condit, *The Port of New York*, vol. I, p. 92; Fitch, *American Building*, p. 170.

38Collins, *Concrete: The Vision of a New Architecture*, pp.57–58; Wight, "The Pioneer Concrete Residence in America"; Condit, *American Building*, pp. 170–71.

39Freyer, ed., *History of Real Estate*, p. 406; Fleming, "Whence the Skyscraper," pp.505–9.

40White and Willensky, *AIA Guide to New York City*, p. 44.

41Gayle and Gillon, Cast-Iron Architecture in New York, p. vi.

42van Leeuwen, *The Skyward Trend of Thought*, p. 93.

43Ibid., p. 71; Condit, *American Building*, p. 116.

44van Leeuwen, The Skyward Trend of Thought, pp. 94, 96; Condit, American Building, p. 116.

45Gayle and Gillon, *Cast-Iron Architecture in New York*, pp. 88–89; *New York Times*, March 1877, p. 1–4.

46Freyer, ed. *History of Real Estate*, p. 53.

47Meadows, *Historic Building Facades*, p. 23.

48Condit, *American Building*, p. 134; Condit, *The Port of New York*, vol. I, p. 92; Fitch, *American Building*, p. 170.

49Author's observations on site.

50Federal Writers Project, *New York Panorama*, p. 216; Condit, *American Building*, p. 117; Condit, *The Port of New York*, vol. I, p. 109.

51Loyrette, *Gustave Eiffel*, p. 100; Condit, *American Building Art: The 19th Century*, p. 46.

52Freitag, *Fireproofing*, p. 13.

53White and Willensky, AIA Guide to New York City, p. 105.

54Birkmire, *Skeleton Construction in Buildings*, p. 11.

55Birkmire, *Skeleton Construction in Buildings*, pp. 9, 11; Goldberger, *The City Observed*, p. 183.

56White and Willensky, *AIA Guide to New York City*, p. 390.

57van Leeuwen, *The Skyward Trend of Thought*, p. 89; King, ed., *King's Handbook of New York City*, p. 824; Gayle and Gillon, *Cast-Iron Architecture in New York*, p. 7; White and Willensky, *AIA Guide to New York City*, p. 31; Freyer, ed., *History of Real Estate*, p. 115.

58White and Willensky, *AIA Guide to New York City*, p. 53.

59Federal Writers Project, *WPA Guide to New York*, p. 99; *History of Architecture and the Building Trades*, vol. II, p. 218; van Leeuwen, *The Skyward Trend of Thought*, pp. 103–6; Kirby et al., *Engineering in History*, p. 322; Freyer, ed., *History of Real Estate*, pp. 465–66.

60"A New System of Fireproof Floor Construction," pp. 434–35.

61Condit, *American Building Art: The l9th Century*, p. 48; Freyer, ed., *History of Real Estate*, pp. 118, 467–68; Fleming, "A Half-Century of the Skyscraper," pp. 634–38; Condit, *American Building*, p. 118; Condit, *The Port of New York*, vol. I, p. 109; Goldberger, *Skyscraper*, p. 27; MacKay, *The Building of Manhattan*, p. 33.

62Freyer, ed., *History of Real Estate*, p. 118.

63Birkmire, *Skeleton Construction in Buildings*, pp. 115–16; Condit, "The Wind Bracing of Buildings," pp. 92–105.

64Birkmire, *Skeleton Construction in Buildings*, pp. 151–55.

65Stern, Gilmartin, and Mellins, *New York 1930*, p. 217; Stern, Gilmartin, and Massengale, *New York 1900*, p. 261; King, *King's Handbook of New York City*, p. 128.

66Birkmire, *Skeleton Construction in Buildings*, pp. 9, 11; King, *King's Handbook of New York City*, p. 220.

67"Concrete," *American Architect and Building News*, pp. 97–100; Birkmire, *Skeleton Construction in Buildings*, p. 222; Fleming, "A Half-Century of the Skyscraper," pp. 634–38.

68*History of Architecture and the Building Trades*, vol. II, p. 257.

69Condit, *American Building*, p. 119; Mujica, *History of the Skyscraper*, p. 56; *History of Architecture and the Building Trades*, vol. II, p. 224.

70"Some Details of the Ireland Building," pp. 127–28; *New York Times*, August 9–17, 1895; *History of Architecture and the Building Trades*, vol. II, p. 254; *Engineering News* editorial, August 15, 1895, p. 104; "Fall of a Building," pp. 111–12; *Engineering News* editorial, November 5, 1896, p. 312.

71Freitag, *Fireproofing*, p. 25; Freitag, *Fire Prevention*, p. 129; "The Efficiency of Modern Fireproof Building Construction," pp. 257–59; "The Effect of Fire on a Partially Fireproofed Building," p. 332.

72Toch, "Condition of the Steel at the Gillander Building," pp. 54–55.

73"The Development of Structural Ironwork Design," pp. 417–18.

74van Leeuwen, *The Skyward Trend of Thought*, p. 109; White and Willensky, *AIA Guide to New York City*, p. 32.

75Collins, *Concrete: The Vision of a New Architecture*, p. 62.

76"The Collapse of a Portion of a New Building in New York City," p. 23; *New York Times*, June 4 and 5, 1897; *History of Architecture and the Building Trades*, vol. II, p. 256; "Fatal Collapse of a Part of a Cast-Iron Building," pp 33–35.

77"The Removal of a Steel Frame Building," pp. 113–14.

78Condit, *The Port of New York*, vol. I, pp. 122–23.

79Birkmire, *Planning and Construction of High Office Buildings*, pp. 189–90; Mujica, *History of the Skyscraper*, pp. 29, 59.

80Diamonstein, *The Landmarks of New York*, p. 215.

81*New York Times*, March 18, 19, and 26, 1899; *Engineering News* editorial, March 23, 1899, p. 185.

82Author's observations on site.

83Freitag, *Fire Prevention*, pp. 183–185, 351.

84Starrett, *Skyscrapers*, p. 41.

85Collins, *Concrete: The Vision of a New Architecture*, p. 62; "A Concrete CourtHouse," p. 8

86Goldberger, *The City Observed*, p. 144.

87Condit, *The Port of New York*, vol. I, p. 260.

88Freitag, *Fire Prevention*, p. 152.

89van Leeuwen, *The Skyward Trend of Thought*, pp. 111–12.

90Ibid.

91Goldberger, *The City Observed*, p. 61.

92Interview with Fred Elsasser, December 1991.

93"Structural Engineering in Apartment House Construction, pp. 354–55; "The Collapse of the Darlington," pp. 217–19; "Concerning the Fall of the Darlington Building"; "Further Notes on the Collapse of the Darlington," pp. 250–51; Fleming, "A Half-Century of the Skyscraper," pp. 634–38; Parsons, "Collapse of a Building During Construction," pp. 454–56; Feld, *Construction Failure*, p. 42.

94van Leeuwen, *The Skyward Trend of Thought*, pp. 111–12.

95Collins, *Concrete: The Vision of a New Architecture*, pp. 88–89.

96Ibid.

97"Reinforced Concrete Pile Foundation," pp. 594–95.

98Stern, Gilmartin, and Massengale, *New York 1900*, p. 57; Collins, *Concrete: The Vision of a New Architecture*, pp. 88–89.

99Collins, *Concrete: The Vision of a New Architecture*, pp. 88–89.

[100]Birkmire, *Skeleton Construction in Buildings*, pp. 9, 11; "A New System of Fireproof Floor Construction," pp. 434–35; Diamonstein, *The Landmarks of New York*, p. 280; Goldberger, *The City Observed*, p. 178.

[101]Gray, "A Case of Terra Cotta Failure," p. 7.

[102]Collins, *Concrete: The Vision of a New Architecture*, pp. 88–89.

[103]Post, "Big Job on Big Exterior in Big Apple," p. 53; Mujica, *History of the Skyscraper*, p. 58.

[104]Birkmire, *Skeleton Construction in Buildings*, pp. 50–51; Goldberger, *The City Observed*, p. 176.

[105]"Tall Buildings of New York," pp. 400–403; Mujica, *History of the Skyscraper*, p. 57.

[106]Cowan, *Science and Building*, p. 91; Mujica, *History of the Skyscraper*, plate xxx.

[107]White and Willensky, *AIA Guide to New York City*, p. 23.

[108]Author's observations on site.

[109]Condit, *The Port of New York*, vol. II, p. 83.

[110]Condit, *American Building*, pp. 186–87; Mujica, *History of the Skyscraper*, p. 58; Condit, "The Wind Bracing in Buildings," pp. 92–105; Stockbridge, "The Interaction Between Exterior Walls," pp. 258–62; MacKay, *The Building of Manhattan*, pp. 38–39; Shopsin, *Restoring Old Buildings for Contemporary Uses*, pp. 110, 114–15.

[111]Condit, *American Building*, pp. 243–44, quoted with permission of The University of Chicago Press.

[112]Author's observations on site.

[113]Condit, *American Building Art: The 20th Century*, p. 30; Condit, *American Building*, p. 192.

[114]Fleming, "A Half-Century of the Skyscraper," p. 634–38.

[115]White and Willensky, *AIA Guide to New York City*, p. 156; Duncan, *Art Deco*, p. 185.

[116]White and Willensky, *AIA Guide to New York City*, p. 218.

[117]Condit, *American Building Art: The 20th Century*, p. 170; Condit, *American Building*, p. 244.

[118]Goldberger, *The City Observed*, p. 147.

[119]Gayle, Look, and Waite, *Metals in America's Historic Buildings*, vol. I, p. 85; Condit, *American Building*, p. 187.

[120]White and Willensky, *AIA Guide to New York City*, p. 170; Cowan, *Science and Building*, p. 290; Comer, *New York City Building Control*, p. 124.

[121]Condit, *American Building*, p. 277.

[122]Stern, Gilmartin, and Mellins, *New York 1930*, p. 107.

[123]White and Willensky, *AIA Guide to New York City*, p. 521.

[124]Goldberger, *On the Rise*, p. 253.

[125]Meadows, *Historic Building Facades*, p. 22; White and Willensky, *AIA Guide to New York City*, p. 247.

[126]Prudon, "Repairing Early Curtain Walls," pp. 123–26; Goldberger, *The Skyscraper*, p. 107, Schaal, *Curtain Walls*, p. 59; White and Willensky, *AIA Guide to New York City*, p. 152; Goldberger, *On the Rise*, p. 253.

[127]"Skyscraper Gets Aluminum Skin"; "Held by Rails"; "A Second Aluminum Faced Skyscraper."

[128]Ibid.

[129]Ibid.

[130]Goldberger, *The City Observed*, p. 141; "Glass Walls Will Show Off Bank's Interior."

[131]White and Willensky, *AIA Guide to New York City*, p. 156; "Miracle at Grand Central."

[132]"Bolting is New, TV Too—Both Meet."

[133]"Girder Will Carry 23 Floors."

[134]Goldberger, *The Skyscraper*, p. 109; "Metal Diamonds Adorn Skyscraper"; "Contract is Awarded for Largest Bolted Skyscraper."

[135]"A Skyscraper Crammed With Innovation."

[136]Byrne, "Skyscraper Sprouts Through Railroad Terminal Tracks."

[137]White and Willensky, *AIA Guide to New York City*, p. 127; Condit, *American Building*, p. 246.

[138]Goldberger, *The City Observed*, p. 155; Schaal, *Curtain Walls*, p. 89.

[139]"Tallest Welded Building in East Topped Out."

[140]Prudon, "Repairing Early Curtain Walls," pp. 123–26; Condit, "The Wind Bracing of Buildings," pp. 92–105.

[141]White and Willensky, *AIA Guide to New York City*, p. 148.

[142]Ibid., p. 76.

[143]Ibid., p. 519.

[144]Ibid., p. 171; Huxtable, *Will They Ever Finish Bruckner Boulevard?*, pp. 98–102.

[145]White and Willensky, *AIA Guide to New York City*, p. 70.

[146]Ibid., p. 174.

[147]Goldberger, *The City Observed*, p. 134.

[148]White and Willensky, *AIA Guide to New York City*, p. 128.

[149]Goldberger, *The City Observed*, p. 200; White and Willensky, *AIA Guide to New York City*, p. 181.

[150]Prudon, "Repairing Early Curtain Walls," pp. 123–26; Goldberger, *The Skyscraper*, p. 128; Condit, *American Building*, p. 198.

[151]Goldberger, *The City Observed*, p. 261.

[152]Ibid., pp. 158–59.

### APPENDIX C

[1]Plummer and Wanner, *Principles of Tile Engineering*, p. 8.

[2]Freyer, ed., *History of Real Estate*, p. 465.

[3]Green, "Cements and Concrete"; Kirby, et al., *Engineering in History*, p. 197.

[4]Waite, "Cinder Concrete Floors," p. 1776.

[5]Hool, *Reinforced Concrete Construction*, vol. II, pp. 163–71.

[6]AISC, *Steel Construction Manual*, 4th ed., p. 7.

[7]Green, "Cements and Concrete"; Sturgis, *Dictionary of Architecture*, vol. I, col. 484–86.

[8]Green, "Cements and Concrete."

[9]Freyer, ed., *History of Real Estate*, p. 465.

### APPENDIX D

[1]Hodgkinson, *Experimental Researches*, pp. 229, 234–39.

[2]Fairbairn, *On the Application of Cast and Wrought Iron*, p. 36.

[3]Ibid.

[4]Ferris, *AISC Iron and Steel Beams*, p. 5.

[5]Ibid.

[6]"The Strength of Columns," *Engineering News*, p. 93.

[7]Ibid.

[8]Kidder, "Strength of Columns," pp. 26–27.

[9]Ibid.

[10]Ibid.

[11]Ferris, *AISC Iron and Steel Beams*, p. 5.

[12]Burr, *Elasticity and Resistance*, pp. 220, 258, 376.

[13]Ibid.

[14]Christie, "Strength and Elasticity of Structural Steel," pp. 254, 260.

[15]Ferris, *AISC Iron and Steel Beams*, p. 5.

[16]Johnson, "On the Strength of Columns," p. 530.

[17]Ibid.

[18]Ibid.

[19]Ibid.

[20]*Manual of the Bouton Foundry*, pp. 11, 14, 65.

[21]Ibid, pp. 149, 150.

[22]Ferris, *AISC Iron and Steel Beams*, p. 5.

[23]*Manual of the Bouton Foundry*, pp. 149, 150.

[24]Johnson, "Proper Tests for Cast Iron," p. 258.

[25]Ferris, *AISC Iron and Steel Beams*, p. 5.

[26]Trautwine, *Civil Engineer's Pocket-Book*, pp. 398, 434e.

[27]Ibid., pp. 400, 434e.

[28]Ibid.

[29]Ferris, *AISC Iron and Steel Beams*, p. 5.

[30]*Building Laws*, 1891, p. 43.

[31]DuBois, *The Strains of Framed Structures*, p. 327.

[32]Ibid.

[33]Birkmire, *Skeleton Construction in Buildings*, p. 75.

[34]Ferris, *AISC Iron and Steel Beams*, p. 5.

[35]Ibid.

[36]"Standard Specifications for Steel," p. 130.

[37]Ibid.

[38]Ibid.

[39]Ferris, *AISC Iron and Steel Beams*, p. 6.

[40]Ibid.

[41]Ibid., p. 5

[42]*Building Code*, 1901, pp. 15, 118–21.

[43]Ibid., pp. 16, 118–21.

[44]Ibid., p. 16.

[45]Ibid., pp. 16, 118–21.

[46]Ibid., p. 115.

[47]Ibid., pp. 118–21.

[48]Ibid.

[49]Ibid.

[50]Ibid.

[51]Ibid.

[52]Ibid.

[53]Ibid.

[54]Ibid.

[55]Ibid.

[56]Ibid.

[57]Ferris, *AISC Iron and Steel Beams*, p. 6.

[58]Ibid.

[59]International Correspondence Schools, *Mechanics' Handbook*, p. 151.

[60]Ibid.

[61]Ibid.

[62]Ferris, *AISC Iron and Steel Beams*, p. 5.

[63]Ibid., p. 6

[64]Ibid.

[65]Hudson, *Deflections and Statically Indeterminate Stresses*, p. 4.

[66]Ibid.

[67]Ibid.

[68]Ibid.

[69]*Pocket Companion*, 1919, pp. 158–64.

[70]Ibid.

[71]Ibid.

[72]Ibid.

[73]Ibid.

[74]Ferris, AISC Iron and Steel Beams, p. 6.

[75]Ibid.

[76]Ibid.

[77]Ibid.

[78]Ibid., p. 5

[79]Hool, *Reinforced Concrete Construction*, vol. I, p. 24.

[80]Ibid.

[81]Merriman, ed., *American Civil Engineers' Handbook*, pp. 325, 359.

[82]Ibid., p. 325

[83]Ibid.

[84]Ibid.

[85]Nolan, ed., *Kidder's Architects and Builders' Handbook*, pp. 376, 412.

[86]Ibid.

[87]Ibid.

[88]Hool and Kinne, *Steel and Timber Structures*, pp. 6, 21.

[89]Ibid.

[90]Ibid.

[91]Ibid.

[92]Ibid.

[93]"Standard Specification for Structural Steel for Buildings," *Steel Construction Manual*, 1st ed., pp. 8–16.

[94]Ibid.

[95]Ibid.

[96]Ibid.

[97]Ibid.

[98]*Ferris, AISC Iron and Steel Beams*, p. 7.

[99]Ibid.

[100]*Building Code*, 1926, pp. 15–21.

[101]Ibid.

[102]Ibid.

[103]Ibid.

[104]Ibid.

[105]Ibid.

[106]Ibid.

[107]Ibid.

[108]Ferris, *AISC Iron and Steel Beams*, p. 7.

[109]Ibid.

[110]Ibid.

[111]Ibid., pp. 7, 9.

[112]Ibid.

[113]Ibid.

[114]"Specification for the Design, Fabrication, and Erection of Structural Steel for Buildings," *Steel Construction Manual*, 2d ed., pp. 227–34.

[115]Ibid.

[116]Ibid.

[117]Ibid.

[118]Ibid.

[119]Ibid.

[120]Ibid., p. 274.

[121]Ibid., 3d ed., pp. 261–78.

[122]Ibid.

[123]Ibid.

[124]Ibid.

[125]Ibid.

[126]Ibid.

[127]Ibid.

[128]Ferris *AISC Iron and Steel Beams*, pp. 7–9.

[129]"Specification for the Design, Fabrication, and Erection of Structural Steel for Buildings," *Steel Construction Manual*, 4th ed., pp. 271–92.

[130]Ibid.

[131]Ibid.

[132]Ibid.

[133]Ibid.

[134]Ibid.

[135]Ibid.

[136]Ibid.

[137]*Building Laws*, 1942, p. 98.

[138]Ibid. pp. 99–104.

[139]Ibid.

[140]Ibid.

[141]Ibid.

[142]Ibid.

[143]Ibid.

[144]Ibid.

[145]"Specification for the Design, Fabrication, and Erection of Structural Steel for Buildings," *Steel Construction Manual*, 5th ed., pp. 277–305.

[146]Ibid.

[147]Ibid.

[148]Ibid.

[149]Ibid.

[150]Ibid.

[151]Ibid.

[152]Ibid.

[153]Ibid.

[154]Ibid.

[155]Ibid.

[156]Ibid.

[157]Ibid.

[158]Ferris, *AISC Iron and Steel Beams*, p. 9.

[159]Ibid.

[160]Urquhart, *Civil Engineering Handbook*, 3d ed., p. 524.

[161]Ibid.

**APPENDIX E**

[1]Hodgkinson, *Experimental Researches*, pp. 229, 234–39, 240–47, 248–57.

[2]"The Strength of Columns," *Engineering News*, p. 93.

[3]Ibid.

[4]Ibid.

[5]Kidder, "Strength of Columns," pp. 26–27.

[6]Ibid.

[7]Ibid.

[8]Ibid.

[9]Ibid.

[10]Ibid.

[11]Ibid.

[12]Ibid.

[13]Birkmire, *Skeleton Construction in Buildings*, p. 33.

[14]Johnson, "On the Strength of Columns," p. 530.

[15]Ibid.

[16]Ibid.

[17]Ibid.

[18]*Manual of the Bouton Foundry*, p. 10.

[19]Ibid.

[20]Birkmire, *Skeleton Construction in Buildings*, p. 34.

[21]Trautwine, *Civil Engineer's Pocket-Book*, p. 439.

[22]Ibid.

[23]Ibid.

[24]Ibid.

[25]Birkmire, *Skeleton Construction in Buildings*, pp. 26–28.

[26]Ibid.

[27]Ibid.

[28]Ibid.

[29]Birkmire, *Planning and Construction of High Office Buildings*, p. 198.

[30]Ibid.

[31]Ibid.

[32]*Building Code*, 1901, p. 112.

[33]Ibid., pp. 116–17.

[34]Ibid.

[35]Ibid.

[36]*Pocket Companion*, 1903, pp. 125, 126.

[37]Ibid., p. 148.

[38]Ibid.

[39]Blakely, *Dimensions, Weights, and Properties*, p.

255.

[40]Ibid.

[41]Ibid.

[42]Johnson, Bryan, and Turneaure, *Modern Framed Structures*, p. 167.

[43]Ibid.

[44]Ibid.

[45]*Pocket Companion*, 1919, pp. 158–64.

[46]Ibid., pp. 294, 320.

[47]Ibid.

[48]Ibid., p. 325.

[49]Ibid., pp. 220–21, 247.

[50]Ibid.

[51]Merriman, ed., *American Civil Engineers' Handbook*, pp. 363–64.

[52]Ibid.

[53]Ibid.

[54]"Standard Specification for Structural Steel for Buildings," *Steel Construction Manual*, 1st ed., pp. 8–16.

[55]Building Code, 1926, pp. 15–21.

[56]Ibid.

[57]Ibid.

[58]"Specification for the Design, Fabrication, and Erection of Structural Steel for Buildings," *Steel Construction Manual*, 2d ed., pp. 227–34.

[59]Building Code, 1936, p. 191.

[60]"Specification for the Design, Fabrication, and Erection of Structural Steel for Buildings," *Steel Construction Manual*, 3d ed., pp. 261–78.

[61]Building Code, 1939, p. 250.

[62]Grinter, *Design of Modern Steel Structures*, p. 184.

[63]"Specification for the Design, Fabrication, and Erection of Structural Steel for Buildings," *Steel Construction Manual*, 4th ed., pp. 271–92.

[64]Building Laws, 1942, p. 98.

[65]Ibid., pps. 99–104.

[66]Ibid.

[67]"Specification for the Design, Fabrication, and Erection of Structural Steel for Buildings," *Steel Construction Manual*, 5th ed., pp. 285–88.

[68]*The Lally Column Catalog*, pp. 7, 10, 21, 30.

[69]"Specification for the Design, Fabrication, and Erection of Structural Steel for Buildings," *Steel Construction Manual*, 6th ed., pp. 5-15–5-20.

[70]"Specification for the Design, Fabrication, and Erection of Structural Steel for Buildings," Steel Construction Manual, 7th ed., pp. 5-3–5-166.

[71]"Specification for the Design, Fabrication, and Erection of Structural Steel for Buildings," *Steel Construction Manual*, 8th ed., pp. 5-3–5-168.

**APPENDIX F**

[1]Waite, ed., *Iron Architecture in New York City*, p. 31.

[2]Freyer, ed., *History of Real Estate*, p. 288.

[3]Ibid.

[4]Gray, "Status for an 1866 Frame House," p. 6.

[5]Ibid.

[6]Building Laws, 1891, pp. 46–47.

[7]Freyer, ed., *History of Real Estate*, p. 288.

[8]Ibid.

**APPENDIX G**

[1]Burr, *A Course on the Stresses*, 1st ed., p. 19.

[2]Ibid.

[3]*Building Laws*, 1891, p. 43.

[4]Ibid.

[5]Ibid.

[6]Birkmire, *Skeleton Construction in Buildings*, pp. 63–64.

[7]Ibid.

[8]Ibid.

[9]Ibid., p. 64.

[10]Birkmire, *Planning and Construction of High Office Buildings*, pp. 119–21.

[11]Ibid.

[12]Ibid.

[13]Ibid.

[14]*Building Code*, 1901, p. 111.

[15]Ibid.

[16]Ibid.

[17]Ibid.

[18]Ibid.

[19]Ibid.

[20]Ibid.

[21]Ibid.

[22]Ibid.

[23]Ibid., pp. 121–22.

[24]*Pocket Companion*, 1923, p. 265.

[25]Ibid.

[26]Ibid.

[27]Ibid.

[28]Ibid.

[29]Ibid.

[30]Ibid.

[31]Ibid.

[32]Ibid.

[33]Hool and Kinne, *Steel and Timber Structures*, p. 30.

[34]Ibid., p. 4.

[35]Ibid.

[36]Ibid.

[37]Ibid.

[38]Ibid.

[39]Ibid.

[40]Ibid.

[41]Ibid.

[42]Ibid.

[43]Ibid.

[44]Ibid.

[45]Ibid.

[46]Ibid.

[47]Ibid.

[48]*Building Code*, 1926, pp. 15–21.

[49]Ibid.

[50]Ibid.

[51]Ibid.

[52]Ibid.

[53]Ibid.

[54]Ibid.

[55]Ibid.

[56]Ibid.

[57]Ibid.

[58]*Building Code*, 1936, p. 191.

[59]Ibid.

[60]Ibid.

[61]Ibid.

[62]Ibid.

[63]Ibid.

[64]Ibid.

[65]Ibid.

[66]Ibid.

[67]*Building Code*, 1939, p. 250.

[68]Ibid.

[69]Ibid.

[70]Ibid.

[71]Ibid.

[72]Ibid.

[73]Ibid.

[74]Urquhart, ed., *Civil Engineering Handbook*, 3d ed., p. 553.

[75]Ibid.

[76]Ibid.

[77]Ibid.

[78]Ibid.

[79]Ibid.

[80]Ibid.

[81]*Building Laws*, 1942, p. 88.

[82]Ibid.

[83]Ibid., p. 89.

[84]Ibid.

[85]Ibid.

[86]Ibid.

[87]Ibid.

[88]Ibid., p. 90.

[89]Ibid.

[90]Ibid.

[91]Ibid.

[92]Ibid.

[93]Ibid.

[94]Ibid.

[95]Ibid., p. 91.

[96]Ibid., p. 92.

**APPENDIX H**

[1]Von Emperger, "The Calculation of Flat Arches," pp. 186–87.

[2]Johnson, "The Ultimate Strength of Concrete Steel Beams," pp. 261–62.

[3]Waite, "Cinder Concrete Floors," p. 1775.

[4]"Second Report of Joint Committee on Concrete and Reinforced Concrete," in Hool, *Reinforced Concrete Construction*, vol. II, pp. 615–59.

[5]Perrine and Strehan, "Cinder Concrete Floor Construction," pp. 523–621.

[6]Ibid.

[7]*Pocket Companion*, 1919, pp. 326–30.

[8]Hool and Kinne, *Structural Members and Connections*, p. 436.

[9]Ibid.

[10]*American Welded Wire Fabric*, pp. 46, 48, 49.

[11]*Building Laws*, 1942, pp. 94–98.

[12]*American Welded Wire Fabric*, pp. 30–32.

[13]Plummer and Wanner, *Principles of Tile Engineering*, p. 383.

# BIBLIOGRAPHY

AISC. *Steel Construction*, 1st ed. New York: American Institute of Steel Construction, 1924.

———. *Steel Construction*, 2d ed. New York: American Institute of Steel Construction, 1934.

———. *Steel Construction*, 3d ed. New York: American Institute of Steel Construction, 1937.

———. *Steel Construction*, 4th ed. New York: American Institute of Steel Construction, 1941.

———. *Steel Construction Manual*, 5th ed. New York: American Institute of Steel Construction, 1947.

———. *Manual of Steel Construction*, 6th ed. New York: American Institute of Steel Construction, 1963.

———. *Manual of Steel Construction*, 7th ed. New York: American Institute of Steel Construction, 1970.

———. *Manual of Steel Construction*, 8th ed. Chicago: American Institute of Steel Construction, 1980.

———. *Manual of Steel Construction: Allowable Stress Design*, 9th ed. Chicago: American Institute of Steel Construction, 1989.

"AISC Develops New Rules for Heavy Shapes." *Modern Steel Construction*, February 1989, 21.

Aitchison, G. "On Iron as a Building Material." *Papers Read at the Royal Institute of British Architects, Session 1863–64*, London: The Royal Institute of British Architects, 1864, 97–107.

Albion, Robert Greenhalgh. *The Rise of New York Port*. New York: Charles Scribner and Sons, 1939.

"Aluminum Exterior." *Engineering News-Record*, October 12, 1950, 28.

"American Portland Cement." *Engineering News*, June 21, 1879, 197–98.

*American Welded Wire Fabric for Concrete Reinforcement*. Pittsburgh, PA: American Steel & Wire Division of United States Steel, 1944.

*Arbed-Rolled Wide Flange Beams*, 4th ed. New York: Trade-Arbed, 1988.

ASCE, *Guideline for Structural Condition Assessment of Existing Buildings*. New York: American Society of Civil Engineers, 1991.

Aus, Gunvald. "Test of Long-Span Terra-Cotta Floor Arches." *Engineering News*, November 7, 1895, 314.

Badger, Daniel. *Illustrations of Iron Architecture, Made by The Architectural Iron Works of the City of New York*. New York: Baker & Godwin, Printers, 1865; New York: Dover Publications, 1981, with a new introduction by Margot Gayle.

Baier, Julius. "Wind Bracing in the St. Louis Tornado, with Special Reference to the Necessity of Wind Bracing for High Buildings." *Transactions of the American Society of Civil Engineers*, vol. XXXVII. New York: American Society of Civil Engineers, 1897.

Bannister, Turpin. "Bogardus Revisited." *Journal of the Society of Architectural Historians*, December 1956, 12–22; March 1957, 11–19.

Belle, John, John Ray Hoke, Jr., and Stephen A. Kliment, eds. *Traditional Details for Building Restoration, Renovation, and Rehabilitation from the 1932–1951 Editions of Architectural Graphic Standards*. New York: John Wiley & Sons, 1991.

*Bethlehem Structural Shapes*. Bethlehem, PA: Bethlehem Steel Company, 1938.

"Big City Lift Slabs." *Engineering News-Record*, June 19, 1958.

"Bill to Allow Building Inspection by A-E's Debated." *Engineering News-Record*, May 6, 1976, 14.

Birkmire, William. *Architectural Iron and Steel, and Its Application in the Construction of Buildings*. New York: John Wiley & Sons, 1892.

———. *Skeleton Construction in Buildings*. New York: John Wiley & Sons, 1894.

———. *The Planning and Construction of High Office Buildings*. New York: John Wiley & Sons, 1898.

Bixby, W. H. "Wind Pressures in Engineering Construction." *Engineering News*, March 14, 1895, 175–84.

Blakely, George H. *Dimensions, Weights, and Properties of Special and Standard Structural Steel Shapes Manufactured by Bethlehem Steel Company*, 1st ed. South Bethlehem, PA: Bethlehem Steel Company, 1907.

"Bolting is New, TV Too—Both Meet." *Engineering News-Record*, April 19, 1956.

Bruce, A. F. "The Transverse Strength of Concrete." *Engineering News*, November 9, 1893, 367–68.

*The Building Code of The City of New York*. New York: Department of Buildings, 1901.

*The Building Code of The City of New York*. New York: Department of Buildings, 1926.

*The Building Code of The City of New York*. New York: Department of Buildings, 1936.

*The Building Code of The City of New York*. New York: Department of Buildings, 1939.

*Building Code of The City of New York*. New York: Gould Publications, 1991.

"Building Going Down?" *Civil Engineering*, September 1974, 32.

"Building Inspection in New York City." *Engineering Record*, July 24, 1897, 165.

*The Building Laws, Relating to the Construction of Buildings in the City of New York*. New York: Willis McDonald & Company, 1891.

*Building Laws of the City of New York*. Vol. I (Building Code: General Provisions and Articles 1–11). New York: Department of Buildings, 1942.

"The Buildings of New York." *Engineering News*, January 5, 1884, 6–7.

Burnham, Alan. "Last Look at a Structural Landmark." *Architectural Record*, September 1956, 273–79.

Burr, William H. *A Course on the Stresses in Bridge and Roof Trusses, Arched Ribs, and Suspension Bridges*. 1st ed. New York: John Wiley & Sons, 1880.

———. *A Course on the Stresses in Bridge and Roof Trusses, Arched Ribs, and Suspension Bridges*, 8th ed. New York: John Wiley & Sons, 1893.

———. *The Elasticity and Resistance of the Materials of Engineering*. New York: John Wiley & Sons, 1883.

Byrne, Donald. "Skyscraper Sprouts Through Railroad Terminal Tracks." *Engineering News-Record*, July 10, 1958.

"Cast Iron *vs.* Wrought Iron and Steel in Building Columns." *Engineering Record*, November 19, 1892, 389–90.

"Chicken Little, the Sky Is Falling." *Engineering News-Record*, December 9, 1991, 82.

Christie, James. "The Strength and Elasticity of Structural Steel, and Its Efficiency in the Form of Beams and Struts." *Transactions of the American Society of Civil Engineers*, vol. XIII. New York: American Society of Civil Engineers, 1884.

Clark, T. M. "Fireproof Building in New York." *Scientific American Architects' and Builders' Edition*, November 1885.

Clark, William Gifford, and J. L. Kingston. *The Skyscraper; A Study in the Economic Height of Modern Office Buildings*. New York: American Institute of Steel Construction, 1930.

Clausen, Meredith L. "Belluschi and the Equitable Building in History." *The Journal of the Society of Architectural Historians*, June 1991, 109–29.

"The Collapse of a Portion of a New Building in New York City." *Engineering Record*, June 12, 1897, 23.

"The Collapse of the Darlington Apartment House in New York City." *Engineering News*, vol. LI, no. 10, March 10, 1904, 217–19.

Collins, George R. "Guastavino y Moreno, Rafael and Guastavino y Esposito, Rafael." In *Macmillan Encyclopedia of Architects*. New York: Macmillan Publishing Co., 1982.

Collins, Peter. *Concrete: The Vision of a New Architecture*. New York: Horizon Press, 1959.

Comer, John P. *New York City Building Control, 1800–1941*. New York: Columbia University Press, 1942.

"Comparative Standard Fireproof Floor Tests of the New York Building Department." *Engineering Record*, September 18, 1897, 337–40; September 25, 1897, 359–63; October 2, 1897, 382–87; October 9, 1897, 402–5.

"Composite Beam Patentee Forgoes Royalties." *Engineering News-Record*, November 23, 1964.

"Concerning the Fall of the Darlington Building." *Engineering News*, March 24, 1904, 281–82.

"Concrete." *American Architect and Building News*, vol. 129, 1926, 97–100.

"A Concrete Court House." *American Architect and Building News*, vol. 75, 1902, 87.

Condit, Carl. *American Building*. 2nd ed. Chicago: University of Chicago Press, 1982.

——. *American Building Art: The 19th Century*. New York: Oxford University Press, 1960.

——. *American Building Art: The 20th Century*. New York: Oxford University Press, 1961.

——. "Bogardus, James." In *Macmillan Encyclopedia of Architects*. New York: Macmillan Publishing Co., 1982.

——. *The Port of New York*. 2 vols. Chicago: University of Chicago Press, 1980.

——. "The Two Centuries of Technical Evolution Underlying the Skyscraper." In Council on Tall Buildings and Urban Habitat, *Second Century of the Skyscraper*. New York: Van Nostrand Reinhold, 1988.

——. "The Wind Bracing of Buildings." *Scientific American*, February 1974, 92–105.

Connor, Frank L. *Labor Costs of Construction*. Chicago: Gillette Publishing Company, 1931.

"Contract is Awarded for Largest Bolted Skyscraper." *Engineering News-Record*, January 26, 1956.

Cook, John Philip. *Composite Construction Methods*. New York: Wiley Interscience, 1977.

"The Corrosion of Reinforcing Metal in Cinder Concrete Floors." *Engineering News*, November 1, 1906, 458, 461; November 22, 1906, 549–50.

"Corruption Clean-up." *Civil Engineering*, March 1975, 69.

Cowan, Henry J. *Science and Building*. New York: John Wiley & Sons, 1978.

Cross, Hardy. "Analysis of Rigid Frames by the Distribution of Fixed End Moments." In *Proceedings of the ASCE*. New York: American Society of Civil Engineers, 1930.

Crowly, Herbert D. "The Use of Terra Cotta in the United States; How it has Increased." *Architectural Record*, July 1905, 86–94.

Cushman, Allerton S., and Henry Gardner. *The Corrosion and Preservation of Iron and Steel*. New York: McGraw-Hill, 1910.

Dencer, F. W. *Detailing and Fabricating Structural Steel*. New York: McGraw-Hill, 1924.

"The Development of Structural Ironwork Design." *Engineering News*, December 24, 1896, 417–18.

Diamonstein, Barbaralee. *The Landmarks of New York*. New York: Harry N. Abrams, 1988.

DuBois, A. Jay. *The Strains of Framed Structures*. New York: John Wiley & Sons, 1892.

Duncan, Alastair. *Art Deco*. New York: Thames and Hudson, 1988.

"Early Bridge Building Pointed Way to Modern Use of Structural Steel." In *Steel Facts*, no. 79. New York: American Iron and Steel Institute, August 1946.

"The Effect of Fire on a Partially Fireproofed Building." *Engineering News*, November 14, 1895, 332.

"The Efficiency of Modern Fireproof Building Construction." *Engineering News*, April 16, 1896, 257–59.

*Engineering News*, editorial on Darlington Apartments, June 15, 1911, 730–31.

*Engineering News*, editorial on floor construction, July 1, 1897, 8.

*Engineering News*, editorial on high buildings, April 6, 1889, 299.

*Engineering News*, editorial on high buildings, December 15, 1892.

*Engineering News*, editorial on high buildings, October 25, 1894, 342.

*Engineering News*, editorial on Hotel Windsor, March 23, 1899, 185.

*Engineering News*, editorial on Ireland Building, August 15, 1895, 104.

*Engineering News*, editorial on Ireland Building, November 5, 1896, 312.

*Evaluation of Reinforcing Steel Systems in Old Reinforced Concrete Structures.* Chicago: Concrete Reinforcing Steel Institute, 1981.

"Experiments in Flexure of Beams." *Engineering News*, March 4, 1876, 76.

"Experiments with a Flitch Plate Girder." *Engineering News*, October 7, 1876, 325.

Faija, Henry. "Portland Cement and Concrete." *Engineering News*, May 12, 1883, 217–18.

Fairbairn, William. *On the Application of Cast and Wrought Iron to Building Purposes.* New York: John Wiley, 1854.

"Falling Masonry Fatally Injures Barnard Student." *New York Times*, May 17, 1979, B3-6.

"Fall of a Building with Cast Iron Columns." *Engineering News*, August 15, 1895, 111–12.

"Fatal Collapse of a Part of a Cast-Iron Column Building in New York." *Engineering Record*, June 12, 1897, 33–35.

The Federal Writers' Project of the Works Progress Administration. *New York Panorama.* New York: Pantheon Books, 1984.

The Federal Writers' Project of the Works Progress Administration. *The WPA Guide to New York.* New York: Pantheon Books, 1982.

Feld, Jacob. *Construction Failure.* New York: John Wiley & Sons, 1968.

Ferris, Herbert W., ed. *AISC Iron and Steel Beams, 1873–1952.* Chicago: American Institute of Steel Construction, 1953.

Ferris, Hugh. *The Metropolis of Tomorrow.* Princeton, NJ: Princeton Architectural Press, 1986.

"Fireproof Floor Construction in New York City." *Engineering News*, February 4, 1897, 72–75.

Fisher, Douglas A. *The Epic of Steel.* New York: Harper & Row, 1967.

Fisher, John W., and Alan W. Pense. "Experience with the Use of Heavy W Shapes in Tension." *AISC Engineering Journal*, 2d quarter 1987, 63–77.

Fitch, James Marston. *American Building: The Historical Forces That Shaped It.* New York: Schocken Books, 1973.

———. *Historic Preservation.* Charlottesville: University Press of Virginia, 1990.

Fleming, Robins. "A Half Century of the Skyscraper." *Civil Engineering*, vol. 4, no. 12, December 1934, 634–38.

———. "For and Against the Skyscraper." *Civil Engineering*, vol. 5, no. 6, June 1935.

———. "Whence the Skyscraper." *Civil Engineering*, vol. 4, no. 10, October 1934, 505–9.

Fox, William H. "The Corrosion of Steel in Reinforced Cinder Concrete." *Engineering News*, May 23, 1907, 569–70.

Freitag, Joseph Kendall. *Fire Prevention and Fire Protection.* 2d ed. New York: John Wiley & Sons, 1921.

———. *The Fireproofing of Steel Buildings.* New York: John Wiley & Sons, 1899.

Freyer, William J. "The New York Building Laws." *Architectural Record*, November 1891, 69–82.

———. "Skeleton Construction." *Architectural Record*, December 1891, 228–35.

Freyer, William J., ed. *History of Real Estate, Building, and Architecture in New York City.* 1898. Reprint. New York: Arno Press, 1967.

Friberg, Bengt F. "Combined Form and Reinforcement for Concrete Slabs." *ACI Journal Proceedings*, May 1954.

"Further Notes on the Collapse of the Darlington Building." *Engineering News*, March 17, 1904, 250–51.

Gayle, Margot, and Edmund Gillon, Jr. *Cast-Iron Architecture in New York.* New York: Dover Publications, 1974.

Gayle, Margot, David W. Look, and John G. Waite. *Metals in America's Historic Buildings: Part I. A Historical Survey of Metals.* Washington, D.C.: Preservation Assistance Division, National Park Service, U.S. Department of the Interior, 1980.

Giedion, Sigfreid. *Space, Time and Architecture.* Cambridge, MA: Harvard University Press, 1944.

Gillette, Leslie H. *American Institute of Steel Construction: The First 60 Years.* Chicago: American Institute of Steel Construction, 1980.

"Girder Will Carry 23 Floors." *Engineering News-Record*, March 8, 1956.

"Glass Walls Will Show Off Bank's Interior." *Engineering News-Record*, January 14, 1954.

Goldberger, Paul. *The City Observed: New York*. New York: Vintage Books, 1979.

———. *On The Rise*. New York: Penguin Books, 1985.

———. *The Skyscraper*. New York: Alfred A. Knopf, 1982.

Gray, Christopher. "A Case of Terra Cotta Failure." *New York Times*, February 2, 1992, real estate sect., 7.

———. "An 1869 Work with a Shaky Future." *New York Times*, June 23, 1991, real estate sect., 6.

———. "Status for an 1866 Frame House." *New York Times*, May 19, 1991, real estate sect., 6.

Green, Bernard R. "Cements and Concrete." *Engineering News*, February 24, 1877, and March 17, 1877.

Green, Robert S. *Design for Welding*. Cleveland, OH: James F. Lincoln Arc Welding Foundation, 1948.

Grinter, Linton E. *Design of Modern Steel Structures*. New York: Macmillan, 1941.

Gupte, Pranay B. "City Is Studying Why Masonry Fell, Killing Student." *New York Times*, May 18, 1979, B1-4.

Gwilt, Joseph. *The Encyclopedia of Architecture*. London: Longmans, Green, 1867. Reprint. New York: Bonanza Books, 1982.

Hayden, Arthur G. "Continuous Frame Design Used for Concrete Highway Bridges." *Engineering News-Record*, January 11, 1923, 73–75.

———. "Rigid Frames in Concrete Bridge Construction." *Engineering News-Record*, April 29, 1926, 686–89.

"Held By Rails." *Engineering News-Record*, June 17, 1954.

"High Tensile Bolts Speed Erection." *Engineering News-Record*, April 14, 1955.

Hinton, Henry L. *Catalog of the National Fireproofing Company*. New York: National Fireproofing Corporation, 1903.

"Historic Terra Cotta-Clad Tower Gets $9-Million Face-Lift." *Engineering News-Record*, July 27, 1978.

"History and Properties of Light-Weight Aggregates." *Engineering News-Record*, April 24, 1919, 802–4.

*History of Architecture and the Building Trades of Greater New York*. 2 vols. New York: Union Historical Company, 1899.

Hitchcock, Henry-Russell. "American Influence Abroad." In *The Rise of an American Architecture*, edited by Edgar Kaufman. New York: Praeger, 1970.

Hodgkinson, Eaton. *Experimental Researches on the Strength and Other Properties of Cast Iron*. 2d ed. London: John Weale, 1861.

Hodgkinson, Eaton, ed. *A Practical Essay on the Strength of Cast Iron and Other Metals by the late Thomas Tredgold, C.E.* 5th ed. London: John Weale, 1861.

Hoerr, John P. *And the Wolf Finally Came: The Decline of the American Steel Industry*. Pittsburgh, PA: University of Pittsburgh Press, 1988.

Hool, George A. *Reinforced Concrete Construction*. Vol. I: Fundamental Principles. New York: McGraw-Hill, 1917.

———. *Reinforced Concrete Construction*. Vol. II: Retaining Walls and Buildings. New York: McGraw-Hill, 1913.

Hool, George A., and Nathan C. Johnson. *Concrete Engineers' Handbook*. New York: McGraw-Hill, 1918.

Hool, George A., and W. S. Kinne. *Foundations, Abutments and Footings*. New York: McGraw-Hill, 1923.

———. *Reinforced Concrete and Masonry Structures*. New York: McGraw-Hill, 1924.

———. *Steel and Timber Structures*. New York: McGraw-Hill, 1924.

———. *Stresses in Framed Structures*. New York: McGraw-Hill, 1923.

———. *Structural Members and Connections*. New York: McGraw-Hill, 1923.

Hudson, Clarence W. *Deflections and Statically Indeterminate Stresses*. New York: John Wiley & Sons, 1912.

Huxtable, Ada Louise. "Harper & Brothers Building—1854, New York." *Progressive Architecture*, February 1957, 153–54.

———. *Will They Ever Finish Bruckner Boulevard?* New York: Collier Books, 1972.

"Inspection Proposal Deserves Support." *Engineering News-Record*, May 6, 1976, 68.

International Correspondence Schools. *The Mechanics' Handbook*. Scranton, PA: International Textbook Company, 1904.

"Investigating the New York Building Department." *Engineering News*, 1899, 285–86.

"Iron and Concrete Adapted to Resist Transverse Strains." *Engineering News*, September 8, 1888, 180–81.

"Iron Castings." *Engineering News*, June 17, 1876, 197.

Jenny, Daniel. "Structural Lightweight Concrete for Composite Design." *AISC Engineering Journal*, October 1965, 122–24.

Jensen, Robert. "Board and Batten Siding and the Balloon Frame: Their Incompatibility in the Nineteenth Century." *Journal of the Society of Architectural Historians*, March 1971, 40–50.

Jewett, Robert. "Solving the Puzzle of the First American Structural Rail-Beam." *Technology and Culture*, July 1969, 371–91.

Johnson, J. B. "Proper Tests for Cast Iron." *Engineering News*, September 14, 1889, 258.

———. "The Ultimate Strength of Concrete-Steel Beams." *Engineering News*, October 21, 1897, 261–62.

Johnson, J. B., C. W. Bryan, and F. E. Turneaure. *The Theory and Practice of Modern Framed Structures*. New York: John Wiley & Sons, 1909.

Johnson, Thomas H. "Comparison of Formulæ for the Strength of Columns." *Engineering News*, January 22, 1888, 482–83.

———. "On the Strength of Columns: Discussing the Experiments Which Have Been Accumulated, and Proposing New Formulas." *Transactions of the American Society of Civil Engineers,* vol. 15. New York: American Society of Civil Engineers, 1886, 517–36.

Joint Committee of the ASCE, ASTM, AREA, Association of American Portland Cement Manufacturers. *Second Report of the Joint Committee on Concrete and Reinforced Concrete*. Reprinted in George Hool, *Reinforced Concrete Construction*, vol. II, 1913.

Kahn, David M. "Bogardus, Fire, and the Iron Tower." *Journal of the Society of Architectural Historians*, October 1976, 186–203.

Kesler, R. "Reinforced Concrete Materials—A Remarkable Heritage." *Journal of the Structural Division of the American Society of Civil Engineers*, April 1977.

Kessner, Thomas. *Fiorello H. LaGuardia and the Making of Modern New York*. New York: Penguin Books, 1991.

Ketchum, Milo S. *The Structural Engineer's Handbook*. 2d ed. New York: McGraw-Hill, 1918.

Kidder, F. E. "Strength of Columns." *Engineering News*, January 17, 1880, 26–27.

King, Moses, ed. *King's Handbook of New York City*. 2d ed. Boston: Moses King, Publisher, 1893.

Kirby, Richard S., and Philip Laursen. *The Early Years of Modern Civil Engineering*. New Haven: Yale University Press, 1932.

Kirby, Richard S., Sidney Withington, Arthur B. Darling, and Frederick G. Kilgour. *Engineering in History*. New York: Dover Publications, 1990.

Kulak, Geoffrey L., John W. Fisher, and John H. A. Struik. *Guide to Design Criteria for Bolted and Riveted Joints*. 2d ed. New York: John Wiley & Sons, 1987.

Kurtz, Charles. "A Review of Concrete-Metal Construction." *American Architect and Building News*, vol. 72, 1901, 61–62.

*The Lally Column Catalog*. Cambridge, MA: Lally Column Company, 1953.

"A Large Building Endangered by Dry Rot." *Scientific American Architects' and Builders' Edition*, November 1887.

Leabu, Victor. "Composite Design with Lightweight Aggregate in Building Projects." *AISC Engineering Journal*, October 1965, 135–44.

Lee, Antoinette J. *Iron, Engineers and Urbanization*. New York: American Society of Civil Engineers. Preprint issued for Annual and National Environmental Engineering Meeting, 1972.

"Lightweight Fireproofing for Steel Framing." *Engineering News-Record*, November 6, 1952.

Liscombe, R. W. "A 'New Era in My Life': Ithiel Town Abroad." *Journal of the Society of Architectural Historians*, March 1991, 5–17.

"Local Laws of the City of New York for the Year 1980, No. 10." *The City Record*, vol. cviii, no. 32169.

Lowe, David, ed. *The Great Chicago Fire*. New York: Dover Publications, 1979.

Loyrette, Henri. *Gustave Eiffel*. Translated by Rachel and Susan Gomme. New York: Rizzoli International Publications, 1985.

McCabe, James Dabney. *New York by Sunlight and Gaslight*. Philadelphia: Hubbard Brothers, 1882.

McCormac, Jack C. *Structural Steel Design*. 3d ed. New York: Harper & Row Publishers, 1981.

McCullough, David. *The Great Bridge*. New York: Avon Books, 1972.

MacKay, Donald A. *The Building of Manhattan*. New York: Harper & Row, 1987.

MacKay, H. M., Peter Gillespie, and C. Leluau. "Report on the

Strength of Steel I-Beams Haunched with Concrete." *Engineering Journal*, Engineering Institute of Canada, August 1923, 365–69.

McKay, Richard C. *South Street: A Maritime History of New York*. New York: Haskell House Publishers, 1971.

*Manual of the Bouton Foundry Company*. Chicago: R. R. Donnelley & Sons, 1887.

"Marble Falls from Skyscraper Facade." *Engineering News-Record*, December 23, 1976, 25.

Marinelli, Janet. "Architectural Glass and the Evolution of the Storefront." *The Old-House Journal*, July/August 1988, 34–43.

Meadows, Robert. *Historic Building Facades: A Manual for Inspection and Rehabilitation*. New York: New York Landmarks Conservancy, 1986.

Merriman, Mansfield, ed. *American Civil Engineers' Handbook*. 4th ed. New York: John Wiley & Sons, 1920.

Merriman, Mansfield, and Henry S. Jacoby. *A Text-Book on Roofs and Bridges, Part I, Stresses in Simple Trusses*. 6th ed. New York: John Wiley & Sons, 1917.

———. *A Text-Book on Roofs and Bridges, Part I, Stresses in Simple Trusses*, 5th ed. New York: John Wiley & Sons, 1899.

———. *A Text-Book on Roofs and Bridges, Part III, Bridge Design*, 5th ed. New York: John Wiley & Sons, 1917.

———. *A Text-Book on Roofs and Bridges, Part IV, Higher Structures*, 3d ed. New York: John Wiley & Sons, 1914.

"Metal Deck, Connectors Pass Composite-Beam Test." *Engineering News-Record*, July 5, 1962.

"Metal Deck Moves Into Bay Area." *Engineering News-Record*, October 22, 1959.

"Metal Diamonds Adorn Skyscraper." *Engineering News-Record*, February, 16, 1956.

"The Metropolitan Concrete and Wire Floor." *Engineering News*, November 14, 1895, 333.

"Miracle at Grand Central." *Engineering News-Record*, July 9, 1956.

Morley, Jane. "Frank Bunker Gilbreth's Concrete System." *Concrete International*, November 1990, 57–62.

Mujica, Francisco. *History of the Skyscraper*. Paris: Archaeology & Architecture Press, 1929. Reprint. New York: DaCapo Press, 1977.

National Trust for Historic Preservation. *All About Old Buildings: The Whole Preservation Catalog*. Edited by Diane Maddex. New York: Pantheon, 1976.

"New Reinforcing Bars." *Engineering News-Record*, July 13, 1950, 21.

"A New System of Fireproof Floor Construction." *Engineering News*, November 9, 1889, 434–35.

"New York City Revises Inspection Bill." *Engineering News-Record*, November 4, 1976, 16.

New York Landmarks Preservation Commission. *SoHo-Cast Iron Historic District; Designation Report*. New York: New York Landmarks Preservation Commission, 1973.

*New York Times*, accounts of fire at 1–5 Bond Street, March 7, 1877, 1-4; March 8, 1977, 4-2.

*New York Times*, accounts of partial collapse of Ireland Building at 3rd Street and West Broadway, August 9, 1895, 1-8; August 10, 1895, 2-3; August 11, 1895, 8-3; August 12, 1895, 8-3; August 13, 1895, 6-1; August 14, 1895, 8-4; August 15, 1895, 9-1; August 16, 1895, 9-6; August 17, 1895, 8-3.

*New York Times*, accounts of partial collapse of Brown Co. soap factory at Twelfth Avenue and 51st Street, June 4, 1897, 1-8; June 5, 1897, 2-7.

*New York Times*, accounts of the Hotel Windsor fire, March 18, 1899, 1-5; March 19, 1899, 1-5; March 26, 1899, 4-1.

Nolan, Thomas, ed. *Kidder's Architects' and Builders' Handbook*. 17th ed. New York: John Wiley & Sons, 1921.

Norton, Charles L. "The Protection of Steel from Corrosion." *Engineering News*, January 14, 1904, 29.

———. "Tests to Determine the Protection Afforded to Steel by Portland Cement Concrete." *Engineering News*, October 23, 1902, 333–34.

"Office Building Assigned Triple Role." *Engineering News-Record*, January 30, 1958.

O'Reilly, Joseph J., ed. *Fire Fighting*. New York: The Chief Publishing Company, 1911.

Page, Philip P., Jr. "New Type of Shear Connector Cuts Costs of Composite Construction." *Engineering News-Record*, May 10, 1956.

Parcel, John, and George Maney. *An Elementary Treatise on Statically Indeterminate Structures*. 2d ed. New York: John Wiley & Sons, 1936.

Parsons, Harry de B. "Collapse of a Building During Construction." *Engineering News*, vol. LI, no. 19, May 12, 1904, 454–56.

Perrine, Harold, and George Strehan. "Cinder Concrete Floor Construction Between Steel Beams." *Transactions of the American Society of Civil Engineers*, vol. LXXIX. New York: American Society of Civil Engineers, 1915, 523–621.

Petersen, J. L. "History and Development of Precast Concrete in the U.S." *Journal of the American Concrete Institute*, 1954, 477–500.

Peterson, Charles. "Inventing the I-beam: Richard Turner, Cooper & Hewett and Others." In *The Technology of Historic American Buildings*, edited by H. Ward Jandl. Washington, D.C.: Foundation for Preservation Technology, 1983.

Philbrick, Philetus H. *Beams and Girders: Practical Formulas for Their Resistance*. New York: Van Nostrand, 1886.

Placzek, Adolf, ed. *Macmillan Encyclopedia of Architects*. New York: Macmillan, 1982.

Plummer, Harry C., and Edwin F. Wanner. *Principles of Tile Engineering: Handbook of Design*. Washington, D.C.: Structural Clay Products Institute, 1947.

*Pocket Companion Containing Useful Information and Tables Appertaining to the Use of Steel*. Pittsburgh, PA: Carnegie Steel Company, 1903.

*Pocket Companion Containing Useful Information and Tables Appertaining to the Use of Steel*. Pittsburgh, PA: Carnegie Steel Company, 1919.

*Pocket Companion for Engineers, Architects and Builders Containing Useful Information and Tables Appertaining to the Use of Steel*. 23d ed. Pittsburgh, PA: Carnegie Steel Company, 1923.

Portland Cement Association. "Concrete Technology: 20th Century Concrete Material and Admixture Advances." Advertisement, *Engineering News-Record*, vol. 226, no. 18, May 6, 1991, C-8.

Post, Nadine M. "Big Squeeze for Lean Landmark." *Engineering News-Record*, vol. 226, no. 2, January 14, 1991.

———. "Big Job on Big Exterior in Big Apple." *Engineering News-Record*, vol. 227, no. 23, December 9, 1991, 53–54.

Potter, Thomas. *Concrete: Its Uses in Building*. 3d ed. New York: D. Van Nostrand Company, 1908.

"Protection of Iron from Corrosion." *Engineering News*, June 27, 1878, 205.

Prudon, Theodore H. M. "Repairing Early Curtain Walls." *Progressive Architecture*, February 1992, 123–26.

Purdy, Corydon. "Steel Skeleton Type of High Building." *Engineering News*, December 12, 1891.

Quimby, Henry. "Wind Bracing in High Buildings." *Engineering Record*, November 19, 1892, 394–95; December 31, 1892, 99; January 14, 1893, 138; January 21, 1893, 161–62; January 28, 1893, 180; February 25, 1893, 260; March 11, 1893, 298–99; March 18, 1893, 320.

Rains, William A. "A New Era in Fire Protective Coatings for Steel." *Civil Engineering*, September 1976, 80–83.

Ramsey, Charles, and Harold Sleeper. *Architectural Graphic Standards*. Facsimile edition of 1st ed. New York: John Wiley & Sons, 1990.

"Reinforced Concrete Pile Foundation for the Latteman Building, Brooklyn, N.Y." *Engineering News*, December 7, 1905, 594–95.

"The Removal of a Steel Frame Building." *Engineering News*, January 29, 1903, 113–14.

Republic Steel advertisement, *Engineering News-Record*, March 20, 1958.

Sabnis, Gajanan M. ed. *Handbook of Composite Construction Engineering*. New York: Van Nostrand, 1979.

Salmon, Charles G., and John E. Johnson. *Steel Structures: Design and Behavior*. 2d ed. New York: Harper & Row Publishers, 1980.

Sante, Luc. *Low Life: Lures and Snares of Old New York*. New York: Farrar, Strauss, Giroux, 1991.

Saurbrey, Alexis. "The 'Ransome Unit System'; A Separately Molded Reinforced Concrete Construction." *Engineering News*, June 15, 1911, 724–27.

Schaal, Rolf. *Curtain Walls: Design Manual*. Translated by Thomas E. Burton. New York: Reinhold Publishing Company, 1962.

"A Second Aluminum-Faced Skyscraper." *Engineering News-Record*, November 12, 1953.

Seeley, Herman B. "The Art of Fireproofing." *Engineering News*, Part I, April 9, 1896, 234–37; Part II, April 16, 1896, 250–52.

"Self Inspection for N.Y.C. Builders?" *Civil Engineering*, April 1976, 94.

Shaw, Esmond. *Peter Cooper and Wrought Iron Beams*. New York: Cooper Union, 1960.

Shedd, Thomas. *Structural Design in Steel*. New York: John Wiley & Sons, 1934.

Shopsin, William C. *Restoring Old Buildings for Contemporary Uses*. New York: Whitney Library of Design, 1986.

Shryock, Joseph. "Comparative Diagrams of Compression Formulas for Steel Struts or Columns." *Lefax Filing System*. Philadelphia: Standard Corporation, 1914.

Silver, Nathan. *Lost New York*. New York: American Legacy Press, 1967.

Singleton, Jack. *Fireproofing Structural Steel*. New York: American Institute of Steel Construction, 1929.

Skempton, A. W. "Evolution of the Steel Frame Building." *The Guilds' Engineer*, vol. X, 1959, 37–51.

"A Skyscraper Crammed with Innovation." *Engineering News-Record*, June 13, 1957.

"Skyscraper Gets Aluminum Skin." *Engineering News-Record*, August 6, 1953.

Smith, William Sooy. "The Modern Tall Building: Corrosion and Fire Dangers." *Cassier's Magazine*, May 1902, 56–60.

"Some Details of the Ireland Building." *Engineering News*, August 22, 1895, 127–28.

"Spandrel Repairs Avert Collapse." *Engineering News-Record*, February 16, 1978.

Spann, Edward K. *The New Metropolis*. New York: Columbia University Press, 1981.

Sprauge, E. T. E. *50-Year Steel Joist Digest*. Richmond, VA: Steel Joist Institute, 1982.

"Spray Fireproofing on Steel Floors." *Engineering News-Record*, April 3, 1958.

Spurr, Henry V. *Wind Bracing: The Importance of Rigidity in High Towers*. New York: McGraw-Hill, 1930.

Squires, Frederick. "Houses at Forest Hills Gardens." *Concrete-Cement Age*, January 1915, 2-8, 53–54.

"Standard Specifications for Steel; Association of American Steel Manufacturers." *Engineering News*, August 27, 1896, 130.

Starrett, Paul. *Changing the Skyline*. New York: Whittlesey House, 1938.

Starrett, William A. *Skyscrapers and the Men Who Build Them*. New York: Charles Scribner's Sons, 1928.

Stern, Robert A. M., Gregory Gilmartin, and John M. Massengale, *New York 1900*. New York: Rizzoli International Publishers, 1983.

Stern, Robert A. M., Gregory Gilmartin, and Thomas Mellins, *New York 1930*. New York: Rizzoli International Publishers, 1987.

Stetina, Henry J. "Choosing the Best Structural Fastener." *Civil Engineering*, November 1963.

Stockbridge, Jerry G. "The Interaction Between Exterior Walls and Building Frames in Historic Tall Buildings." In Council on Tall Buildings and Urban Habitat, *Developments in Tall Buildings 1983*. Stroudsburg, PA: Hutchinson Ross Publishing Company, 1983.

Stokes, I. N. Phelps. *The Iconography of Manhattan Island 1498–1909*, vol. 5, 1926. Reprint. New York: Arno Press, 1967.

"The Strength of Columns." *Engineering News*, April 21, 1877, 93.

"Structural Engineering in Apartment House Construction." *Engineering News*, vol. LI, no. 15, April 14, 1904, 353–54.

Sturgis, Russell, ed. *Sturgis' Illustrated Dictionary of Architecture and Building*. 3 vols. New York: Macmillan, 1902. Reprint. New York: Dover Publications, 1989.

Sutherland, Hale, and Raymond Reese. *Introduction to Reinforced Concrete Design*. 2d ed. New York: John Wiley & Sons, 1943.

Swank, James M. *History of the Manufacture of Iron in All Ages, and Particularly in the United States from Colonial Times to 1891*. 2d ed. Philadelphia: The American Iron and Steel Association, 1892.

"Tall Buildings of New York." *Scientific American*, December 5, 1908, 400–403.

"Tallest Welded Building in East Topped Out." *Civil Engineering*, July 1960.

Task Committee on Structural Connections of the Committee on Metals of the Structural Division of the American Society of Civil Engineers and the Research Council on Riveted and Bolted Structural Joints. *Bibliography on Bolted and Riveted Joints*. New York: American Society of Civil Engineers, 1967.

Thomson, John W. *Cast Iron Buildings: Their Construction and Advantages, by James Bogardus*. New York: J. W. Harrison, 1856. Reprinted in *The Origins of Cast Iron Architecture in America*. Introduction by Walter Knight Sturges. New York: DaCapo, 1970.

Thornton, Charles, Udom Hungsprucke, and Robert DeScenza. "Vertical Expansion of Vintage Buildings." *Modern Steel Construction*, vol. 31, no. 6, June 1991.

Thornton, Charles, Jagdish Prasad, Robert DeScenza, and Joseph Lieber. "Economic Upgrading of Vintage Buildings." *Modern Steel Construction*, vol. 31, no. 7, July 1991.

Toch, Maximillian. "The Condition of the Steel of the Gillander Building." *Engineering News*, July 14, 1910, 54–55.

Toprac, A. A. "Strength of Three New Types of Composite Beams." *AISC Engineering Journal*, January 1965, 21–30.

Torroja, Eduardo. *The Philosophy of Structures*. Translated by J. J. Polivka and Milos Polivka. Berkeley: University of California Press, 1958.

Trautwine, John C. *The Civil Engineer's Pocket-Book*. 14th ed. New York: John Wiley & Sons, 1889.

Trestain, Tom, and Jacques Rousseau, "Technics Topics: Steel Stud / Brick Veneer Walls." *Progressive Architecture*, February 1992, 113–16.

Trimble, Brian. "Brick Institute Recommendations." *Progressive Architecture*, February 1992, 117–18.

Troup, Emile W. J. "Fire-Resistant Design of Interior Structural Steel." In Council on Tall Buildings and Urban Habitat, *Developments in Tall Buildings 1983*. Stroudsburg, PA: Hutchinson Ross Publishing Company, 1983.

Turneaure, F. E., ed. *Johnson's Materials of Construction, Rewritten by M. O. Withey and James Aston*. New York: John Wiley & Sons, 1919.

Turneaure, F. E., and E. R. Maurer. *Principals of Reinforced Concrete Construction*. 4th ed. New York: John Wiley & Sons, 1932.

"Two Killed as Tenement Collapses." *Engineering News-Record*, August 18, 1977.

"United Engineering Center." *Civil Engineering*, July 1960.

Urquhart, Leonard C., ed. *Civil Engineering Handbook*. 3d ed. New York: McGraw-Hill, 1950.

van Leeuwen, Thomas A. P. *The Skyward Trend of Thought*, Cambridge, MA: MIT Press, 1988.

Von Emperger, Frederick. "The Calculation of Flat Arches." *Engineering News*, March 19, 1896, 186–87.

Waite, Guy B. "Cinder Concrete Floors." *Transactions of the American Society of Civil Engineers*, vol. 77. New York: ASCE, 1914, 1773–1823.

Waite, John G., ed. *Iron Architecture in New York City*. New York State Historic Trust and the Society for Industrial Archeology, 1972.

Waite, John G., and Margot Gayle. *The Maintenance and Repair of Architectural Cast Iron*. New York: New York Landmarks Conservancy, 1991.

Walker, Frank R. *The Building Estimator's Reference Book*. 9th ed. Chicago: Frank R. Walker Company, 1940.

Wang, Chu-Kia, and Charles G. Salmon. *Reinforced Concrete Structures*. 4th ed. New York: Harper & Row, 1985.

Wang, Marcy Li, Isao Sakamoto, and Bruce L. Bassler, eds. *Cladding: Council on Tall Buildings and Urban Habitat Committee 12A*. New York: McGraw-Hill, 1992.

Ward, W. E. "Béton in Combination with Iron as a Building Material." *Transactions of the American Society of Mechanical Engineers*, 1883, 388–403.

Weiskopf, S. C. unpublished design aids written for own use, 1916.

Weisman, Winston. "A New View of Skyscraper History." In *The Rise of an American Architecture*, edited by Edgar Kaufmann. New York: Praeger Publishers, 1970.

White, Norval, and Elliot Willensky. *AIA Guide to New York City*. 2nd ed. New York: Collier Books, 1978.

White, William. "Ironwork: Its Legitimate Uses and Proper Treatment." In *Papers Read at the Royal Institute of British Architects, Session 1865–1866*. London: Royal Institute of British Architects, 1866.

"Why the Finest New Buildings Have Q-floor." Advertisement, *Engineering News-Record*, April 14, 1955.

Wight, Peter B. "The Pioneer Concrete Residence in America." *Architectural Record*, vol. 25, 1909, 359–63.

———. "Origin and History of Hollow Tile Fire Proof Floor Construction." *The Brickbuilder*, March 1897, 53–54; April 1897, 73–75; May 1897, 98–99.

Wildt, Roger. "Fire Protection of Structural Steel Framing." *Civil Engineering*, September 1966.

Williams, Clifford D., and Ernest C. Harris, *Structural Design in Metals*. New York: The Ronald Press Company, 1949.

Wilson, A. C. "Wind Bracing with Knee-Braces or Gusset-Plates." *Engineering Record*, September 5, 1908, 272–74.

Winter, George, and Arthur H. Nilson. *Design of Concrete Structures*. 9th ed. New York: McGraw-Hill, 1979.

Winterton, K. "5000 B.C.–1962 A.D., A Brief History of Welding Technology." *Welding and Metal Fabrication*, November 1962, 438–42; December 1962, 488–93; February 1963, 71–76.

## PHOTO CREDITS

The photographs on the following pages appear courtesy of the United States History, Local History & Genealogy Division, The New York Public Library, Astor, Lenox and Tilden Foundations: 2, 14, 19, 20 (top), 21, 22, 24 (top), 26, 31, 32, 44, 46, 50, 51, 52, 53, 54, 55, 60, 80, 122.

All other photographs and drawings are by the author.

## ACKNOWLEDGMENTS

Grateful acknowledgment is made to:

The University of Chicago Press for permission to use the quotations on pages 52, 58, and 165 from *American Building*, 2d ed., by Carl Condit, copyright © 1968, 1982 by The University of Chicago Press, and the quotation on page 51 from *The Port of New York*, vol. I, by Carl Condit, copyright © 1980 by The University of Chicago Press.

McGraw-Hill, Inc. for permission to use material on page 204 from *Civil Engineering Handbook*, 3d ed., by Leonard Urquhart, copyright © 1934, 1940, 1950 by The McGraw-Hill Book Company, Inc.

American Institute of Steel Construction for permission to use material from *Steel Construction*, 4th ed., *Steel Construction Manual*, 5th ed., *Manual of Steel Construction*, 6th, 7th, and 8th eds., and *AISC Iron and Steel Beams*, 1873–1952, edited by Herbert W. Ferris.

# INDEX

high-strength steel in, 145
historical development, 90–91
jack arch, 31, 171
masonry wall building codes, 24
metal decks, 97, 132–136, 139,
    141–142
modern steel construction, 149
modern steel design, 141–142
national standardization, 87–88
present considerations, 89, 112–113
steel beams, 58
terra-cotta systems, 42, 58–62, **59**, **61**,
    **62**
traditional wood construction, 13
web openings, 149–150
wrought-iron beams, 21, 31, 38, 58, 69
*see also* tile arch floor
Ford Foundation Building, 126, 167
Forty Fourth Street, 164
Forty Second Street, 125
    McGraw-Hill Building (old), 166
    Mobil Building, 166
Forty Sixth Street, **24**
Forty Third Street, 126, 167
Foster, Richard, 167
foundations, 10
    in Ireland Building failure, 48
foundries, 28, 29
Fourth Avenue, 160
Friberg, Bengt, 132
Frost Associates, 168
full moment connections, 146–147
Fulton Street, 19

Gaynor, John P., 160
General Fireproofing Company, 97
General Post Office (old), 59, **60**, 161
German Winter Garden, 160
Gilbert, Bradford, 52, 162
Gilbert, Cass, 165
Gildemeister, Charles, 159
Gillander Building, 163
Gilman, Arthur, 21, 161
Gilsey Building, 159
girders
    built-up, 71–72, 89
    definition, 171
    double, 71
    equivalent strength section comparison,
        **72**, 73
    live load reduction, 155
    plate, 71–72
    tension-rod, 29, 83–84
    wide-flange beams, 73
    wrought composite, 31
glass-fiber reinforced concrete (GFRC),
    126, **128**, 130
glass walls, 121–124, 127, 130
Goethals, George, 165
Gottlieb, A. S., 165
grain elevator, 54, 160
Granco Steel products, 134
Grand Central Depot, 121–122, **122**, 161
Grand Central Station, 163

Grand Central Terminal, 165
Greene Street, **39**
Greenwich Street, **18**, **42**
Gregory Industries, 144
Grey, Henry, 72
Grey rolling mill, 72
Gropius, Walter, 167
Gustavino Timbrel Vault, **94**, 94–95, 105
gypsum, 133, 137, 139

Hard-burned Hollow Tile, 94
Hardenbergh, Henry, 163, 165
Harper & Brothers Building, 31, 38, 58,
    160
Harrison, Wallace, 166
Harrison & Abramovitz, 166
Hatfield, R. F., 161
Haughwout Building, **30**, 30, 160
Havemeyer Building, 51, 81, 162
Havemeyer reinforcing bar, 108
Hayden Planetarium, 166
H columns, 73, 74, 75
headed shear studs, 144, 171
header course, 85–86
helices, 143–144
Henry Street, Brooklyn, **14**, 162
Herringbone reinforcing bar, 108
Herter Brothers, 162
Hewitt, Abraham, 31, 69
high-strength steel, 145, 153
    connections, 145–146, 171
Hildenbrand, Wilhelm, 122, 161
Hodgkinson, Eaton, 13, 69
hollow walls, 25–26, 119
Home Insurance Building, 50, **53**
Hood, Godley & Foulihoux, 166
Hoppin Company, 159
Hotel New Netherland, 162
Hotel Savoy, 163
house construction, 13
Howells & Stokes, 165
Hubert, Pirsson & Co., 162
Huckel, Samuel, Jr., 163
Hugh O'Neill Store, **29**
Hunt, Richard Morris, 21, 57, 161
Hy-rib floor system, 97

I-beams, 171
    alternate shapes, 70–71
    in built-up girders, 71–72, 89
    development of, 31, 58, 69–70
    as lintels, 85
    moment of inertia, 71
    present considerations, 88–89
    section comparison, **73**
    specifications, 69–70
    standardization, 70
    steel, 58
Inland Bar, 108
intumescent paint, 138
Ireland Company building, 47–49, 163
iron construction
    acceptance of, by architects and

engineers, 30
column size, **22**
development of structural engineering
    profession, 13
early foundries, 28, 29
early use, 21
emergence of rolled-iron technology, 31,
    38, 83
historical developments, 10
pioneering architects, 13
with tile arch floors, 21, 22
*see also* cast-iron construction; cast-iron
    facades; steel construction;
    wrought iron

Jackson, James, 28, 160
Jackson Building, 162
Janes foundry, 28
Johnson, George, 13, 53–54, 58, 160
Johnson, Philip, 167
Johnson's formula, 77
joist, definition, 171
joists, concrete, 110–112, **111**
joists, open-web, 83–84, **84**, 172
joists, wood
    in brick rowhouse, 25
    in cast-iron buildings, 31–32
    deflection criteria, 15
    embedded in brick walls, 15, 16–17
    fire cut, 16, **17**, 170
    fire risk, 57
    framing around stairs, 67
    hanger construction, 15, 15–16
    iron building ties and, 17
    masonry wall stability, 16–17
    materials, 15
    metal box enclosures, 17
    mortise and tenon connections, 15, 16,
        26
    potential defects, 26–27
    stress design, 14–15, 16

Kahn & Jacobs, 166, 167
Kawneer Company, 122
K-bracing, 171
Keep Building, 37, 38, 57, 163
Keller George, 164
Kellum, John, 13, 160
Kendall, George, 21
Kessler, S. J., 167
Keystone Bridge Company, 161
Kilburn, Henry F., 163
Kimball & Thompson, 163
King, Gamaliel, 13, 160
King Contracting, 161
Kips Bay Plaza, 167
Komendant, August, 167
Kreischer, Balthasar, 58
Kroehl, Julius, 53, 160

labor costs, 11
    in cast-ironconstruction, 56–57